Learning and Analytics in Intelligent Systems

Volume 24

The main aim of the series is to make available a publication of books in hard copy form and soft copy form on all aspects of learning, analytics and advanced intelligent systems and related technologies. The mentioned disciplines are strongly related and complement one another significantly. Thus, the series encourages cross-fertilization highlighting research and knowledge of common interest. The series allows a unified/integrated approach to themes and topics in these scientific disciplines which will result in significant cross-fertilization and research dissemination. To maximize dissemination of research results and knowledge in these disciplines, the series publishes edited books, monographs, handbooks, textbooks and conference proceedings.

Indexed by EI Compendex.

More information about this series at https://link.springer.com/bookseries/16172

Maria Virvou · George A. Tsihrintzis ·
Lakhmi C. Jain
Editors

Advances in Selected Artificial Intelligence Areas

World Outstanding Women in Artificial Intelligence

Springer

Editors
Maria Virvou
Department of Informatics
University of Piraeus
Piraeus, Greece

George A. Tsihrintzis
Department of Informatics
University of Piraeus
Piraeus, Greece

Lakhmi C. Jain
KES International
Shoreham-By-Sea, UK

ISSN 2662-3447 ISSN 2662-3455 (electronic)
Learning and Analytics in Intelligent Systems
ISBN 978-3-030-93051-6 ISBN 978-3-030-93052-3 (eBook)
https://doi.org/10.1007/978-3-030-93052-3

This Springer imprint is published by the registered company Springer Nature Switzerland AG
The registered company address is: Gewerbestrasse 11, 6330 Cham, Switzerland

Foreword

This book, edited by Profs. Maria Virvou, George A. Tsihrintzis and Lakhmi C. Jain, highlights highly significant research led by *women* in selected areas of Artificial Intelligence (AI) and Machine Learning (ML). This work is laudable, and its importance is hard to exaggerate. AI and ML applications are nearly ubiquitous, and researchers need to be reflective of the societal and ethical impact of their work. This collection of contributions presents the state-of-the-art research and projects led by outstanding women in various aspects of AI and its applications toward inspiring other women to get involved in this very important discipline which will reshape our society in years to come.

The chapters in this volume serve to illustrate recent trends and paradigms in AI[1] such as explainable AI with the goal of building explanatory models, to overcome shortcomings of automatic decision-making especially when dealing with crucial, ethical or legal issues (C. Zanni-Merk and A. Jeannin-Girardon); trustworthy human-AI collaboration which highlights emergent multi-faceted role of AI in human-machine interactions (L. Razmerita, A. Brun and T. Nabeth); pushing the limits of the no free lunch theorem developments by designing general-purpose classification systems (A. Lumini, L. Nanni and S. Brahnam); exploring linguistic intelligence as the root of computational reasoning (D. L. De Luise); and feature selection in this era of big data (V. Bolón-Canedo, A. Alonso-Betanzos, L. Morán-Fernández and B. Cancela).

[1] Sheela Ramanna, Lakhmi Jain, Robert Howlett, (Eds.) Emerging Paradigms in Machine Learning. Springer, Berlin 2013, ISBN: 978-3-642-28698-8.

Enabled by faster and powerful computers, we are witnessing a fascinating range of AI applications impacting our physical health, environment, mental well-being and digital identity. Data-driven findings in these areas are crucial in assisting policy makers and governments, and the following papers highlight these issues: state-of-the-art approaches to solving problems of biometric security through AI methods (M Gavrilova, I. Luchak, T. Sudhakar and S. N. Tumpa); adaptive neural networks that embed genomic knowledge for precision medicine such as cancer detection (S. Belciug); advanced deep learning methods to analyze offensive behavior in cyberspaces via abusive content identification in social media data (R Nayak and H. S. Ba); and exploring challenges and methods in smoke detection in outdoor spaces crucial in early fire alarms (M. N. Favorskaya).

The central problem of our age is how to act decisively in the absence of certainty.[2] Fuzzy sets,[3] rough sets,[4,5] and evolutionary and probabilistic reasoning[6] are all methodologies fundamentally important in AI applications as they bridge the gap between approximate and crisp forms of knowledge structures. The papers in this volume serve to illustrate the broad range of machine learning tasks using these methodologies for solving real-world problems: feature selection and reduction methods with rough sets (U. Stanczyk), fuzzy clustering (M. Sato-Ilic), effective hybrid genetic algorithms (J. C. Siala and O. Harrabi), and theory and philosophy of Bayesian networks (D. E. Holmes).

The editors and contributors are to be congratulated on producing a book that covers a variety of challenging problems in AI and ML led by outstanding women scientists who can serve as role models for a new generation of scientists and leaders, reflecting what a more inclusive AI community should look like. On a personal note, as a professor working in AI and machine learning, my own role model has been my

---

[2] Bertrand Russell (1950). An Inquiry into Meaning and Truth. George Allen and Unwin, London; W.W. Norton, New York.

[3] L.A. Zadeh, Fuzzy sets, Information and Control 8 (1965) 338–353.

[4] Zdzisław Pawlak and Andrzej Skowron, (2007). Rudiments of rough sets. Information Sciences Journal, Elsevier, 177, 3-27.

[5] James Peters and Andrzej Skowron (2007). Zdzislaw Pawlak life and work (1926-2006), Information Sciences Journal, Elsevier 177, 1-2.

[6] Pearl J (1988). Probabilistic Reasoning in Intelligent Systems. In: Networks of Plausible Inference. Morgan Kaufmann Publishers, San Francisco.

mother B. V. Saroja, a retired Professor in Mathematics, from College of Engineering, Osmania University, India, from whom I draw inspiration.

Winnipeg, Canada
October 2021

Sheela Ramanna

Dr. Sheela Ramanna is a Full Professor and past Chair of the Applied Computer Science Department. She is the co-founder of the ACS graduate studies program at the University of Winnipeg. She received a Ph.D. in Computer Science from Kansas State University, U.S.A. and a B.S. in Electrical Engineering and M.S. in Computer Science and Engineering from Osmania University, India. She serves on the Editorial Board of Springer Transactions on Rough Sets (TRS) Journal, Elsevier Engineering Applications of AI Journal, KES Journal and Advisory Board of the International Journal of Rough Sets and Data Analysis. She is the Managing Editor of the TRS and is a Senior Member of the IRSS (Intl. Rough Set Society). She has co-edited a book with L. C. Jain and R. Howlett on Emerging Paradigms in Machine Learning published in 2013 by Springer. She has served as Program Co-Chair for IJCRS 2021 track of IFSA/EUSFLAT 2021, MIWAI 2013, RSKT 2011, RSCTC 2010 and JRS2007. Her research is funded by NSERC Discovery and Collaborative Grants and MITACS Accelerate Programs. She has received more than $1,150,000 in research funding since 1992. She has published over 60 peer-reviewed articles in the past 6 years and has given keynote and invited talks at international conferences. The focus of her research is in fundamental and applied research in machine learning and granular computing. Her current interests include (i) Tolerance-based granular computing techniques (fuzzy sets, rough sets and near sets) with applications in social networks, natural language processing, computer vision and audio signal processing, (ii) topological data analysis and (iii) deep learning applications.

Preface

As new technological challenges are perpetually arising, Artificial Intelligence research interests are focusing on the incorporation of improvement abilities into machines in an effort to make them more efficient and more useful. Recent reports indicate that the demand for scientists with Artificial Intelligence skills significantly exceeds the market availability and that this shortage will intensify further in the years to come. A potential solution includes attracting more women into the field, as women currently make up only 26 percent of Artificial Intelligence positions in the workforce. The present book serves a dual purpose: On the one hand, it sheds light on the very significant research led by *women* in the areas of Artificial Intelligence, in the hopes of inspiring other women to follow studies in that area and get involved in related research. On the other hand, it highlights the state-of-the-art and current research in selected Artificial Intelligence areas and applications.

The book consists of an editorial note and an additional thirteen (13) chapters, all authored by *invited* women-researchers who work on various Artificial Intelligence areas and stand out for their significant research contributions. Each draft chapter was reviewed by two independent reviewers, whose comments and suggestions were incorporated into the finalized chapter versions.

In more detail, the chapters in the book are organized into three parts, namely (i) Advances in Artificial Intelligence Paradigms (five (5) chapters), (ii) Advances in Artificial Intelligence Applications (five (5) chapters), and (iii) Recent Trends in Artificial Intelligence Areas and Applications (three (3) chapters).

This research book is directed toward professors, researchers, scientists, engineers, and students in Artificial Intelligence-related disciplines. It is also directed toward readers who come from other disciplines and are interested in becoming versed in some of the most recent Artificial Intelligence-based technologies. An extensive list of bibliographic references at the end of each chapter guides the readers to probe further into the Artificial Intelligence areas of interest to them.

We are grateful to the authors and reviewers for their excellent contributions and visionary ideas. We are also thankful to Springer for agreeing to publish this book in its Learning and Analytics in Intelligent Systems series. Last, but not the least, we are grateful to the Springer staff for their excellent work in producing this book.

Piraeus, Greece Maria Virvou
Piraeus, Greece George A. Tsihrintzis
Sydney, Australia Lakhmi C. Jain

Contents

Chapter 1
Introduction to Advances in Selected Artificial Intelligence Areas

Maria Virvou, George A. Tsihrintzis, and Lakhmi C. Jain

Abstract As new technological challenges are perpetually arising, Artificial Intelligence research interests are focusing on the incorporation of improvement abilities into machines in an effort to make them more efficient and more useful. Recent reports indicate that the demand for scientists with Artificial Intelligence skills significantly exceeds the market availability and that this shortage will intensify further in the years to come. A potential solution includes attracting more women into the field, as women currently make up only 26 percent of Artificial Intelligence positions in the workforce. The present book serves a dual purpose: On one hand, it sheds light on the very significant research led by *women* in areas of Artificial Intelligence, in hopes of inspiring other women to follow studies in the area and get involved in related research. On the other hand, it highlights the state-of-the-art and current research in selected Artificial Intelligence areas and applications.

1.1 Editorial Note

Artificial Intelligence, in all its aspects, approaches and methodologies, aims at incorporating *improvement abilities* into machines, i.e. at developing mechanisms, methodologies, procedures and algorithms that allow machines to become more efficient and more useful when performing specific tasks, either on their own or with the help of a supervisor/instructor [1].

M. Virvou (✉) · G. A. Tsihrintzis (✉)
Department of Informatics, University of Piraeus, 185 34 Piraeus, Greece
e-mail: mvirvou@unipi.gr

G. A. Tsihrintzis
e-mail: geoatsi@unipi.gr

L. C. Jain
Liverpool Hope University, Hope Park, UK

KES International, Shoreham-by-Sea, UK

M. Virvou et al. (eds.), *Advances in Selected Artificial Intelligence Areas*,
Learning and Analytics in Intelligent Systems 24,
https://doi.org/10.1007/978-3-030-93052-3_1

In recent years, Artificial Intelligence has (re-)emerged as a field of very active and intense research worldwide, exhibiting stunning worldwide pace of growth, fascinating success rates and tremendous impact on science, technology and society. Indeed, new research is continuously reported on both theoretical advances and innovative tools and methodologies. Additionally, new application areas are under continuous exploration and success stories are constantly announced. Consequently, these advances of Artificial Intelligence areas and paradigms and its technological applications have already affected many aspects of life, including the workplace, people's homes and both professional and social human relationships and interactions. In fact, the impact of Artificial Intelligence has been so strong that Artificial Intelligence is characterized as one (perhaps the main) *disruptive technology* [2].

On the other hand, new technological challenges are perpetually arising, which appear interrelated and are often involved in disciplines other than Artificial Intelligence [3–5]. This presses Artificial Intelligence to play an even more important role in the years to come. An example of this trend is the Internet of Things (IOT), which forms one of the pillars of new technologies, such as smart homes [6], smart cities [7–9], smart industries [10] or smart healthcare [11, 12] of the, so-called, 4th Industrial Revolution [13] and Society 5.0 [14, 15].

While advances are taking place, the availability of scientists with Artificial Intelligence skills is clearly insufficient [16, 17] and the corresponding demand is increasing despite the COVID-19 pandemic [18]. Indeed, even the most recent reports clearly indicate that the market demand significantly exceeds the market availability [19]. In this context, the Deloitte report [20] is particularly interesting, as it points out that the shortage in Artificial Intelligence talent could be overcome if more women got involved as "a 2020 World Economic Forum report, however, found that women make up only 26 percent of data and AI positions in the workforce, while the Stanford Institute for Human-Centered AI's 2021 *AI Index Report* found that women make up just 16 percent of tenure-track faculty focused on AI globally." Clearly, a potential solution to the shortage of scientists with Artificial Intelligence skills includes attracting more women into the field [21]. Indeed, efforts are being made in this direction, including formation of supporting communities [22] or identifying and presenting women at the forefront of Artificial Intelligence evolution who can act as role models [23].

The purpose of the present book is two-fold: On one hand, we would like to highlight the very significant research led by *women* in areas of Artificial Intelligence and, thus, inspire other women to follow studies in the area and get involved in related research. On the other hand and as Artificial Intelligence is currently a discipline of exciting new research and applications, we would like to highlight the state-of-the-art and current research in selected Artificial Intelligence areas and applications.

All chapters in the book were *invited* from a small number of world-outstanding women-researchers, but may have been co-authored. A biographical sketch of the (leading) woman-researcher is included at each chapter end.

In more detail, the book consists of the current editorial chapter and an additional thirteen (13) chapters, organized into three parts, as follows:

The *first part* of the book consists of 5 chapters devoted to *Advances in Artificial Intelligence Paradigms*.

More specifically, Chap. 2, by Verónica Bolón-Canedo, Amparo Alonso-Betanzos, Laura Morán-Fernández and Brais Cancela, is entitled "Feature Selection: From the Past to the Future." The authors analyze the paramount need for feature selection, briefly review the most popular feature selection methods and some typical applications and discuss new challenges facing researchers in the Big Data era.

Chapter 3, by Urszula Stańczyk, is entitled "Application of Rough Set-Based Characterisation of Attributes in Feature Selection and Reduction." The author presents new research on application of reduction-based characterisation of features, employed to support classification by selected inducers working outside rough set domain.

Chapter 4, by Mika Sato-Ilic, is entitled "Advances in Fuzzy Clustering used in Indicator for Individuality." The author describes the advances of fuzzy clustering integrated to convex clustering for extracting differences of subjects based on data consisting of objects, variables, and subjects.

Chapter 5, by Alessandra Lumini, Loris Nanni, and Sheryl Brahnam, is entitled "Pushing the Limits against the No Free Lunch Theorem: Towards building General-purpose (GenP) Classification Systems." The authors present their research on building general purpose classification systems that require little to no parameter tuning for performing competitively across a range of tasks within a domain or with specific data types, such as images, that span across several fields.

Chapter 6, by Dawn E. Holmes, is entitled "Bayesian Networks: Theory and Philosophy." The author explores the theory of Bayesian networks with particular reference to Maximum Entropy Formalism and also discusses objective Bayesianism together with some brief remarks on applications.

The *second part* of the book consists of 5 chapters devoted to *Advances in Artificial Intelligence Applications*.

More specifically, Chap. 7, by Marina Gavrilova Iryna Luchak, Tanuja Sudhakar and Sanjida Nasreen Tumpa, is entitled "Artificial Intelligence in Biometrics: Uncovering Intricacies of Human Body and Mind." The authors provide a comprehensive overview of the current progress and state-of-the-art approaches to solving problems of biometric security through artificial intelligence methods, including both classical machine learning paradigms and novel deep learning architectures.

Chapter 8, by Margarita N. Favorskaya, is entitled "Early Smoke Detection in Outdoor Space: State-of-the-Art, Challenges and Methods." The author follows the evolution of research approaches from conventional image processing and machine learning methods based on the motion, semi-transparence, color, shape, texture and fractal features to deep learning solutions using various deep network architectures.

Chapter 9, by Richi Nayak and Hee Sook Baek, is entitled "Machine Learning for Identifying Abusive Content in Text Data." The authors review various types of machine learning techniques, including the currently popular deep learning methods, that can be used in the analysis of social media data for identifying abusive content.

Chapter 10, by Jouhaina Chaouachi Siala and Olfa Harrabi, is entitled "Toward Artificial Intelligence tools for solving the Real World Problems: Effective Hybrid

Genetic Algorithms Proposal." The authors demonstrate how the design of an Interactive Decision Support System integrating a hybrid genetic algorithm during the optimization process is a performing approach to providing optimal solutions. The authors apply their approach to two NP-Hard problems, namely the University Course Timetabling problem (UCT) and the Solid Waste collection Problem (SWP), with good results.

Chapter 11, by Smaranda Belciug, is entitled "Artificial Neural Networks for Precision Medicine in Cancer Detection." The author presents several novel adaptive neural networks that embed genomic knowledge into their architecture, increasing their medical diagnosis performance and computational speed, while decreasing computational cost.

Finally, the *third part* of the book consists of 3 chapters devoted to *Recent Trends in Artificial Intelligence Areas and Applications*.

More specifically, Chap. 12, by Cecilia Zanni-Merk and Anne Jeannin-Girardon, is entitled "Towards the Joint Use of Symbolic and Connectionist Approaches for Explainable Artificial Intelligence." The authors focus on the joint use of symbolic and connectionist artificial intelligence with the aim of improving explainability.

Chapter 13, by Daniela López De Luise, is entitled "Linguistic Intelligence as a Root for Computing Reasoning." The author considers language processing as a tool for reasoning, understanding and mimics.

Finally, Chap. 14, by Liana Razmerita, Armelle Brun and Thierry Nabeth, is entitled "Collaboration in the Machine Age: Trustworthy Human-AI Collaboration." The authors provide insights in the state of the art of AI developments in relation to Human-AI collaboration.

1.2 Book Summary and Future Volumes

In this book, we present some very significant advances in selected Artificial Intelligence areas and applications, while at the same time attempting to inspire more women to get involved in studies and research in this area via highlighting the very significant research led by women around the world. The book is directed towards professors, researchers, scientists, engineers and students in Artificial Intelligence and Computer Science-related disciplines. It is also directed towards readers who come from other disciplines and are interested in becoming versed in some of the most recent advances in the field of Artificial Intelligence. We hope that all of them will find it useful and inspiring in their works and researches.

On the other hand, societal demand continues to pose challenging problems, which require ever more advanced theories and ever more efficient tools, methodologies, systems and Artificial Intelligence-based technologies to be devised to address them. Thus, the readers may expect that additional related volumes will appear in the future.

References

1. E. Rich, K. Knight, S.B. Nair, *Artificial Intelligence*, 3rd edn. (McGraw Hill, 2010)
2. https://corporatefinanceinstitute.com/resources/knowledge/other/disruptive-technology/
3. G.A. Tsihrintzis, D.N. Sotiropoulos, L.C. Jain (Eds.), *Machine Learning Paradigms – Advances in Data Analytics*, vol. 149. Intelligent Systems Reference Library Book Series (Springer, Berlin, 2018)
4. G.A. Tsihrintzis, M. Virvou, E. Sakkopoulos, L.C. Jain (Eds.), *Machine Learning Paradigms - Applications of Learning and Analytics in Intelligent Systems*, vol. 1. Learning and Analytics in Intelligent Systems Book Series (Springer, Berlin, 2019)
5. M. Virvou, E. Alepis, G.A. Tsihrintzis, L.C. Jain (Eds.), *Machine Learning Paradigms – Advances in Learning Analytics*, vol. 158. Intelligent Systems Reference Library Book Series (Springer, Berlin, 2020)
6. H. Jiang, C. Cai, X. Ma, Y. Yang, J. Liu, Smart home based on WiFi sensing: a survey. IEEE Access **6**, 13317–13325 (2018)
7. R. Du, P. Santi, M. Xiao, A.V. Vasilakos, C. Fischione, The sensable city: a survey on the deployment and management for smart city monitoring. IEEE Commun. Surv. Tutor. **21**(2),1533–1560 (2019)
8. B.P.L. Lau, S.H. Marakkalage, Y. Zhou, N.U. Hassan, C. Yuen, M. Zhang, U.X. Tan, A survey of data fusion in smart city applications. Inf. Fusion **52**, 357–374 (2019)
9. Z. Mahmood (Ed.), *Smart Cities – Development and Governance Frameworks*. Computer Communications and Networks (Springer, Berlin, 2018)
10. X. Liu, J. Cao, Y. Yang, S. Jiang, CPS-based smart warehouse for industry 4.0: a survey of the underlying technologies. Computers **7**(1), 13 (2018)
11. M.M. Dhanvijay, S.C. Patil, Internet of things: a survey of enabling technologies in healthcare and its applications. Comput. Netw. **153**, 113–131 (2019)
12. Q. Cai, H. Wang, Z. Li, X. Liu, A survey on multimodal data-driven smart healthcare systems: approaches and applications. IEEE Access **7**, 133583–133599 (2019)
13. K. Schwabd, The fourth industrial revolution - what it means and how to respond. Foreign Affairs (2015), https://www.foreignaffairs.com/articles/2015-12-12/fourth-industrial-revolution, Accessed 12 Dec 2015
14. From Industry 4.0 to Society 5.0: the big societal transformation plan of Japan, https://www.i-scoop.eu/industry-4-0/society-5-0/
15. Society 5.0, https://www8.cao.go.jp/cstp/english/society5_0/index.html
16. https://www.bruegel.org/2020/08/europe-has-an-artificial-intelligence-skills-shortage/
17. https://www.zdnet.com/article/artificial-intelligence-skills-shortages-re-emerge/
18. https://www.linkedin.com/pulse/after-us-covid-19-outbreak-ai-job-growth-slows-zhichun-jenny-ying/
19. https://www.forbes.com/sites/forbestechcouncil/2021/06/10/16-tech-roles-that-are-experiencing-a-shortage-of-talent/?sh=639f6ca93973
20. https://www2.deloitte.com/us/en/pages/consulting/articles/state-of-women-in-ai-today.html
21. https://hyperight.com/expert-pov-how-can-we-solve-the-ai-skills-shortage/
22. https://www.womeninai.co/
23. https://www.forbes.com/sites/robtoews/2020/12/13/8-leading-women-in-the-field-of-ai/?sh=4a6a5b3c5c97

Maria Virvou was born in Athens, Greece. She received a B.Sc. Degree in Mathematics from the National and Kapodistrian University of Athens, Greece, a M.Sc. Degree in Computer Science from the University College London (UCL), U.K. and a Ph.D. Degree in Computer Science and Artificial Intelligence from the University of Sussex, U.K. Her postgraduate and doctoral studies were funded by a scholarship obtained from the Greek State Scholarship Foundation.

She is currently a FULL PROFESSOR, HEAD OF THE DEPARTMENT, DIRECTOR OF POST- GRADUATE STUDIES and DIRECTOR OF THE SOFTWARE ENGINEERING LAB in the Department of Informatics, University of Piraeus, Greece. She is AUTHOR/CO-AUTHOR of over 350 research papers published in international journals, books and conference proceedings and of 7 books and monographs in Computer Science. She has been EDITOR of over 20 collections of papers in conference proceedings or books, published by major academic publishers, such as IEEE, Springer and IOS Press. She is currently CO-FOUNDER and CO-EDITOR-IN-CHIEF of the Springer book series "Learning and Analytics in Intelligent Systems". She has also been EDITOR-IN-CHIEF of the SpringerPlus Journal (Springer) for the whole area of Computer Science. Additionally, she has been an ASSOCIATE EDITOR of many other journals. She has been GENERAL CO-CHAIR of the yearly conference series of International Conference on Information, Intelligence, Systems and Applications (IISA 2013–2021), technically-sponsored by IEEE, which aims at promoting research in the area of interactive multimedia and major applications. She has been the GENERAL CHAIR / PROGRAM CHAIR of over twenty-five (25) International Conferences. She has been the PRINCIPAL INVESTIGATOR or CO-INVESTIGATOR of numerous national / international research projects. She has supervised 12 Ph.D. alumni and many of them currently hold academic positions in Universities. She has been teaching under-graduate and post-graduate courses in Educational Software, Software Engineering and Mobile Software, Human Computer Interaction, Programming Languages and Compilers, Software Personalization Technologies, User Modeling, Adaptive Tutoring Systems.

Prof.-Dr. Virvou has been a recipient of many best paper awards in international conferences. She has been an invited keynote speaker for many international conferences. She has received an honorary award by UNESCO in honour and recognition of her outstanding scholarly achievements and contributions to the field of Computer Science. She received the 1st Research Award from the Research Centre of the University of Piraeus for high quality international journal publications among Faculty Members of the University of Piraeus for the years 2004–2005 and 2005–2006 respectively. According to Microsoft Academic Search exploring entity analytics of 262,751,231 Authors she has been ranked as top 1st author in the Computer Science area of EDUCATIONAL SOFTWARE regarding citations and 2nd regarding publications. In addition, she has been ranked as top 1st author in publications in the Computer Science areas of MOBILE AUTHORING TOOLS, ADAPTIVE TUTORING, BI-MODAL AFFECTIVE COMPUTING, INTELLIGENT HELP, AUTOMATIC REASONING HELP, EDUCATIONAL GAME SOFTWARE, VIRTUAL REALITY EDUCATIONAL GAME, RUP LIFE CYCLE SOFTWARE. She has been ranked as 2nd top author regarding publications in the areas of SOFTWARE PERSONALIZATION STUDENT MODELING, INTELLIGENT HELP SYSTEMS FOR UNIX USERS, KNOWLEDGE ENGINEERING AFFECTIVE SYSTEMS, top 4th author in the area of UML INTELLIGENT COLLABORATIVE LEARNING and top 5th in the whole area of USER MODELING. Moreover she has been ranked among the top 40 researchers worldwide in publications for the Computer Science area of MULTIMEDIA (out of 979.432 publications, 8.626.011 citations, 1.600.000 authors), among the top 50 authors in the area of USER INTERFACE (out of 176.156 publications, 3.462.738 citations) and among the top 65 authors in HUMAN COMPUTER INTERACTION (out of 538.396 publications, 6.497.244 citations). She is among the top 2% of the most influential scientists worldwide in the area of ARTIFICIAL INTELLIGENCE, according to Ioannidis, J. P., Boyack, K. W., & Baas, J. (2020) "Updated science-wide author databases of standardized citation indicators", PLoS Biology, 18(10), e3000918.

Part I
Advances in Artificial Intelligence Paradigms

Chapter 2
Feature Selection: From the Past to the Future

Verónica Bolón-Canedo, Amparo Alonso-Betanzos, Laura Morán-Fernández, and Brais Cancela

Abstract Feature selection has been widely used for decades as a preprocessing step that allows for reducing the dimensionality of a problem while improving classification accuracy. The need for this kind of technique has increased dramatically in recent years with the advent of Big Data. This data explosion not only has the problem of a large number of samples, but also of big dimensionality. This chapter will analyze the paramount need for feature selection and briefly review the most popular feature selection methods and some typical applications. Moreover, as the new Big Data scenario offers new opportunities to machine learning researchers, we will discuss the new challenges that need to be faced: from the scalability of the methods to the role of feature selection in the presence of deep learning, as well as exploring its use in embedded devices. Beyond a shadow of doubt, the explosion in the number of features and computing technologies will point to a number of hot spots for feature selection researchers to launch new lines of research.

Keywords Feature selection · Dimensionality reduction · Big data · Big dimensionality

---

Part of the content of this chapter was previously published in *Knowledge-Based Systems* (https://doi.org/10.1016/j.knosys.2015.05.014, https://doi.org/10.1016/j.knosys.2020.105885, https://doi.org/10.1016/j.knosys.2019.105326), *Knowledge and Information Systems* (https://doi.org/10.1007/s10115-012-0487-8), and *Information Fusion* (https://doi.org/10.1016/j.inffus.2018.11.008).

V. Bolón-Canedo (✉) · A. Alonso-Betanzos · L. Morán-Fernández · B. Cancela
Department of Computer Science and Information Technologies, Centro de Investigación CITIC, Universidade da Coruña, Campus de Elviña s/n, 15071 A Coruña, Spain
e-mail: veronica.bolon@udc.es

A. Alonso-Betanzos
e-mail: amparo.alonso.betanzos@udc.es

L. Morán-Fernández
e-mail: laura.moranf@udc.es

B. Cancela
e-mail: brais.cancela@udc.es

M. Virvou et al. (eds.), *Advances in Selected Artificial Intelligence Areas*, Learning and Analytics in Intelligent Systems 24,
https://doi.org/10.1007/978-3-030-93052-3_2

2.1 Introduction

Driven by recent advances in algorithms, computing power, and big data, artificial intelligence (AI) has made substantial breakthroughs in recent years. In particular, machine learning has had great success because of its impressive ability to automatically analyze large amounts of data. However, the advent of big data has brought not only a large number of samples, but also ultra-high dimensionality. Feature selection (FS) is the process of selecting the relevant features and discarding the irrelevant or redundant ones. There are considerable noisy and useless features that are often collected or generated by different sensors and methods, which also occupy a lot of computational resources. Therefore, feature selection performs a crucial role in the framework of machine learning of removing nonsense features and preserving a small subset of features to reduce the computational complexity.

There are several applications in which it is necessary to find the relevant features: in bioinformatics (e.g. to identify a few key biomolecules that explain most of an observed phenotype [1]), in respect to the fairness of decision making (e.g. to find the input features used in the decision process, instead of focusing on the fairness of the decision outcomes [2]), or in nanotechnology (e.g. to determine the most relevant experimental conditions and psychicochemical features to be considered when making a nanotoxicology risk assessment [3]). A shared aspect of these applications is that they are not pure classification tasks. In fact, an understanding of which features are relevant is as important as accurate classification, as these features may provide us with new insights into the underlying system.

There are also scenarios in which feature selection has been successfully applied in order to improve subsequent classification, such as DNA microarray analysis, image classification, face recognition, text classification, etc. However, the advent of big data has raised unprecedented challenges for researchers. This chapter outlines hot spots in feature selection research, aimed at encouraging scientific community to seek and embrace new opportunities and challenges that have recently arisen.

The remainder of this chapter is organized as follows: Sect. 2.2 introduces the need for feature selection and Sect. 2.3 describes a brief history about feature selection. Then, Sect. 2.4 defines the most popular feature selection techniques and, finally, Sect. 2.5 outlines the hot spots in feature selection and open research lines.

2.2 The Need for Feature Selection

In recent years, most enterprises and organizations have stored large amounts of data in a systematic way, but without a clear idea of its potential usefulness. In addition, the growing popularity of the Internet has generated data in many different formats (text, multimedia, etc.), and from many different sources (systems, sensors, mobile devices, etc.). To be able to extract useful information from all these data, we require new analysis and processing tools. Most of these data have been generated in the last few years—as we continue to generate quintillions of bytes daily [4]. Big data—large volumes and ultrahigh dimensionality—is now a recurring feature of various machine learning application fields, such as text mining and information retrieval [5]. Weinberger et al. [6], for instance, conducted a study of a collaborative email-spam filtering task with 16 trillion unique features, whereas the study by Tan et al. [5] was based on a wide range of synthetic and real-world datasets of tens of million data points with $\mathcal{O}(10^{14})$ features. The growing size of datasets raises an interesting challenge for the research community; to cite Donoho et al. [7] "our task is to find a needle in a haystack, teasing the relevant information out of a vast pile of glut".

Ultrahigh dimensionality implies massive memory requirements and a high computational cost for training. Generalization capacities are also undermined by what is known as the "curse of dimensionality". According to Donoho et al. [7], Bellman coined this colorful term in 1957 to describe the difficulty of optimization by exhaustive enumeration on product spaces [8]. This term refers to various phenomena that arise when analyzing and organizing data in high-dimensional spaces (with hundreds or thousands of dimensions) that do not occur in low-dimensional settings. A dataset is usually represented by a matrix where the rows are the recorded instances (or samples) and the columns are the attributes (or features) that represent the problem at hand. In order to tackle the dimensionality problem, the dataset can be summarized by finding "narrower" matrices that in some sense are close to the original. Since these narrower matrices have a smaller number of samples and/or features, they can be used much more efficiently than the original matrix. The process of finding these narrow matrices is called *dimensionality reduction*.

Feature extraction is a dimensionality reduction technique that addresses the problem of finding the most compact and informative set of features for a given problem so as to improve data storage and processing efficiency. Feature extraction is decomposed into the steps of construction and selection. Feature construction methods complement human expertise in converting "raw" data into a set of useful features using preprocessing transformation procedures such as standardization, normalization, discretization, signal enhancement, local feature extraction, etc. Some construction methods do not alter space dimensionality, while others enlarge it, reduce it or both. It is crucial not to lose information during the feature construction stage; Guyon and Elisseeff [9] recommend that it is always better to err on the side of being too inclusive rather than run the risk of discarding useful information. Adding features may seem reasonable but it comes at a price: the increase in the dimensionality of patterns incurs the risk of losing relevant information in a sea of possibly irrelevant, noisy or redundant features. The goal of feature selection methods is to reduce the number of initial features so as to select a subset that retains enough information to obtain satisfactory results.

In a society that needs to deal with vast quantities of data and features in all kinds of disciplines, there is an urgent need for solutions to the indispensable issue of feature selection. To understand the challenges that researchers face, the next section will briefly describe the origins of feature selection.

2.3 History of Feature Selection

Feature selection is defined as the process of detecting the relevant features and discarding the irrelevant and redundant ones with the goal of obtaining a subset of features that accurately describe a given problem with a minimum degradation of performance [10]. Theoretically, having a large number of input features might seem desirable, but the *curse of dimensionality* is not only an intrinsic problem of high-dimensionality data, but more a joint problem of the data and the algorithm being applied. For this reason, researchers began to select features in a pre-processing phase in an attempt to convert their data into a lower-dimensional form.

The first research into feature selection dates back to the 1960s [11]. Hughes [12] used a general parametric model to study the accuracy of a Bayesian classifier as a function of the number of features, concluding as follows: "Measurement selection, reduction and combination are not proposed as developed techniques. Rather, they are illustrative of a framework for further investigation". Since then, research into feature selection has posed many challenges, with some researchers highly skeptical of progress; in "Discussion of Dr. Miller's paper" [13], for example, RL Plackett stated: "If variable elimination has not been sorted out after two decades of work assisted by high-speed computing, then perhaps the time has come to move on to other problems". In the 1990s, notable advances were made in feature selection used to solve machine learning problems [14–16]. Nowadays, feature selection is acknowledged to play a crucial role in reducing the dimensionality of real problems, as evidenced by the growing number of publications on the issue [9, 10, 17, 18].

Given its ability to enhance the performance of learning algorithms, feature selection has attracted increasing interest in the field of machine learning, in processes such as clustering [19, 20], regression [21, 22] and classification [15, 23], whether supervised or unsupervised.

2.4 Feature Selection Techniques

There are a vast number of feature selection techniques, and several ways in which they could be classified. One quite commonly used is to distinguish three major different groups of techniques depending on the existing relationship between the feature selection algorithm and the inductive learning method used to infer a model, namely [9]:

- *Wrappers*, which involve optimizing a predictor as a part of the selection process. These methods iteratively select or delete a feature or set of features using the accuracy of the classification model. Because of this, usually these methods tend to be computationally intensive.
- *Filters*, which rely on the general characteristics of training data and carry out the feature selection process as a pre-processing step with independence of the induction algorithm. Thus, instead of the error rate used by wrappers, filters employ

measures (such as mutual information or Pearson product-moment correlation coefficient), that are chosen to capture the relevance of the feature set while being fast to compute. Thus, filters are usually less computationally intensive than wrappers.

- *Embedded methods*, which perform feature selection in the process of training and are usually specific to given learning machines.

In Table 2.1 a summary of the most relevant pros and cons characteristics of the three groups of feature selection methods mentioned above is shown, as well as some examples of each.

Within filters, one can also distinguish between *univariate* and *multivariate* methods. Univariate methods (such as is the case of InfoGain) are fast and scalable, but ignore feature dependencies. On the other hand, multivariate filters (such as CFS, INTERACT, etc.) model feature dependencies, but at the cost of being slower and less scalable than univariate techniques.

Besides this classification, feature selection methods can also be divided according to two approaches: *individual evaluation* and *subset evaluation* [24]. Individual evaluation is also known as feature ranking and works by assigning weights to each individual feature according to its degree of relevance. Moreover, subset evaluation delivers candidate feature subsets based on a certain search strategy. Evaluation of the candidate subsets requires a scoring metric that grades each subset of features and compares their result with the previous best one with respect to this measure. While individual evaluation is not able to remove redundant features as they are likely to have similar rankings, the subset evaluation approach can handle feature redundancy with feature relevance. However, methods in this framework can suffer from an inevitable problem caused by searching through feature subsets required in the subset generation step, as exhaustive search is generally not feasible, and some stopping criterion needs to be used.

Some of the most well-known methods in each of the models shown in Table 2.1 are briefly described below.

2.4.1 Filter Methods

- **Correlation-based Feature Selection** (CFS) is a multivariate algorithm that ranks feature subsets according to a correlation based heuristic evaluation function [25]. The bias of the evaluation function is toward subsets that contain features that are highly correlated with the class label and uncorrelated with each other (independent). Irrelevant features should be ignored because they will have low correlation with the class. On the other hand, redundant features should be screened out because they will be highly correlated with one or more of the remaining features. The acceptance of a feature will depend on the extent to which it predicts classes in areas of the instance space not already predicted by other features.

Table 2.1 Feature selection techniques

Method	Advantages	Disadvantages	Examples
Filter 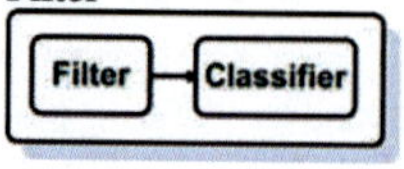	Classifier's independent	No interaction with the classifier	Consistency-based CFS
	Lower computational cost than wrappers	May select redundant variables	INTERACT
	Fast execution		$\mathcal{M}_d$
	Good generalization ability		Information Gain mRMR
	Robust to overfitting		ReliefF
Embedded 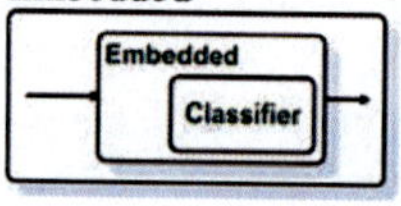	Interaction with the classifier	Classifier-dependent selection	FS-Percepton SVM-RFE
	Lower computational cost than wrappers		Lasso
	Captures feature dependencies		Elastic Net
Wrapper 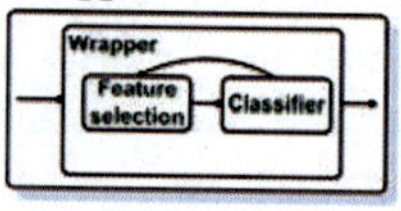	Interaction with the classifier	Computationally intense	Wrapper-C4.5 Wrapper-SVM
	Captures feature dependencies	Risk of overfitting	
	High accuracy	Classifier-dependent selection	

- The **Consistency-based Filter** [26] is a multivariate method which selects subsets of the features evaluating its interest by the level of consistency in the class values when the training instances are projected onto the subset of attributes.
- **Information Gain** [27] is one of the most common attribute evaluation methods. This univariate filter provides an ordered ranking of all the features and then a threshold is required.
- The **Fast correlated-based filter** (FCBF) [28] is based on a measure named symmetrical uncertainty (SU), defined as the ratio between the information gain (IG) and the entropy (H) of two features. This filter method was designed for high-dimensionality data and is effective in removing both irrelevant and redundant features, although it fails in detecting the interaction between features.

- The **INTERACT** algorithm [29] is a subset filter based on the same measure of symmetrical uncertainty (SU) as FCBF, but includes also the consistency contribution, which is an indicator about how significantly the elimination of a feature will affect consistency. The algorithm consists of two major parts. In the first part, the features are ranked in descending order based on their SU values. In the second part, features are evaluated one by one starting from the end of the ranked feature list. If the consistency contribution of a feature is less than an established threshold, the feature is removed, otherwise it is selected. The authors stated that this method can handle feature interaction, and efficiently selects relevant features.
- **ReliefF** [30] is an extension of the original Relief algorithm [31]. The original Relief works by randomly sampling an instance from the data and then locating its nearest neighbors from the same and opposite class. The values of the attributes of the nearest neighbors are compared to the sampled instance and used to update relevance scores for each attribute. The hypothesis is that an attribute that is useful should differentiate between instances from different classes and have the same value for instances from the same class. ReliefF adds the ability of dealing with multiclass problems and is also more robust and capable of dealing with incomplete and noisy data. Furthermore it has low bias, includes interaction among features and may capture local dependencies which other methods miss.
- The **mRMR** (minimum Redundancy Maximum Relevance) method [32] is a well-known method that selects features that have the highest relevance with the target class and are also minimally redundant, i.e., selects features that are maximally dissimilar to each other. Both optimization criteria (Maximum-Relevance and Minimum-Redundancy) are based on mutual information. The method owes its popularity to its accuracy, although it is computationally expensive, and then some optimizations have been presented [33] so as to permit its use in the present big data/high dimensional scenarios.
- The $\boldsymbol{M_d}$ filter [34] is an extension of mRMR which uses a measure of monotone dependence (instead of mutual information) to assess relevance and irrelevance. One of its contributions is the inclusion of a free parameter (λ) that controls the relative emphasis given on relevance and redundancy. When λ is equal to zero, the effect of the redundancy disappears and the measure is based only on maximizing the relevance. On the other hand, when λ is equal to one, it is more important to minimize the redundancy among variables. The authors of the method state that $\lambda = 1$ performs better than other λ values.

2.4.2 Embedded Methods

- **SVM-RFE** (Recursive Feature Elimination for Support Vector Machines) was introduced by Guyon in [35]. This embedded method performs feature selection by iteratively training a SVM classifier with the current set of features and removing the least important feature indicated by the SVM. Thus, SVM-RFE is an iteration process of the backward removal of features.

- **FS-P** (Feature Selection - Perceptron) [36] is an embedded method based on a perceptron (a type of artificial neural network that can be seen as the simplest kind of feedforward neural network: a linear classifier) trained in the context of supervised learning. The interconnection weights are used as indicators of which features could be the most relevant and provide a ranking for them.
- **Lasso** (Least Absolute Shrinkage and Selection Operator) [37] is a powerful method that performs simultaneously two main tasks: regularization and feature selection. It does so by putting a constraint on the sum of the absolute values of the model parameters, that has to be less than a fixed value (upper bound). In order to do so the method applies a shrinking (regularization) procedure where it penalizes the coefficients of the regression variables shrinking some of them to zero. Those variables that still have a non-zero coefficient after the shrinking process are selected to be part of the model, as they are deemed to be relevant.
- **Elastic Net** [38] is a regularized regression method that linearly combines the L1 and L2 penalties of the Lasso and Ridge regression [39] methods, respectively, with the aim of overcoming some of the limitations of the Lasso model for feature selection, which are of two types: first, Lasso selects at most n variables before it saturates. Second, when there is a group of highly correlated variables, LASSO tends to select one variable from a group and ignore the others. The Elastic Net adds a quadratic part (used in Ridge regression, or Tikhonov regularization) to the penalty employed in Lasso. This quadratic penalty term makes the loss function strongly convex, and therefore it has a unique minimum. The method works in two steps: first it finds the ridge regression coefficients, and then it does a LASSO type shrinkage.
- **Concrete autoencoders** [40] are a variant to the Lasso approach. Instead of applying a regularization term to a linear classifier, they propose to use a more complex architecture, relying the FS to a matrix mask (number of features vs number of selected features) that is attached to the input data. Contrary to Lasso, this approach requires to establish the number of features in advance. Additionally, they can work on both supervised and unsupervised problems. More recently, the End-to-End Feature Selection (E2E-FS) [41] algorithm is able to provide a more accurate solution by only using a vector mask the size of the number of total features.

2.4.3 Wrapper Methods

Wrappers evaluate attribute sets by using a learning algorithm as a black box, as part of the evaluation is the estimation of the accuracy of the classifier that guides the search for good feature subset candidates [16, 42]. Classifiers that are frequently used are those shown in Table 2.1: SVM, C4.5, Naive-Bayes, etc. An exhaustive search is most of the times unfeasible, as the size of the search space is quadratic on the number of features, and thus heuristic search is performed (mainly hill climbing or best-first search).

Cross validation is used to estimate the accuracy of the learning scheme for a set of attributes, and this cross-validation might be repeated several times, depending on the size of the datasets to be evaluated. The algorithm may start with the empty set of attributes and search forward, adding attributes until performance does not improve further (forward selection), or on the contrary backward elimination might be carried out, starting with the full set of features. Forward selection approaches are usually faster, as building classifiers is faster when few features are present.

2.5 What Next in Feature Selection?

Ongoing advances in computer-based technologies have enabled researchers and engineers to collect data at an increasingly fast pace. To address the challenge of analyzing these data, feature selection becomes an imperative preprocessing step that needs to be adapted and improved to be able to handle high-dimensional data. We have highlighted the need for feature selection but, in the new big data scenario, an important number of challenges are emerging, representing current hot spots in feature selection research.

2.5.1 Scalability

In the new era of big data, machine learning methods need to be able to deal with the unprecedented scale of data. Analogous to big data, the term "big dimensionality" has been coined to refer to the unprecedented number of features arriving at levels that are rendering existing machine learning methods inadequate [4].

The widely-used UCI Machine Learning Repository [43] indicates that, in the 1980s, the maximum dimensionality of data was only about 100. By the 1990s, this number had increased to more than 1500 and, nowadays, to more than 3 million. In the popular LIBSVM Database [44] the maximum dimensionality of the data was about 62000 in the 1990s, increasing to some 16 million in the 2000s and to more than 29 million in the 2010s.

The problem is that most existing learning algorithms were developed when dataset sizes were much smaller, but nowadays different solutions are required for the case of small-scale versus large-scale learning problems. Small-scale learning problems are subject to the usual approximation-estimation trade-off, but this trade-off is more complex in the case of large-scale learning problems, not only because of accuracy but also due to the computational complexity of the learning algorithm. Moreover, since most algorithms were designed under the assumption that the dataset would be represented as a single memory-resident table, these algorithms are useless when the entire dataset does not fit in the main memory. Dataset size is therefore one reason for scaling up machine learning algorithms. However, there are other settings where a researcher could find the scale of a machine learning task daunting [45], for instance:

- Model and algorithm complexity: A number of high-accuracy learning algorithms either rely on complex, non-linear models, or employ computationally expensive subroutines.
- Inference time constraints: Applications that involve sensing, such as robot navigation or speech recognition, require predictions to be made in real time.
- Prediction cascades: Applications that require sequential, interdependent predictions have a highly complex joint output space.
- Model selection and parameter sweeps: Tuning learning algorithm hyper-parameters and evaluating statistical significance require multiple learning executions.

Scalability is defined as the impact of an increase in the size of the training set on the computational performance of an algorithm in terms of accuracy, training time and allocated memory. Thus the challenge is to find a trade-off among these criteria—in other words, to obtain "good enough" solutions as "fast" as possible and as "efficiently" as possible. As explained before, this issue becomes critical in situations in which there are temporal or spatial constraints as happens with real-time applications dealing with large datasets, unapproachable computational problems requiring learning and initial prototyping requiring rapidly implemented solutions.

Similarly to instance selection, which aims at discarding superfluous, i.e., redundant or irrelevant, samples [46], feature selection can scale machine learning algorithms by reducing input dimensionality and therefore algorithm run-time. However, when dealing with a dataset containing a huge number of both features and samples, the scalability of the feature selection method also assumes crucial importance. Since most existing feature selection techniques were designed to process small-scale data, their efficiency is likely to be downgraded, if not reduced totally, with high-dimensional data. Figure 2.1 shows run-time responses to modifications to the number of features and samples for four well-known feature selection ranker methods applied to the SD1 dataset, a synthetic dataset that simulates DNA microarray data [47].

In this scenario, feature selection researchers need to focus not only on the accuracy of the selection but also on other aspects, such as stability and scalability. Broadly speaking, although most classical univariate feature selection approaches (with each feature considered separately) have an important advantage in terms of scalability, they ignore feature dependencies and thus potentially perform less well than other feature selection techniques. Multivariate techniques, in contrast, may improve performance, but at the cost of reduced scalability [48]. The scalability of a feature selection method is thus crucial and deserves more attention from the scientific community. One of the solutions commonly adopted to deal with the scalability issue is to distribute the data into several processors, discussed in the following section.

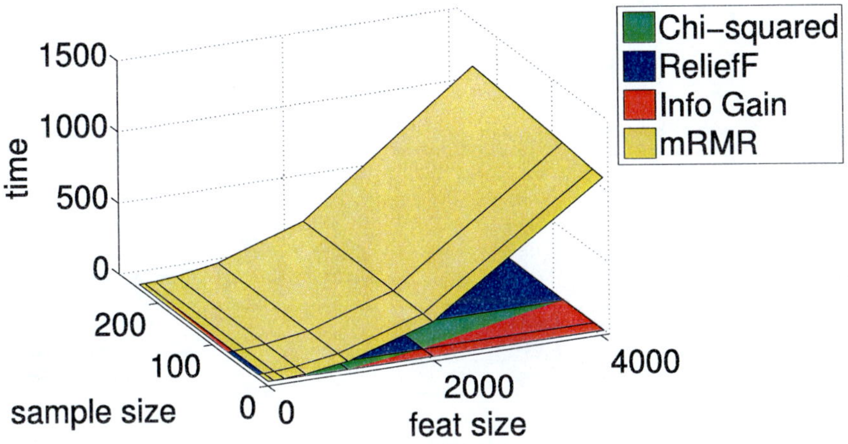

Fig. 2.1 Run-time scalability in response to modifications in the number of features and samples for four feature selection methods applied to the SD1 dataset

2.5.2 Distributed Feature Selection

Traditionally, feature selection is applied in a centralized manner, i.e., a single learning model is used to solve a given problem. However, since nowadays data may be distributed, feature selection can take advantage of processing multiple subsets in sequence or concurrently. There are several ways to distribute a feature selection task [49]:

(i) **The data is together in one very large dataset**. The data can be distributed on several processors, an identical feature selection algorithm can be run on each and the results combined.

(ii) **The data may be in different datasets in different locations** (e.g., in different parts of a company or even in different cooperating organizations). As for the previous case, an identical feature selection algorithm can be run on each and the results combined.

(iii) **Large volumes of data may be arriving in a continuous infinite stream in real time**. If the data is all streaming into a single processor, different parts can be processed by different processors acting in parallel. If the data is streaming into different processors, they can be handled as above.

(iv) **The dataset is not particularly large but different feature selection methods need to be applied to learn unseen instances and combine results** (by some kind of voting system). The whole dataset may be in a single processor, accessed by identical or different feature selection methods that access all or part of the data.

This last approach, known as *ensemble learning*, has recently been receiving a great deal of attention [50], and will be described in more detail in Sect. 2.5.3.

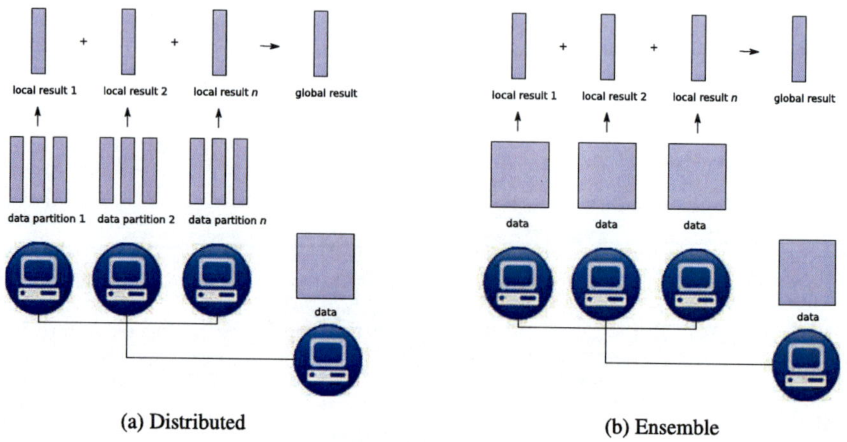

Fig. 2.2 Partitioned feature selection scenarios

Figure 2.2 shows different partitioned feature selection scenarios. Figure 2.2a represents the situation described in (i): the original data is distributed between several processors and local results are combined in a final result. Figure 2.2b represents the situation described in (iv): the data is replicated on different processors, local results are obtained as a consequence of applying different feature selection methods and, again, local results are combined into a global result.

As mentioned, most existing feature selection methods are not expected to scale efficiently when dealing with millions of features; indeed, they may even become inapplicable. A possible solution might be to distribute the data, run feature selection on each partition and then combine the results. The two main approaches to partitioned data distribution are by features (vertically) or by samples (horizontally). Distributed learning has been used to scale up datasets that are too large for batch learning by samples [51–53]. While distributed learning is not common, there have been some developments regarding data distribution by features [54, 55]. One proposal is a distributed method where data partitioning is both vertical and horizontal [56]. Another is a distributed parallel feature selection method that can read data in distributed form and perform parallel feature selection in symmetric multiprocessing mode via multithreading and massively parallel processing [57]. However, when dealing with big dimensionality datasets, researchers, of necessity, have to partition by features. In the case of DNA microarray data, the small sample size combined with big dimensionality prevent the use of horizontal partitioning. However, the previous mentioned vertical partitioning methods do not take into account some of the particularities of these datasets, such as the high redundancy among features, as is done in the methods described by Sharma et al. [58] and Bolón-Canedo et al. [59], the latter at a much lower computational cost.

Several paradigms for performing distributed learning have emerged in the last decade. MapReduce [60] is one such popular programming model with an associated implementation for processing and generating large data sets with a parallel, distributed algorithm on a cluster. Hadoop, developed by Cutting and Cafarella in 2005 [61], is a set of algorithms for distributed storage and distributed processing of very large datasets on computer clusters; it is built from commodity hardware and has a processing part based on MapReduce. Developed more recently, is Apache Spark [62], a fast, general engine for large-scale data processing, popular among machine learning researchers due to its suitability for iterative procedures. Developed within the Apache Spark paradigm was MLib [63], created as a scalable machine learning library containing algorithms. Although it already includes a number of learning algorithms such as SVM and naive Bayes classification, k-means clustering, etc., as yet, it includes no feature selection algorithms. This poses a challenge for machine learning researchers, as well as offering an opportunity to initiate a new line of research.

Another open line of research is the use of graphics processing units (GPUs) to distribute and thus accelerate calculations made in feature selection algorithms. With many applications to physics simulations, signal processing, financial modelling, neural networks, and countless other fields, parallel algorithms running on GPUs often achieve up to 100x speedup over similar CPU algorithms. The challenge now is to take advantage of GPU capabilities to adapt existing state-of-the-art feature selection methods to be able to cope effectively and accurately with millions of features.

2.5.3 *Ensembles for Feature Selection*

Classification has been the most common machine learning scenario for the use of ensemble paradigms [64]. Nevertheless, the idea of ensemble learning is not only applicable to classification, but it can also be used to improve other machine learning disciplines, such as feature selection. Nowadays, in areas such as genomics and many others of bioinformatics, image classification, face recognition, or text mining,among many others, it is common to have to deal with datasets containing a very large number of features. This is undoubtedly an interesting challenge, as said before, because most classical machine learning methods cannot deal efficiently with high dimensionality. There exists a vast body of feature selection methods in the literature, as seen above, including filters based on distinct metrics, embedded and wrapper methods using different induction algorithms [4]. The proliferation of feature selection algorithms, however, has not brought about a general methodology allowing for the intelligent selection from existing algorithms. Thus, in order to make a correct choice, not only users need to know the domain well, but also they are expected to understand technical details of available algorithms.

A solution for the problem can be the ensemble paradigm applied to feature selection. In this way, combining the output of several feature selectors, the performance can usually be improved and the user is released from having to chose a single method. The increase in performance does not come only from having several models, as it is also the case with classification ensembles, but also from the diversity of the feature

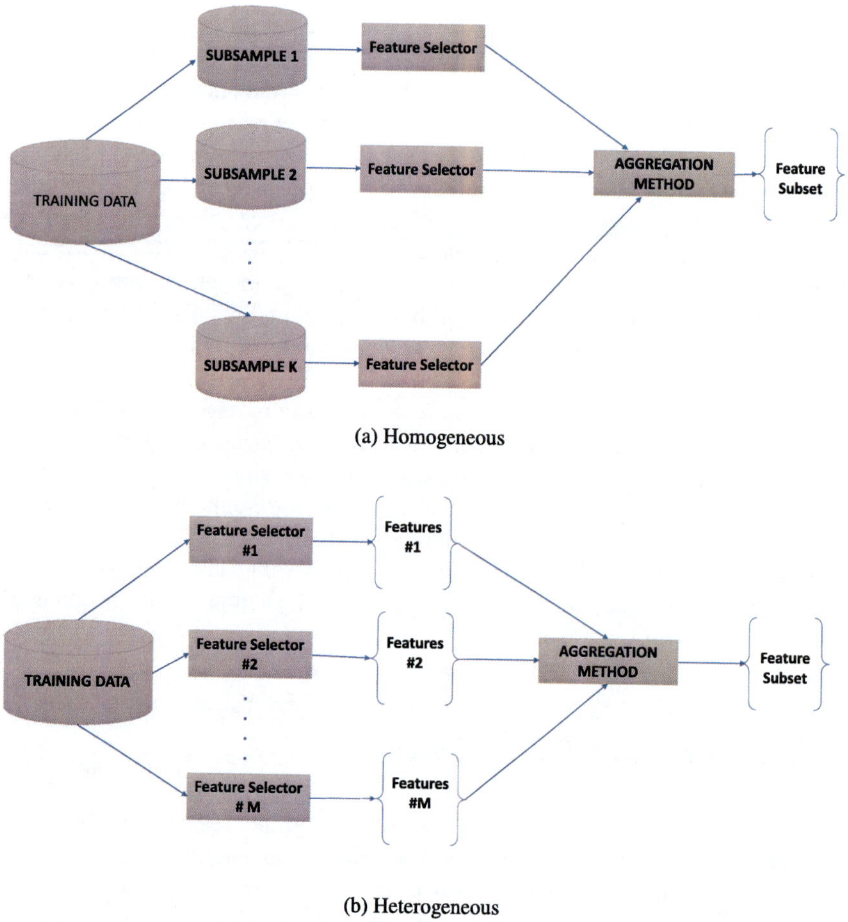

Fig. 2.3 Homogeneous and Heterogeneous feature selection ensembles

subsets obtained. Ensembles for feature selection can be broadly classified, as can be seen in Fig. 2.3, into homogeneous (if they use the same base feature selector) and heterogeneous (if different feature selectors are employed), respectively (as seen before in previous subsection).

In general, when designing a feature selection ensemble, there are several aspects that need to be taken into account:

- The individual feature selection methods to be used. Obviously, employing more than one FS method has a computational cost. For this reason, ensembles containing filters and embedded methods are more popular than those with wrappers. Each individual methods has its advantages and limitations, and thus they should be selected carefully for the ensemble to guarantee diversity while increasing the regularity of the feature selection process. Available metrics for stability [65, 66] and diversity [67] can help us choose the appropriate individual methods.

- Not only the type, but also the number of different methods to use is a variable to be estimated. Although studies on the matter have been performed for the determination of the size of classification ensembles, this issue is an open research line for feature selection ensembles, and most of the times only statistical tests are used for determining their size, always maintaining the balance between complexity, diversity and stability of the process.
- The number and size of the different training sets to use. The situation is analogous as in the variable above, and although some studies have been performed for classification ensembles, there are no reported studies on the size of the optimal training sets for feature selection ensembles, although some authors have studied the consequences of distributing the training set regarding the number of features and using ranker methods [68].
- The aggregation (also named combination) method to use. Different methods are available for combining label predictions, subsets of features or rankers [69]. Their behavior has been studied scarcely for ensembles for feature selection [70].
- The threshold method to use if the individual feature selection methods of the ensemble are rankers, that is, if the methods return an ordered list of all features involved in the problem. For most studies, the thresholds chosen are based on a fixed percentage of retained features, for example 25%, 50%, or a fixed number of top features [71, 72]. Other authors have tried to derive a threshold based on different metrics, as in [70, 73].

Ensembles for feature selection are relatively recent however, their appearance being caused for the need of more accurate, robust and stable feature selection. This is a preprocessing step that, if before the Big data era was already relevant, nowadays has converted in essential for Machine Learning pipelines.

2.5.4 *Visualization and Interpretability*

In recent years, several dimensionality reduction techniques for data visualization and preprocessing have been developed. However, although the aim may be better visualization, most techniques have the limitation that the features being visualized are transformations of the original features [74–76]. Thus, when model interpretability is important, feature selection is the preferred technique for dimensionality reduction.

A model is only as good as its features, for which reason features have played and will continue to play a preponderant role in model interpretability. Users have a twofold need for interpretability and transparency in feature selection and model creation processes: (i) they need more interactive model visualizations where they can change input parameters to better interact with the model and visualize future scenarios and (ii) they need more interactive feature selection process where, using interactive visualizations, they are empowered to iterate through different feature subsets rather than be tied to a specific subset chosen by an algorithm.

Some recent works describe using feature selection to improve the interpretability of models obtained in different fields. One example is a method for the automatic and iterative refinement of a recommender system, in which the feature selection step selects the best characteristics of the initial model in order to automatically refine it [77]. Another is the use of feature selection to improve decision trees—representing agents simulating personnel in an organization so as to model sustainability behaviours—through an expert review of their theoretical consistency [78]. Yet another is a generative topographic mapping-based data visualization approach that estimates feature saliency simultaneously as the visualization model is trained [79]. Krause et al. [80] describe a tool in which visualization helps users develop a predictive model of their problem by allowing them to rank features (according to predefined scores), combine features and detect similarities between dimensions.

However, data is everywhere, continuously increasing, and heterogeneous. We are witnessing a form of Diogenes syndrome referring to data: organizations are collecting and storing tonnes of data, but most do not have the tools or the resources to access and generate strategic reports and insights from their data. Organizations need to gather data in a meaningful way, so as to evolve from a data-rich/knowledge-poor scenario to a data-rich/knowledge-rich scenario. The challenge is to enable user-friendly visualization of results so as to enhance interpretability. The complexity implied by big data applications also underscores the need to limit the growth in visualization complexity. Thus, even though feature selection and visualization have been dealt with in relative isolation from each other in most research to date, the visualization of data features may have an important role to play in real-world high dimensionality scenarios. However, it is also important to bear in mind that, although visualization tools are increasingly used to interpret and make complex data understandable, the quality of associated decision making is often impaired due to the fact that the tools fail to address the role played by heuristics, biases, etc. in human-computer interactive settings. Therefore, interactive tools similar to that described by Krause et al. [80] are an interesting line of research.

2.5.5 *Instance-Based Feature Selection*

While classic FS algorithms select the most important features that will help the classifier make a good prediction, Instance-based feature selection (Ib-FS) aims to infer which features are most relevant for each specific input case. The advantage of this approach is that it customizes feature relevance information for each sample.

Let us use an example to demonstrate how Ib-FS can provide insightful instance-level information. Let $\mathbf{X} = \{x_a, x_b, x_c, x_d, x_e\}$ be a random uniform datum ($\mathbf{X} \in [-1, 1]$) and let

Table 2.2 Instance-level results. For each sample, the saliency function successfully provides information about which features are most relevant

$X = \{x_a, x_b, x_c, x_d, x_e\}$	Y	Significant variables	Ib-FS expected output
$\{-0.5, -0.5, -0.5, -0.5, 0.5\}$	1	$\{x_a, x_b, x_c\}$	$\{\mathbf{9.0}, \mathbf{5.9}, \mathbf{8.7}, 0.7, 0.8\}$
$\{-0.5, 0.5, -0.5, 0.5, 0.5\}$	−1	$\{x_a, x_b, x_c\}$	$\{\mathbf{7.4}, \mathbf{6.8}, \mathbf{6.3}, 2.5, 0.1\}$
$\{0.5, 0.5, -0.5, -0.5, -0.5\}$	1	$\{x_a, x_d, x_e\}$	$\{\mathbf{6.3}, 0.9, 0.1, \mathbf{5.1}, \mathbf{5.7}\}$
$\{0.5, 0.5, 0.5, -0.5, 0.5\}$	−1	$\{x_a, x_d, x_e\}$	$\{\mathbf{9.9}, 2.0, 2.4, \mathbf{3.7}, \mathbf{5.7}\}$

$$\mathbf{Y} = \begin{cases} \text{sign}(x_b) \times \text{sign}(x_c) & \text{if } x_a < 0 \\ \text{sign}(x_d) \times \text{sign}(x_e) & \text{otherwise} \end{cases} \tag{2.1}$$

be the expected output. The output will always depend on three variables: one fixed variable (x_a) and two variables that depend on the fixed variable value. While all five variables are, by definition, needed to successfully train the classifier, each instance only takes into account three of them.

Table 2.2 shows some examples about how an ideal Ib-FS algorithm should work. The method should only select, for each instance, the 3 features that are needed to provide a correct output.

The first attempts to solve this problem were based in Saliency [81], which aims to establish, for each input, the degree of importance each feature has in the final output. Later, the Attention Models [82] provide a similar but more accurate solution by using a neural network that outputs the significance probability of each feature. Unfortunately, these techniques, although being widely used in the literature, have limitations in terms of explainability, as no dimensional reduction can be performed.

However, some promising works were already made to overcome this issue. The Saliency-based Feature Selection (SFS) method [83] aims to use the Saliency technique to select the most relevant features. In a different way, the Learning to Explain (L2X) algorithm [84] uses an autoencoder-like architecture that outputs a subset of selected features. INVASE [85] merges three different networks (predictor, baseline and selector) to perform the selection. The challenge now is to merge these selection techniques into complex architectures, aiming to not sacrifice accuracy for the sake of explainability.

2.5.6 *Reduced-Precision Feature Selection*

With the advent and standardization of wireless connectivity paradigms and the cost reduction of electronic components, the number and diversity of Internet of Things devices has exploded over the last decade [86]. Wearable computing has made successful and significant forays in fitness domains, health care, fashion and entertainment, among other application areas. These devices are usually employed as local systems, and their fundamental requirements are to work with little computing power and small memories. However, these requirements become challenging since emerging computing devices are not just sensor devices: they must perform sophisticated computation, collect and aggregate data for propagation to the cloud, and respond in real time to user requests. This data must be fed on a machine learning system to analyze information and make decisions. Unfortunately, limitations in the computational capabilities of resource-scarce devices inhibit the implementation of the most current machine learning algorithms on them. Then, the data must be sent to a remote computational infrastructure. However, an interest in a different paradigm based on Edge Computing has emerged. Edge computing aims at changing this passive situation improving efficiency by allowing the nodes of the network or the very own devices to analyze the generated data.

The process of feature selection is typically performed on a machine using high numerical representation, i.e. double-precision floating point calculations (64 bits). Using a more powerful general purpose processor provides significant benefits in terms of speed and capability to solve more complex problems. But this capability does not come without cost; a conventional microprocessor can require a substantial amount of off-chip support hardware, memory, and often a complex operating system [87]. In contrast to up-to-date computers, these requirements are often not met by embedded systems, low energy computers or integrated solutions that need to optimize the used hardware resources. The majority of the existing approaches available investigated the effect of reduced precision in neural networks and deep learning [88–90]. In some cases, the whole algorithm may be run in reduced precision; in others, it may be possible to limit the use of high precision to critical parts of the algorithm. Turning other machine learning algorithms, however, has not received the same amount of attention, as is the case with feature selection.

In information theoretic feature selection, the main challenge is to estimate the mutual information, one of the most common measures of dependence used in machine learning, for which it is necessary to estimate the probability distributions. Internally, it counts the occurrences of values within a particular group (i.e. its frequency). Thus, based on Tschiatschek et al. [91]'s work for approximately computing probabilities, Morán-Fernández et al. [92] investigated mutual information with limited number of bits by considering this measure with reduced precision counters. Instead a 64-bit resolution, a fixed-point representation was used. Also, since mutual information parameters are typically represented in the logarithmic domain, a lookup table was used to determine the logarithm of the probability of a particular event. The lookup table is indexed in terms of number of occurrences of an event (individual counters) and the total number of events (total counter) and stores values for the logarithms in the desired reduced precision representation. To limit the maximum size of the lookup table and the bit-width required for the counters, a maximum integer number was assumed. Experimental results over several synthetic and real datasets—through different feature selection methods based on this measure—showed that 16 bits were sufficient to return the same feature ranking than that of 64-bit representation. This opens the door to its use in other feature selection algorithms based on mutual information and as preprocessing step of some low precision classifiers, when implementing them in embedded systems for on-device analysis. Having on-device machine learning has some tremendous profits regarding privacy, reliability, efficient use of network bandwidth and power saving.

References

1. H. Climente-González, C. Azencott, S. Kaski, M. Yamada, Block HSIC Lasso: model-free biomarker detection for ultra-high dimensional data. Bioinformatics **35**(14), i427–i435 (2019)
2. N. Grgic-Hlaca, M.B. Zafar, K.P. Gummadi, A. Weller, Beyond distributive fairness in algorithmic decision making: Feature selection for procedurally fair learning. AAAI **18**, 51–60 (2018)
3. I. Furxhi, F. Murphy, M. Mullins, A. Arvanitis, C.A. Poland, Nanotoxicology data for in silico tools: a literature review. Nanotoxicology 1–26 (2020)
4. Y. Zhai, Y. Ong, I.W. Tsang, The emerging "big dimensionality". IEEE Comput. Intell. Mag. **9**(3), 14–26 (2014)
5. M. Tan, I.W. Tsang, L. Wang, Towards ultrahigh dimensional feature selection for big data. J. Mach. Learn. Res. **15**, 1371–1429 (2014)
6. K. Weinberger, A. Dasgupta, J. Langford, A. Smola, J. Attenberg, Feature hashing for large scale multitask learning, in *Proceedings of the 26th Annual International Conference on Machine Learning* (2009), pp. 1113–1120
7. D.L. Donoho et al., High-dimensional data analysis: the curses and blessings of dimensionality, in *AMS Math Challenges Lecture* (2000), pp. 1–32
8. R. Bellman, *Dynamic Programming* (Princeton UP, Princeton, NJ, 1957), p. 18
9. I. Guyon, A. Elisseeff, An introduction to variable and feature selection. J. Mach. Learn. Res. **3**, 1157–1182 (2003)
10. I. Guyon, *Feature Extraction: Foundations and Applications*, vol. 207 (Springer, Berlin, 2006)
11. B. Bonev, *Feature Selection Based on Information Theory* (Universidad de Alicante, 2010)
12. G. Hughes, On the mean accuracy of statistical pattern recognizers. IEEE Trans. Inf. Theory **14**(1), 55–63 (1968)
13. A.J. Miller, Selection of subsets of regression variables. J. R. Stat. Society. Ser. (Gen.) 389–425 (1984)
14. A.L. Blum, P. Langley, Selection of relevant features and examples in machine learning. Artif. Intell. **97**(1), 245–271 (1997)
15. M. Dash, H. Liu, Feature selection for classification. Intell. Data Anal. **1**(3), 131–156 (1997)
16. R. Kohavi, G.H. John, Wrappers for feature subset selection. Artif. Intell. **97**(1), 273–324 (1997)
17. H. Liu, H. Motoda, *Computational Methods of Feature Selection* (CRC Press, 2007)
18. Z.A. Zhao, H. Liu, *Spectral Feature Selection for Data Mining* (Chapman & Hall/CRC, 2011)
19. C. Boutsidis, P. Drineas, M.W. Mahoney, Unsupervised feature selection for the k-means clustering problem, in *Advances in Neural Information Processing Systems* (2009), pp. 153–161
20. V. Roth, T. Lange, Feature selection in clustering problems, in *Advances in Neural Information Processing Systems* (2003)
21. R. Leardi, A. Lupiáñez González, Genetic algorithms applied to feature selection in pls regression: how and when to use them. Chemom. Intell. Lab. Syst. **41**(2), 195–207 (1998)
22. D. Paul, E. Bair, T. Hastie, R. Tibshirani, "Preconditioning" for feature selection and regression in high-dimensional problems. Ann. Stat. 1595–1618 (2008)
23. M. Pal, G.M. Foody, Feature selection for classification of hyperspectral data by SVM. IEEE Trans. Geosci. Remote Sens. **48**(5), 2297–2307 (2010)
24. L. Yu, H. Liu, Efficient feature selection via analysis of relevance and redundancy. J. Mach. Learn. Res. 1205–1224 (2004)

25. M.A. Hall, *Correlation-Based Feature Selection for Machine Learning*. PhD thesis, University of Waikato, Hamilton, New Zealand (1999)
26. M. Dash, H. Liu, Consistency-based search in feature selection. J. Artif. Intell. **151**(1–2), 155–176 (2003)
27. A.M. Hall, L.A. Smith, Practical feature subset selection for machine learning. J. Comput. Sci. **98**, 4–6 (1998)
28. L. Yu, H. Liu, Feature selection for high-dimensional data: a fast correlation-based filter solution, in *Proceedings of The Twentieth International Conference on Machine Learning, ICML* (2003), pp. 856–863
29. Z. Zhao, H. Liu, Searching for interacting features, in *Proceedings of 20th International Joint Conference on Artificial Intelligence, IJCAI* (2007), pp. 1156–1161
30. I. Kononenko, Estimating attributes: analysis and extensions of relief, in *Proceedings of European Conference on Machine Learning, ECML*. Lecture Notes in Computer Science (Lecture Notes in Artificial Intelligence) (1994), pp. 171–182
31. K. Kira, L. Rendell, A practical approach to feature selection, in *Proceedings of the 9th International Conference on Machine Learning, ICML* (1992), pp. 249–256
32. H. Peng, F. Long, C. Ding, Feature selection based on mutual information criteria of maxdependency, max-relevance, and min-redundancy. IEEE Transactions on Pattern Analysis and Machine Intelligence **27**(8), 1226–1238 (2005)
33. S. Ramírez-Gallego, I. Lastra, D. Martínez-Rego, V. Bolón-Canedo, J.M Benítez, F. Herrera, A. Alonso-Betanzos, Fast-mRMR: fast minimum redundancy maximum relevance algorithm for high-dimensional big data. Int. J. Intell. Syst. **32**, 134–152 (2017)
34. S. Seth, J.C. Principe, Variable selection: a statistical dependence perspective, in *Proceedings of the International Conference of Machine Learning and Applications* (2010), pp. 931–936
35. I. Guyon, J. Weston, S. Barnhill, V. Vapnik, Gene selection for cancer classification using support vector machines. Mach. Learn. **46**, 389–422 (2002)
36. M. Mejía-Lavalle, E. Sucar, G. Arroyo, Feature selection with a perceptron neural net, in *Proceedings of the International Workshop on Feature Selection for Data Mining* (2006), pp. 131–135
37. R. Tibshirani, Regression shrinkage and selection via the lasso. J. R. Stat. Soc.: Ser. B **58**(1), 267–288 (1996)
38. H. Zou, T. Hastie, Regularization and variable selection via the elastic net. J. R. Stat. Soc.: Ser. B **67**(2), 301–320 (2005)
39. D.W. Marquardt, R.D. Snee, Ridge regression in practice. Am. Stat. **29**(1), 1–20 (1975)
40. M.F. Balin, A. Abid, J.Y. Zou, Concrete autoencoders: differentiable feature selection and reconstruction, in *International Conference on Machine Learning* (2019), pp. 444–453
41. B. Cancela, V. Bolón-Canedo, A. Alonso-Betanzos, E2E-FS: an end-to-end feature selection method for neural networks. arXiv e-prints (2020)
42. E. Frank, M.A. Hall, I.H. Witten. *Data Mining: Practical Machine Learning Tools and Techniques* (Morgan Kaufmann, 2016)
43. D. Dua, C. Graff, UCI machine learning repository (2017)
44. C.C. Chang, C.J. Lin, LIBSVM: a library for support vector machines. ACM Trans. Intell. Syst. Technol. (TIST) **2**(3), 27 (2011)
45. R. Bekkerman, M. Bilenko, J. Langford, *Scaling Up Machine Learning: Parallel and Distributed Approaches* (Cambridge University Press, Cambridge, 2011)
46. J.A. Olvera-López, J.A. Carrasco-Ochoa, J.F. Martínez-Trinidad, J. Kittler, A review of instance selection methods. Artif. Intell. Rev. **34**(2), 133–143 (2010)
47. D. Rego-Fernández, V. Bolón-Canedo, A. Alonso-Betanzos, Scalability analysis of mRMR for microarray data, in *Proceedings of the 6th International Conference on Agents and Artificial Intelligence* (2014), pp. 380–386
48. A. Alonso-Betanzos, V. Bolón-Canedo, D. Fernández-Francos, I. Porto-Díaz, N. Sánchez-Maroño, Up-to-Date feature selection methods for scalable and efficient machine learning, in *Efficiency and Scalability Methods for Computational Intellect* (IGI Global, 2013), pp. 1–26
49. M. Bramer, *Principles of Data Mining*, vol. 180 (Springer, Berlin, 2007)

50. L.I. Kuncheva, C.J. Whitaker, Measures of diversity in classifier ensembles and their relationship with the ensemble accuracy. Mach. Learn. **51**(2), 181–207 (2003)
51. P.K. Chan, S.J. Stolfo, Toward parallel and distributed learning by meta-learning, in *AAAI Workshop in Knowledge Discovery in Databases* (1993), pp. 227–240
52. V.S. Ananthanarayana, D.K. Subramanian, M.N. Murty, Scalable, distributed and dynamic mining of association rules. High Perform. Comput. HiPC **2000**, 559–566 (2000)
53. G. Tsoumakas, I. Vlahavas, Distributed data mining of large classifier ensembles, in *Proceedings Companion Volume of the Second Hellenic Conference on Artificial Intelligence* (2002), pp. 249–256
54. S. McConnell, D.B. Skillicorn, Building predictors from vertically distributed data, in *Proceedings of the 2004 Conference of the Centre for Advanced Studies on Collaborative research* (IBM Press, 2004), pp. 150–162
55. D.B. Skillicorn, S.M. McConnell, Distributed prediction from vertically partitioned data. J. Parallel Distrib. Comput. **68**(1), 16–36 (2008)
56. M. Banerjee, S. Chakravarty, Privacy preserving feature selection for distributed data using virtual dimension, in *Proceedings of the 20th ACM International Conference on Information and Knowledge Management* (ACM, 2011), pp. 2281–2284
57. Z. Zhao, R. Zhang, J. Cox, D. Duling, W. Sarle, Massively parallel feature selection: an approach based on variance preservation. Mach. Learn. **92**(1), 195–220 (2013)
58. A. Sharma, S. Imoto, S. Miyano, A top-r feature selection algorithm for microarray gene expression data. IEEE/ACM Trans. Comput. Biol. Bioinform. **9**(3), 754–764 (2011)
59. V. Bolón-Canedo, N. Sánchez-Maroño, A. Alonso-Betanzos, Distributed feature selection: an application to microarray data classification. Appl. Soft Comput. **30**, 136–150 (2015)
60. J. Dean, S. Ghemawat, MapReduce: simplified data processing on large clusters. Commun. ACM **51**(1), 107–113 (2008)
61. Apache Hadoop. http://hadoop.apache.org/. Accessed January 2021
62. Apache Spark. https://spark.apache.org. Accessed January 2021
63. MLib / Apache Spark. https://spark.apache.org/mllib. Accessed January 2021
64. L.I. Kuncheva, *Combining Pattern Classifiers: Methods and Algorithms* (Wiley, New York, 2013)
65. S. Nogueira, G. Brown, Measuring the stability of feature selection with applications to ensemble methods, in *Proceedings of the International Workshop on Multiple Classifier Systems* (2015), pp. 135–146
66. L.I. Kuncheva, A stability index for feature selection, in *Proceedings of the 25th IASTED International Multiconference Artificial intelligence and applications* (2007), pp. 421–427
67. B. Seijo-Pardo, Porto-Díaz, V. Bolón-Canedo, A. Alonso-Betanzos. Ensemble feature selection, homogeneous and heterogeneous approaches. Knowl.-Based Syst. **114**, 124–139 (2017)
68. V. Bolón-Canedo, K. Sechidis, N. Sánchez-Maroño, A. Alonso-Betanzos, G. Brown, Exploring the consequences of distributed feature selection in DNA microarray data, in *Proceedings 2017 International Joint Conference on Neural Networks (IJCNN)* (2017), pp. CFP17–US–DVD
69. V. Bolón-Canedo, A. Alonso-Betanzos, Ensembles for feature selection: a review and future trends. Inf. Fusion **52**, 1–12 (2019)
70. B. Seijo-Pardo, V. Bolón-Canedo, A. Alonso-Betanzos, On developing an automatic threshold applied to feature selection ensembles. Inf. Fusion **45**, 227–245 (2019)
71. V. Bolón-Canedo, N. Sánchez-Maroño, A. Alonso-Betanzos, A review of feature selection methods on synthetic data. Knowl. Inf. Syst. **34**, 483–519 (2013)
72. V. Bolón-Canedo, N. Sánchez-Maroño, A. Alonso-Betanzos, An ensemble of filters and classifiers for microarray data classification. Pattern Recognit. **45**(1), 531–539 (2012)
73. J. Rogers, S. Gunn, Ensemble algorithms for feature selection. *Deterministic and Statistical Methods in Machine Learning*. Lecture Notes in Computer Science, vol. 3635 (2005), pp. 180–198
74. P. Flach, *Machine Learning: The Art and Science of Algorithms that Make Sense of Data* (Cambridge University Press, Cambridge, 2012)

75. S. Shalev-Shwartz, S. Ben-David, *Understanding Machine Learning: From theory to algorithms* (Cambridge University Press, Cambridge, 2014)
76. K. Bunte, M. Biehl, B. Hammer, A general framework for dimensionality-reducing data visualization mapping. J. Neural Comput. **24**, 771–804 (2012)
77. P. Castells A. Bellogín, I. Cantador, A. Ortigosa (2010) Discerning relevant model features in a content-based collaborative recommender system, in *Preference Learning*, ed. by J. Färnkranz, E. Hällermeier (Springer, Berlin, 2010), pp. 429–455
78. N. Sánchez-Maroño, A. Alonso-Betanzos, O. Fontenla-Romero, C. Brinquis-Núñez, J.G. Polhill, T. Craig, A. Dumitru, R. García-Mira, An agent-based model for simulating environmental behavior in an educational organization. Neural Process. Lett. **42**(1), 89–118 (2015)
79. D.M. Maniyar, I.T. Nabney, Data visualization with simultaneous feature selection, in *2006 IEEE Symposium on Computational Intelligence and Bioinformatics and Computational Biology, 2006. CIBCB'06* (IEEE, 2006), pp. 1–8
80. J. Krause, A. Perer, E. Bertini, Infuse: interactive feature selection for predictive modeling of high dimensional data. IEEE Trans. Vis. Comput. Graph. **20**(12), 1614–1623 (2014)
81. K. Simonyan, A. Vedaldi, A. Zisserman, Deep inside convolutional networks: visualising image classification models and saliency maps (2013), arXiv:1312.6034
82. D. Bahdanau, K. Cho, Y. Bengio, Neural machine translation by jointly learning to align and translate, in *International Conference on Learning Representations (ICLR)* (2015)
83. B. Cancela, V. Bolón-Canedo, A. Alonso-Betanzos, J. Gama, A scalable saliency-based feature selection method with instance-level information. Knowl.-Based Syst. **192**, 105326 (2020)
84. J. Chen, L. Song, M. Wainwright, M. Jordan, Learning to explain: an information-theoretic perspective on model interpretation, in *International Conference on Machine Learning* (2018), pp. 883–892
85. J. Yoon, J. Jordon, M. van der Schaar, Invase: instance-wise variable selection using neural networks, in *International Conference on Learning Representations* (2018)
86. S. Ray, J. Park, S. Bhunia, Wearables, implants, and internet of things: the technology needs in the evolving landscape. IEEE Trans. Multi-Scale Comput. Syst. **2**(2), 123–128 (2016)
87. P. Koopman, Design constraints on embedded real time control systems (1990)
88. B. Jacob, S. Kligys, B. Chen, M. Zhu, M. Tang, A. Howard, H. Adam, D. Kalenichenko, Quantization and training of neural networks for efficient integer-arithmetic-only inference, in *Proceedings of the IEEE Conference on Computer Vision and Pattern Recognition* (2018), pp. 2704–2713
89. N. Wang, J. Choi, D. Brand, C. Chen, K. Gopalakrishnan, Training deep neural networks with 8-bit floating point numbers, in *Proceedings of the 32nd International Conference on Neural Information Processing Systems* (2018), pp. 7686–7695
90. X. Zhang, X. Zhou, M. Lin, J. Sun, Shufflenet: an extremely efficient convolutional neural network for mobile devices, in *Proceedings of the IEEE Conference on Computer Vision and Pattern Recognition* (2018), pp. 6848–6856
91. S. Tschiatschek, F. Pernkopf, Parameter learning of Bayesian network classifiers under computational constraints, in *Joint European Conference on Machine Learning and Knowledge Discovery in Databases* (Springer, 2015), pp. 86–101
92. L. Morán-Fernández, K. Sechidis, V. Bolón-Canedo, A. Alonso-Betanzos, G. Brown, Feature selection with limited bit depth mutual information for portable embedded systems. Knowl.-Based Syst. **197**, 105885 (2020)

Author Biographies

Verónica Bolón-Canedo received her B.S. (2009) and Ph.D. (2014) degrees in Computer Science from the University of A Coruña (Spain). After a postdoctoral fellowship in the University of Manchester, UK (2015), she is currently an Associate Professor in the Department of Computer Science of the University of A Coruña. She received the Best Thesis Proposal Award (2011) and the Best Spanish Thesis in Artificial Intelligence Award (2014) from the Spanish Association of Artificial Intelligence (AEPIA). She has extensively published in the area of machine learning and feature selection. On these topics, she has co-authored two books, seven book chapters, and more than 100 research papers in international conferences and journals. Her current research interests include machine learning, feature selection and big data. She serves as Secretary of the Spanish Association of Artificial Intelligence and is member of the Young Academy of Spain.

Amparo Alonso-Betanzos (Vigo, 1961) is Full Professor in the area of Computer Science and Artificial Intelligence at CITIC-University of A Coruña (UDC), where she coordinates the LIDIA group (Artificial Intelligence R&D Laboratory). Her research lines are the development of Scalable Machine Learning models, and Reliable and Explainable Artificial Intelligence, among others. She has a a PhD in Physics (1988) from the University of Santiago de Compostela, and has been a Postdoctoral Fellow at the Medical College of Georgia, USA (1988-90), where she worked on the development of Expert Systems for medical applications. She has published more than 200 articles in journals and international conferences, and books and book chapters, participating in more than 30 competitive European, national and local research projects. She is President of the Spanish Association of Artificial Intelligence since 2013. She is a member of the "Reserve List" of the High-Level Expert Group on Artificial Intelligence (AI HLG), of the European Commission since 2018. She has participated as a member of the GTIA, Working Group on Artificial Intelligence, of the Ministry of Science, Innovation and Universities (MINCIU), which collaborated in the drafting of the Spanish R&D&I Strategy in Artificial Intelligence in 2018, and is currently a member of the Advisory Council on Artificial Intelligence of the Government of Spain. She is also a Senior Member of IEEE and ACM.

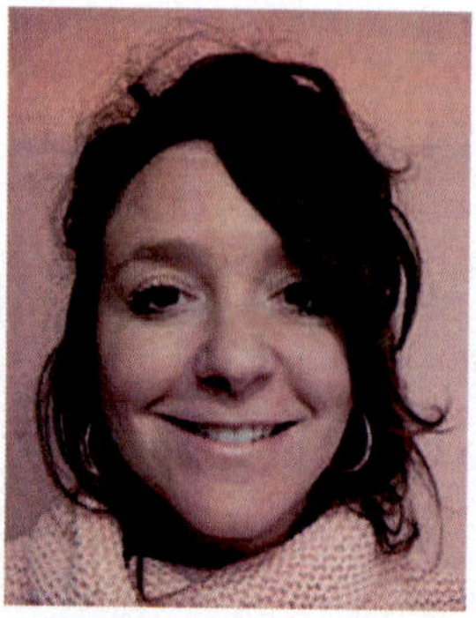

Laura Morán-Fernández received her B.S. (2015) and Ph.D. (2020) degrees in Computer Science from the University of A Coruña (Spain). She is currently an Assistant Lecturer in the Department of Computer Science and Information Technologies of the University of A Coruña. She received the Frances Allen Award (2021) from the Spanish Association of Artificial Intelligence (AEPIA). Her research interests include machine learning, feature selection and big data. She has co-authored three book chapters, and more than 15 research papers in international journals and conferences.

Chapter 3
Application of Rough Set-Based Characterisation of Attributes in Feature Selection and Reduction

Urszula Stańczyk

Abstract Quality of predictions depends heavily on features that are chosen for a classification system to rely on. It is one of the reasons why approaches, focused on feature selection and reduction, play a significant role in data mining. Among all available attributes, these should be detected that are of the highest relevance and importance for a given task. This objective can be achieved by an application of one of feature ranking algorithms. Some of data exploration methods have their own inherent mechanisms dedicated to feature reduction, and decision reducts, defined within rough set theory, offer such option. The chapter presents research on application of reduct-based characterisation of features, employed to support classification by selected inducers working outside rough set domain. The problem to be solved comes from the field of stylometry. It is the study of writing styles with the main task of authorship attribution, while using characteristic features not of qualitative, but quantitative type.

Keywords Feature selection · Ranking · Decision reduct · Stylometry · Authorship attribution · Classification · Rough set theory

3.1 Introduction

The standard aims of classification include obtaining such definitions for concepts to be recognised that lead to acceptably reliable predictions. The descriptions of detected patterns should be detailed enough to enhance their understanding and interpretability of discovered knowledge on one hand, while on the other they need to be sufficiently general to support the operation of labelling for unknown data samples [1]. These two

U. Stańczyk (✉)
Faculty of Automatic Control, Electronics and Computer Science, Department of Graphics, Computer Vision and Digital Systems, Silesian University of Technology, Akademicka 16, 44-100 Gliwice, Poland
e-mail: urszula.stanczyk@polsl.pl

M. Virvou et al. (eds.), *Advances in Selected Artificial Intelligence Areas*, Learning and Analytics in Intelligent Systems 24,
https://doi.org/10.1007/978-3-030-93052-3_3

objectives pull into two opposite directions within the process of feature selection, which is crucial for quality of accuracy for any inducer that is applied to a given task [2].

It may seem that increasing the number of available characteristic features should always work to an advantage, as more attributes bring also more knowledge about described concepts, but in fact, it is quite the opposite. The excessive variables not only increase computational costs and lengthen the required processing time, but also hinder recognition, which is one of the reasons why feature reduction approaches play an important role in data mining. Among all available attributes, these that are of the highest relevance for the task should be detected [3, 4]. This goal can be achieved by an application of one of ranking algorithms, which assign certain scores to all variables and lead to their ordering [5].

Feature selection mechanisms can be independent on data exploration approaches, or inherent to knowledge mining methods [6]. Decision (or relative) reducts are defined within rough set theory [7, 8]. They indicate such minimal subsets of attributes, which are necessary to preserve the predictive power of rule classifiers that can be induced from data. Reducts allow for characterisation of available features, and this discovered knowledge can be employed to support classification not only for rough set processing, but also by other inducers [9].

The chapter presents research on application of reduct-based characterisation of features, in the form of their weighting and ranking, in the process of sequential backward reduction. Ranking-driven selection of attributes was executed for four popular inducers, employed in the tasks of authorship attribution [10], which belongs to stylometric analysis of texts [11]. In the stylometric domain, the writing styles are studied through characteristic features not of qualitative, but quantitative type.

The results from the performed experiments show that attribute rankings, based on inferred decision reducts, enabled to discard significant numbers of attributes, while at least maintaining but also increasing classification accuracy. This validated the usefulness of the proposed methodology.

The content of the chapter is organised as follows. Section 3.2 provides descriptions for all elements of theoretical background, such as approaches to feature selection and reduction, rough set theory and decision reducts, stylometric analysis of texts, and discretisation of variables. Section 3.3 includes details of the framework of performed experiments, and Sect. 3.4 contains comments on the obtained results. Section 3.5 concludes the chapter and indicates directions for future research.

3.2 Background and Related Works

The research works reported in this chapter were dedicated to application of attribute rankings, based on some elements from rough set theory, to feature reduction of selected inducers, employed in the stylometric task of authorship attribution. The following sections briefly describe relevant background and related works.

3.2.1 Estimation of Feature Importance and Feature Selection

The features, which characterise objects and patterns present in the input data, influence the process of classification in greatly varying degrees [12]. Some of them are crucial to correct predictions, others can be simply supporting, while still others superfluous, entirely useless, or even detrimental. Their importance for any given task depends on their numbers, relationships among them, their types and nature, but also on classification systems used for predictions. Feature selection starts with domain knowledge and extraction of information from raw data [13], but this can be insufficient to make the aspect of dimensionality acceptable and feasible. Therefore, dedicated reduction approaches relay on evaluation of significance for variables and their groups, which can be executed in many ways.

The relative relevance of attributes can be estimated through *filtering* algorithms, applied typically within the initial data pre-processing and preparation stage. They are often statistics-oriented, and can measure informative content of variables through such notions as entropy, information gain, or gain ratio [14, 15]. Filters have the advantage of being universal, applicable in any domain, but in consequence it is difficult, or sometimes even impossible, to fine-tune them to specifics of some task.

Conditioning the quality of features on some chosen properties of employed inducers results in a *wrapper* approach to feature selection. It provides opportunities for closer tailoring to model characteristics, but at the cost of certain loss of generality, as a change of the learner will most likely produce entirely different evaluation of relevance for attributes.

Some of data mining methods possess their own inherent mechanisms aiming at dimensionality reduction, which are used as a part of learning processes. These techniques are called *embedded*, as they cannot be separated from data exploration, and constitute the third category of feature selection algorithms.

Estimation of attribute importance can result in their ordering and construction of a *ranking*. The ranking shows which features play the most significant role, and these with lower weights for a task. Some of ranking algorithms assign non-zero weights to all variables [16], while others can entirely reject selected features, when they are considered as redundant or irrelevant. When the employed ranker works independently on an inducer, used later for knowledge mining from the selected features, then this ranker is used as a filter, even if by itself it can be categorised as a wrapper or embedded mechanism.

Based on calculation of some measures, heuristic algorithms, or greedy procedures [17], the search for attributes to select can be executed in either *forward* or *backward* fashion [18]. The former starts with the empty set of variables, which is next gradually expanded, by adding either single elements or their groups. The latter approach begins proceedings with the entire set of available features, and then some of them are discarded relying on chosen criteria.

Regardless of the search direction, the selection process can be stopped when some set conditions are satisfied, such as a threshold number of attributes to be taken under consideration, or observed power of a classifier operating with the chosen variables. Also the entire path can be analysed, with the aim of detection of trends in performance, related to selected subsets of features.

3.2.2 *Rough Sets and Decision Reducts*

Rough set theory (RST) was invented by Pawlak, as an approach to mining uncertain data, with granular structure of the universe U [19]. The atoms of information correspond to equivalence classes of objects that cannot be discerned based on available knowledge on them. The *indiscernibility* relation is one of fundamental notions of RST in its classical version. It imposes granularity on the input space, and requires categorical descriptions of objects. When knowledge is insufficient for precise definitions of sets [20], then they are described by corresponding *approximations*. The lower approximation contains objects that certainly belong to the set, while the upper approximation includes objects that can belong to the set.

Decision table (DT) is a special form of an information system, used for data representation. Columns of a table give values of attributes. Their set $A = \{a_1, \ldots, a_m\}$ is divided into two groups: condition attributes $C = \{c_1, \ldots, c_n\}$ and decision attributes $D = \{d_1, \ldots, d_k\}$. When there are several decision attributes they are called criteria, but in many tasks a single decision attribute is distinguished. Then $D = \{d\}$, and $A = C \bigcup D = C \bigcup \{d\}$, and the values of d are called decisions. Each row of the decision table describes an object of the universe, $x \in U$, by providing information how values of condition attributes lead to a certain class label or decision. Decision tables, by knowledge stored in them, offer specific quality of predictions, which needs to be protected, but not necessarily all available condition attributes are required to guarantee it.

Decision reduct (or relative reduct) is such minimal subset of condition attributes, $R \subseteq C$, which enables unambiguous recognition of all objects in DT with different decisions [21]. For any pair of objects in the universe, $x, y \in U$, with $d(x) \neq d(y)$, an object x is discerned from y, based on attributes included in R. It makes a decision reduct a mechanism dedicated to dimensionality reduction, specific to rough set domain.

From the perspective of knowledge representation, reduct cardinality (called also its *length* or *size*) is its most important property, and for obvious reasons small reducts are preferred over large ones [22]. Unfortunately, even for relatively small decision tables, many reducts with various lengths can exist, and the process of inferring them could be computationally expensive, unless some heuristics are involved. The choice of one subset of attributes, among many alternatives of the same cardinality, is not trivial, and based on them classifiers can have vastly different predictive powers [23].

When all reducts are found, their intersection is called a *core*, and, if non-empty, it contains attributes critical for ensuring the expected prediction quality. On the other hand, when some variable is not included in any of reducts, then it can be rejected as entirely irrelevant for recognition of objects. Most often, however, the core is empty, and the union of reducts results in obtaining the entire set of attributes.

Whether all reducts are found, or only their subset, with the aim at limiting computational costs, the set of reducts can be considered as a new form of representation of knowledge on attributes, discovered in the process of reduct induction. This source can be further investigated and exploited in feature weighting, and construction of attribute rankings, useful not only in rough set domain or for rule classifiers inferred from data, but for other types of inducers as well.

3.2.3 *Reduct-Based Feature Characterisation*

When some set of decision reducts is obtained, it can be analysed not only with respect to its cardinality, but also lengths of reducts, and how often individual condition attributes occur in reducts with specific sizes. As it can be argued that variables, included in higher numbers of small reducts, are more important than features occurring only in few larger reducts, a detailed study of the reduct set can be used in construction of a weighting factor for attributes. This factor can refer to the entire set of reducts, found through exhaustive search, but also to some subset, which is much more easily obtained through heuristic algorithms [24].

Let $S_{Red} = \{Red_1, \ldots, Red_P\}$ be a set of decision reducts, with lengths l varying from some minimum l_{min} to maximum l_{max}. The number of elements included in a set is returned by $card$ function, in particular $card(S_{Red}) = P$. For a given condition attribute c, let $RED(S_{Red}, c)$ denote such subset of decision reducts from the set S_{Red}, where all reducts include attribute c. And let $RED(S_{Red}, c, l)$ be the subset of reducts, including the given attribute c and characterised by specific length l.

The weighting factor W_F for an attribute c, and based on a set of decision reducts S_{Red}, is defined as follows:

$$W_F(S_{Red}, c) = \sum_{i=l_{min}}^{l_{max}} \frac{card\,(RED(S_{Red}, c, i))}{card(S_{Red}) \cdot i}. \tag{3.1}$$

When some attribute is absent in the analysed set of reducts, that is none of reducts includes this particular attribute, the corresponding value of W_F factor equals zero, which is its minimal value. The maximal value would be obtained in the case of such set of reducts, where all elements would be of one and the same length l (then $l_{min} = l_{max} = l$), and also all these reducts would include the considered attribute. Then the maximal value of W_F factor is $1/l$.

For relatively small sets of decision reducts, it is possible that for some attributes the corresponding values of the weighting factor can be the same, but for sets with higher cardinalities it is highly unlikely. Therefore, the returned values can be used for ordering of variables and their ranking.

In the constructed ranking top positions are reserved for features with high values of W_F factor, characteristic to attributes that occur in the highest numbers of reducts, and these reducts have small sizes. At the lowest ranking positions these variables are placed, for which the weighting factor returned smallest values. It means that they are included relatively rarely, and mostly in reducts with high cardinalities, or even not included in any reducts in the considered set. Actually, the attributes completely absent in reducts can be rejected, and the ranking formed only for these that are present at least once.

3.2.4 Stylometry as an Application Domain

The era of stylometric studies began with the statement, considered at the time as revolutionary, that such concept as a writing style can be described not only in qualitative but also quantitative terms, that it can be measured [25]. Of course a style is such complex notion that even formulation of its precise and universal definition proves problematic. However, it can be estimated, and stylometric characteristic features enable description of stylistic profiles for authors. These profiles can be then used for characterisation of writers, comparisons of writing styles, and also for authorship attribution, which is considered as the most important task in the domain of stylometric analysis of texts [26].

Reliable recognition of authorship requires access to several text samples, which are explored to learn stylistic patterns characteristic for writers, their individual linguistic preferences and habits. The knowledge discovered in mining textual data can then be measured against samples of text of unknown or questionable authorship. With such way of processing, the problem of authorship attribution is approached as a classification task, with authors designating classes, and stylometric descriptors employed as attributes that characterise text samples [27].

The linguistic features used in textual analysis can be of following types [28]:

- lexical—return such statistics as frequencies of occurrence of letters, characters, words, phrases, or averages of word lengths, or sentence lengths, also their distributions;
- syntactic—describe patterns and structure of sentence formulation, indicated by employed punctuation marks [29];
- structural—capture elements of formatting and layout—headings, signatures, organisation into paragraphs, for electronic format of text documents also specific fonts, embedded pictures or links;
- content-specific—include words or phrases, which are of higher importance or relevance in a domain of interest.

The choice of a particular group of descriptors is one of great consequence, and to some extent it depends on a number, lengths, and genre of available texts, as, for example, to distinguish authors of blogs, chat posts, or short text messages entirely different markers would be efficient than in the case of analysis of literary works such as poems, or novels [30].

The methods employed for authorship attribution tasks typically come from two fields, either statistics or artificial intelligence domain [31, 32]. In the former case a language model, showing probabilities of transitions from one letter or character into others, and application of Markov chains of some order can be used [33]. Then the matrices of transitions are constructed for texts of known authorship and they are compared with the one obtained from the text that is questioned. Artificial neural networks can be trained to recognise authors, or sets of decision rules can be inferred based on textual data [15]. This last approach has also the advantage of representation of learned knowledge in such transparent form that enables extension of domain knowledge, and enhanced understanding of detected patterns.

3.2.5 Continuous Versus Nominal Character of Input Features

The concepts to be recognised can be described by features with either discrete or continuous values. The latter type causes certain limitations on methods and techniques that can be used for exploration of data, as not all inducers are capable of operation on real-valued attributes. In such case available variables can be transformed by one of *discretisation* approaches [34]. Then the whole ranges of attributes values are divided with selected cut-points into some number of intervals called *bins*, and their finite number is used for categorical representation of input domain, making it granular. Discretisation brings reduction of information, therefore it should always be treated with certain caution.

Beside enabling application of learners working only with nominal features, discretisation process can bring also other advantages. It results in reduction of data, can cause enhanced predictions (when compared to continuous domain), reduced structures of constructed data learning models, which simplifies interpretability, and increased generality of captured patterns [35]. These arguments give reasons for inclusion of discretisation algorithms in the initial data pre-processing stage. It is also possible to firstly explore data, and then to discretise learned patterns, but such reversed order of processing is much less common [36].

The transformation of input domain is performed either in *unsupervised* or *supervised* manner [37]. The former category contains methods, which disregard information on classes in construction of intervals for attribute values. As they do not easily adapt to data characteristics such as distributions of data points in the input space, typically these methods have rather bad reputation.

Supervised approaches are widely considered as superior. They pay close attention to classes, by conditioning the process of validation of possible cut-points on their informative content with respect to recognition, which can be measured by calculated entropy, as in the popular algorithm by Fayyad and Irani [38]. The procedure considers all variables independently, with global perspective on their values, which are analysed in top-down fashion, with splitting intervals, as opposed to the alternative attitude of bottom-up, with merging of bins.

The transformation process of Fayyad and Irani method starts with sorting all values of an attribute, and finding its minimal and maximal values. Then one interval is constructed to represent the whole range of values of the discretised variable. Next some candidate cut-points are determined, which could be used for splitting the interval. Optimality of these cut-points is evaluated, searching for the one that results in the minimal entropy for the split range of values. The processing continues until the stopping criterion, referring to Minimal Description Length principle, is met. This means that for some variables single bins can be returned to represent all their values, when attributes are found as irrelevant and with negligible influence on class distinction [39]. While working in discrete domain, such single bin attributes can be removed from considerations.

3.3 Setup of Experiments

The objective of the research presented in this chapter was evaluation of the effectiveness of a ranking based on decision reducts, in the process of feature reduction for other inducers, operating outside rough set domain, and a study of trends in classifiers performance, observed for such framework.

The proposed methodology consisted of the following stages:

- preparation of input data,
- construction of input datasets,
- discretisation,
- induction of decision reducts,
- weighting of attributes and construction of their rankings,
- ranking-driven reduction of features for selected inducers,
- evaluation of performance for classifiers,
- comparative analysis of results.

The details of these stages are explained in this section, and the obtained results commented in the next one.

3.3.1 Preparation of Input Data and Datasets

In the research selected literary works of four widely known writers were analysed, namely Edith Wharton, Mary Johnston, James O. Curwood, and Jack London. Due to the fact that stylistic profiles of authors of the same gender show closer similarity [40], which could influence observations, these writers were put in pairs. This led to construction of correspondingly female writer dataset, and male writer dataset. Each dataset constituted an example of a binary classification task.

To increase the numbers of available data samples, the chosen novels (available for on-line reading and download thanks to Project Gutenberg[1]), to be subjected to stylometric analysis, were divided into many smaller parts of comparable size. Next, over these text chunks the frequencies of occurrence were calculated for the pre-selected set of 24 stylometric descriptors. They combined 22 lexical and 2 syntactic markers, as follows:

after, almost, any, around, before, but, by, during, how, never, on, same, such, that, then, there, though, until, what, whether, within, who, a comma, a semicolon.

These descriptors were chosen based on the list of the most frequently used words in English language, and most often employed punctuation marks. The usage of such common elements of language is so ingrained in sentence formulation that it enables to capture individuality of stylistic preferences for authors. It is also such characteristic that is almost impossible to imitate, as long as patterns are observed over sufficiently high number of text samples [27].

This initial preparation of data returned datasets with continuous-valued features, each including a training set and two test sets, to be used for evaluation of classifier performance. Samples present in a set were based on such novels, for which in other sets in this dataset there were no representatives. In order to minimise the number of factors possibly influencing predictions, also balance of classes for all sets was ensured [41].

Since rough set processing in the classic version requires categorical attributes, in the next step all sets were independently discretised with supervised Fayyad and Irani algorithm, implemented in WEKA environment [42]. For some of attributes only single bins were found for representation in discrete domain. For female writers 20 variables were left, and four (almost, during, though, within) rejected. For male writers 22 attributes remained, while two (during, never) were discarded.

3.3.2 Decision Reducts Inferred

The prepared two discrete datasets were subjected to rough set processing with Rough Set Exploration System (RSES) [43]. For both female and male writer datasets, all decision reducts were inferred using exhaustive algorithm. The resulting sets of reducts greatly varied in their overall cardinalities, and how they characterised individual attributes, which is shown in Table 3.1.

For female writers the lengths of reducts ranged from 4 to 11, and for male writers from 5 to 12 included attributes. For the former, the attribute "that" was the least often used for construction of reducts, as it occurred only 57 times, whereas the maximum was detected for a comma, used in 730 out of all 914 reducts. On the other hand, the same feature was employed most rarely for male writers, although comparatively

---

[1] https://www.gutenberg.org/.

Table 3.1 Occurrence of attributes in reducts of specific cardinality

Female writer dataset								Attribute	Male writer dataset								
	Reduct cardinality								Reduct cardinality								
Total	4	5	6	7	8	9	10	11		5	6	7	8	9	10	11	12	Total
569	0	3	30	117	161	165	53	40	after	0	3	27	198	505	1102	1524	358	3717
									almost	0	4	107	372	1100	2121	1639	327	5670
396	0	2	35	55	130	83	47	44	any	0	4	95	461	978	2124	2102	364	6128
265	0	2	7	25	75	89	37	30	around	0	10	203	464	764	1855	1503	314	5113
325	0	0	25	77	88	74	33	28	before	0	3	86	373	757	1493	1596	387	4695
533	0	8	22	123	140	145	47	48	but	0	2	48	264	685	1761	1625	373	4758
356	0	8	45	73	80	103	28	19	by	1	32	182	470	1044	2160	1338	179	5406
278	0	1	14	50	76	64	46	27	how	1	28	188	321	610	1454	1520	394	4516
301	0	0	23	57	68	73	28	52	never									
377	1	20	89	93	38	88	41	7	on	1	7	171	588	1475	2844	1886	238	7210
412	0	5	41	105	132	66	22	41	same	0	3	75	401	1036	2027	1820	367	5729
375	0	0	20	82	105	110	35	23	such	0	3	106	432	1008	1885	1518	255	5207
57	0	1	7	3	5	18	16	7	that	0	3	109	526	1256	2370	2012	409	6685
362	0	4	46	88	119	56	21	28	then	0	13	164	505	926	1852	1757	375	5592
462	0	2	15	85	102	145	59	54	there	0	22	266	673	1302	3023	1997	279	7562
									though	1	30	200	563	1289	2239	1446	243	6011
298	0	12	65	92	54	26	12	37	until	0	9	167	513	1171	2398	1799	296	6353
357	0	2	37	72	77	90	27	52	what	0	2	28	269	1056	2362	1727	289	5733
165	0	0	0	26	62	46	16	15	whether	0	1	39	296	799	1394	1566	365	4460
									within	0	2	41	221	417	1046	1633	412	3772
295	1	16	17	47	72	89	30	23	who	0	17	208	530	1048	2173	2104	437	6517
350	1	2	47	98	83	78	28	13	;	0	2	64	409	941	2025	1810	347	5598
730	1	22	117	228	221	111	24	6	,	1	40	429	1207	1334	252	15	0	3278
914	1	22	117	228	236	191	65	54	Total	1	40	429	1257	2389	4196	3267	584	12163

this minimum of 3278 was a rank higher. The most frequently used attribute (7562 from the total of 12163) for this dataset was "there".

Apart from single-bin variables, all other features were included in some number of reducts, and the core of reducts was the empty set. None of variables occurred in reducts as many times as some other feature, and even when they were included in similar numbers of decision reducts, then these reducts were different with respect to their cardinalities. Which spoke for highly individual characterisation of attributes by these statistics based on reducts.

3.3.3 Rankings of Attributes Based on Reducts

Analysis of the sets of inferred decision reducts, and the previously defined weighting factor for attributes, led to construction of their rankings, displayed in Table 3.2. The difference in the number of included features for female and male writer datasets results from supervised dicretisation procedures applied to data, which, as commented above, caused assigning single intervals to represent all values of some variables. In discrete domain these 1-bin attributes had constant value in all data samples contained in the studied decision tables, thus brought zero informative content, and were never considered as candidates for being a part of some reduct. Thus only the remaining variables were included in the presented rankings, as for them the weighting factor returned non-zero values.

Despite using largely the same set of attributes for both datasets, the two rankings differed, as they were based on local properties of the analysed sets of reducts. In the calculation of the score assigned to variables, reducts were weighted by their cardinalities, with preference for smaller ones, as they offer higher dimensionality reduction. Therefore, the previously observed maxima and minima of occurrence for features did not translate directly into their placement at the highest or lowest ranking positions. Not even inclusion in shortest reducts could guarantee high positions for attributes. Although obviously both elements had some influence on the overall score, only analysis executed with taking into account the entire set of considered reducts and their properties led to the final ordering of variables.

3.3.4 Classification Systems Employed

The usefulness of the obtained rankings of attributes for their reduction, was observed for classification systems working outside rough set domain. Four popular inducers were chosen, namely Naive Bayes, k-Nearest Neighbours, PART, and J-48, all implemented in WEKA workbench [42]. Different foundations of these classifiers enabled a wider scope for observations on the influence of ranking-driven reduction of features on the resulting performance [44]. All four systems are capable of working with both continuous and discrete attributes [45].

Table 3.2 Rankings of attributes based on the weighting factor

Ranking position	Female writer dataset	Male writer dataset
1	,	there
2	after	on
3	but	that
4	there	who
5	same	until
6	on	though
7	any	any
8	then	same
9	such	almost
10	;	what
11	by	then
12	what	;
13	until	by
14	before	such
15	who	around
16	never	but
17	how	before
18	around	how
19	whether	whether
20	that	,
21		within
22		after

Naive Bayes (NB) relies on Bayes theorem of conditional probability [46]. It assumes that features describing recognised concepts are completely independent, and in the same degree contribute to correct predictions. All correlations between attributes are treated as non-existent. In most cases this probabilistic classifier works surprisingly well, and its performance is often used as a kind of benchmark for comparisons of obtained results.

k-Nearest Neighbours (kNN) belongs to so-called lazy classifiers, which have no separate learning stage, and the knowledge stored in samples is inferred at the moment when data is queried to find labels for new objects. An object is assigned to such class that is voted for by the majority of its closest neighbours. This closeness is measured by some kind of distance metric, such as Euclidean or Hamming distance, and k specifies the size of the considered neighbourhood. In the performed experiments the value of k was equal 1.

Both J-48 and PART refer to C4.5 algorithm [47], and rely on divide-and-conquer approach. The former constructs either an unpruned or pruned decision tree and in top-down fashion passes samples from the root node and then branches to reach

Table 3.3 Performance of inducers [%] for continuous and discrete domains and the entire set of available attributes

	NB			k-NN			PART			J-48		
	Test1	Test2	AvgT	Test1	Test2	AvgT	Test1	Test2	AvgT	Test1	Test2	AvgT
Dataset	Continuous domain											
Female writers	95.56	91.11	93.34	96.67	94.44	95.56	93.33	93.33	93.33	86.67	92.22	89.45
Male writers	88.89	93.33	91.11	93.33	92.22	92.78	90.00	86.67	88.34	84.44	83.33	83.89
	Discrete domain											
Female writers	90.00	81.11	85.56	95.56	91.11	93.34	91.11	95.56	93.34	90.00	88.89	89.45
Male writers	94.44	93.33	93.89	94.44	92.22	93.33	75.56	92.22	83.89	50.00	50.00	50.00

leaf nodes for final decisions, while PART rule learner induces rules based on the best leaves of the partial decision tree [48].

All four classifiers were used with their default settings, without any fine-tuning of their parameters. As in the research none of classes was preferred (both were considered as of the same importance), classification accuracy was selected as the evaluation measure for inducer performance [49], understood as the number of correctly labelled samples divided by their total number. It was estimated with two test sets, for which the results were presented individually, but also the averages were calculated (denoted as AvgT).

The performance of inducers while working with the entire sets of continuous and discrete variables is given in Table 3.3. As it can be observed, for same cases discretisation resulted in worsened performance (in particular for J-48 and male writer dataset), while in others predictions were improved.

The power of classifiers for the entire set of attributes in discrete domain was treated as a reference point for comparisons with predictions observed in the process of gradual feature reduction. The results are commented in the next section of the chapter.

3.4 Obtained Results of Feature Reduction

The process of feature selection and reduction for studied classifiers was performed backwards, that is starting with the entire set of 24 discrete features. Firstly, single-bin attributes were rejected, which resulted in leaving 20 variables for female writer dataset, and 22 for male writers. Next, one by one, the lowest ranking features were discarded, and the performance for all classifiers evaluated for gradually decreasing numbers of remaining attributes. The process of feature reduction can be executed

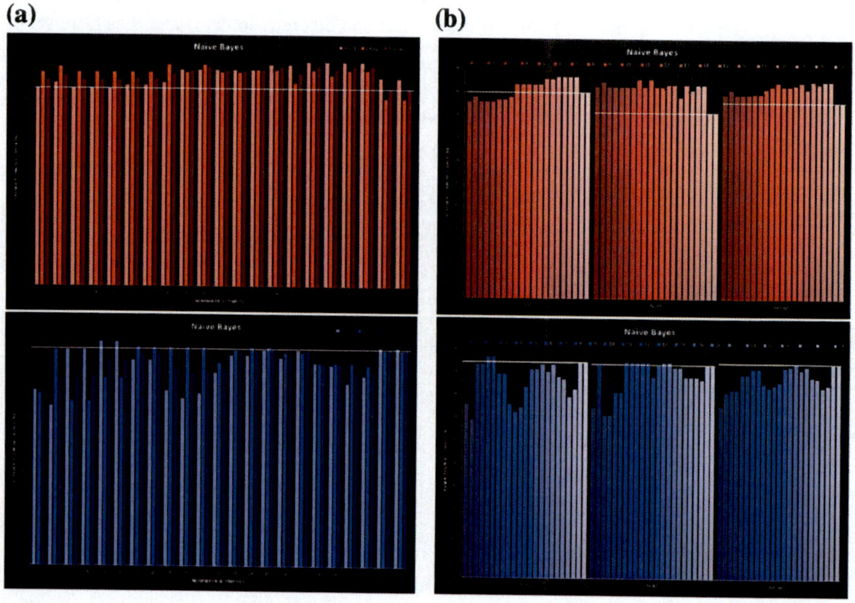

Fig. 3.1 Classification accuracy [%] of Naive Bayes classifiers in the perspective of the number of attributes considered: **a** for all sets at the same time, **b** within a particular set. The top row shows results for female writers and the bottom row for male writer dataset

till such conditions are met as the number of variables left under considerations, or specific level of expected predictions. In the experiments performed it was continued until the set of attributes to reject was exhausted, and only single features were left.

The results were presented in Fig. 3.1 for Naive Bayes, in Fig. 3.2 for *k*-NN, in Fig. 3.3 for PART, and in Fig. 3.4 for J-48 classifier. The performance was shown separately for the two test sets that were employed for evaluation, and also the averaged classification accuracy was displayed. The level of predictions for the entire set of available attributes was used as a point of reference, and in each chart it was shown with the included horizontal lines. For Naive Bayes and *k*-NN classifiers the performance in discrete domain was worse for female writers, but for male writer dataset it was improved. For PART and J-48 discretisation caused decreased predictions for male writer dataset, while for female writers the results for both domains were close.

While in discrete domain, for female writers reduction of features for Naive Bayes, shown in Fig. 3.1, for the most part resulted in increased classification accuracy for the two test sets treated separately, but also with respect to their average. The trend started with the maintained power, then increased, and then slow decrease followed. For male writers the possible gains were less significant with respect to prediction level, where only slight increase could be detected, but still there were cases of maintained performance while discarding a noticeable proportion of features.

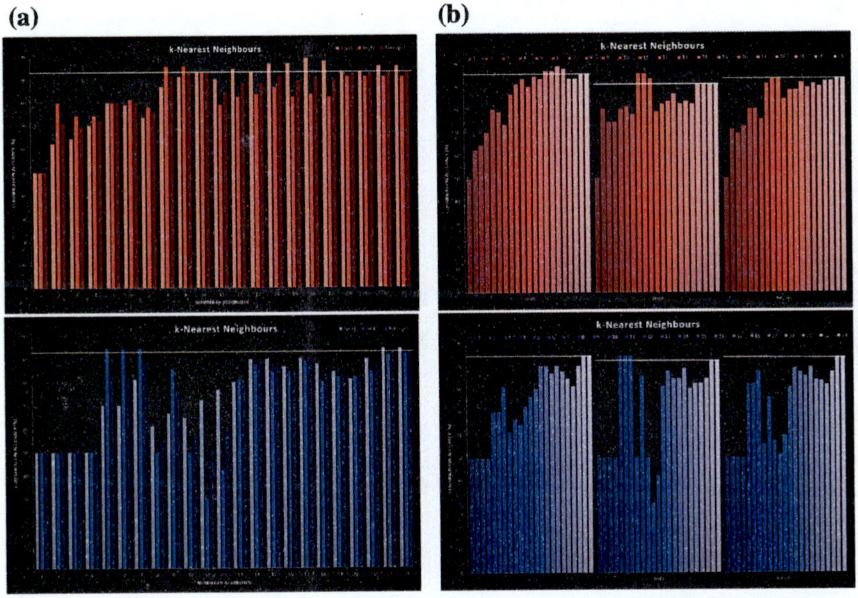

Fig. 3.2 Classification accuracy [%] of k-Nearest Neighbours classifiers in the perspective of the number of attributes considered: **a** for all sets at the same time, **b** within a particular set. The top row shows results for female writers and the bottom row for male writer dataset

For k-NN classifier for both female and male writers not so many attributes could be discarded without undermining the power of inducers, which was shown in Fig. 3.2. For individually treated test sets there were some interesting cases, for both datasets around half of variables could be rejected, but when average over test sets was calculated, only for female writers the same statement was valid. But for the most part reduction of attributes caused degraded performance.

Results of feature reduction for PART were shown in Fig. 3.3. For both datasets the similar trends could be observed: at the beginning of the process the performance stayed at the original level, or some increase was detected, then some decrease in classification accuracy was observed, only to increase once again, and then fall again with more and more rejected attributes.

The trends in performance for the entire process of feature reduction for J-48 classifier were displayed in Fig. 3.4. For male writer dataset, transformation of attributes from continuous-valued into discrete led to compromised performance, thus feature reduction brought only increase in predictions, even when only a single attribute was left. For female writers when there were some enhanced results, the difference was on the smaller size, but significant numbers of attributes could be safely discarded.

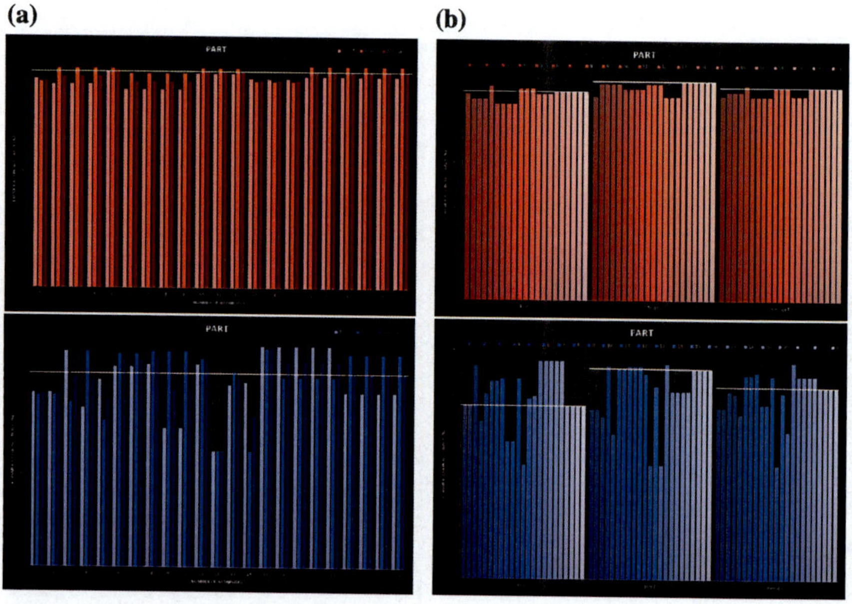

Fig. 3.3 Classification accuracy [%] of PART classifiers in the perspective of the number of attributes considered: **a** for all sets at the same time, **b** within a particular set. The top row shows results for female writers and the bottom row for male writer dataset

The best results of feature reduction for all classifiers were presented in Table 3.4, in terms of the two factors, typically considered as the most important ones, classification accuracy and the number of considered attributes. They could be treated as optimisation criteria, indicating dimensions in search of Pareto points. In each case firstly the highest overall classification accuracy was detected (at least equal to the one for the entire set of available variables), and the corresponding number of variables under consideration for this case was listed, separately for two test sets, and for the average as well.

For individual test sets it was possible to reduce attributes to as few as 2 or 5 for J-48 classifier, or PART for female writers, even the averages for these two inducers enabled such significant reduction of features. For male writers generally there were fewer cases of improved results, in particular for k-NN only the entire set of attributes guaranteed maintaining the averaged quality of predictions. Still, for J-48 classifiers even as few as 3 or 4 attributes, or 5 for NB, gave enhanced power.

(a) (b)

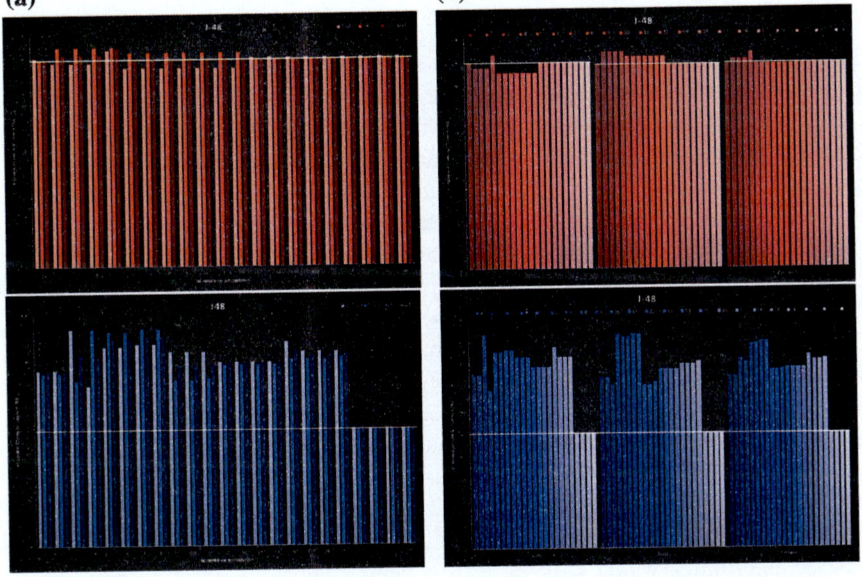

Fig. 3.4 Classification accuracy [%] of J-48 classifiers in the perspective of the number of attributes considered: **a** for all sets at the same time, **b** within a particular set. The top row shows results for female writers and the bottom row for male writer dataset

Table 3.4 The best results of feature reduction for classifiers, (a) the highest performance [%], observed for (b) the smallest sets of considered attributes

	Female writer dataset						Male writer dataset					
	Test1		Test2		AverageT		Test1		Test2		AverageT	
Classifier	(a)	(b)	(a)	(b)	(a)	(b)	(a)	(b)	(a)	(b)	(a)	(b)
NB	93.33	9	95.56	8	95.00	16	96.67	5	94.44	7	93.89	15
k-NN	96.67	14	95.56	8	93.34	9	94.44	22	94.44	5	93.33	22
PART	93.33	5	95.56	16	93.89	5	95.56	15	94.44	15	90.56	8
J-48	93.33	5	94.44	2	93.89	5	93.33	3	93.33	4	90.00	7

Since for all classifiers some attributes could be discarded without worsening the performance, when compared to the one for the entire set of variables, these experimental results validated the proposed methodology of feature reduction based on characteristics of decision reducts inferred from data, even when rankings constructed by weighting of attributes were employed for classification systems operating outside rough set domain.

3.5 Conclusions

The chapter presents research dedicated to the problem of feature selection and reduction, which can greatly influence the efficiency of data mining approaches. In the proposed methodology selection of attributes was driven by a ranking, obtained by weighting features with the defined factor. The calculated scores depended on characterisation of variables by sets of inferred decision reducts, which constitute a mechanism dedicated to dimensionality reduction, embedded in rough set processing. The explored datasets were examples of authorship attribution problem, which belongs to stylometric analysis of texts.

The experiments showed the usefulness of obtained rankings for classifiers with different foundations than rough set theory. The backward rejection of attributes while following the rankings resulted in several cases of significant reduction of the number of considered features while at least maintaining the predictive powers of inducers, or even increasing classification accuracy.

In the research works reported, all decision reducts were found through the exhaustive algorithm, which is rather computationally expensive. Within future research paths other approaches to reduct generation will be tested, such as for example genetic algorithms, which will result in obtaining only subsets of reducts, but at significantly lower computational costs. The rankings based on subsets of reducts will be compared with those based on the entire reduct sets, and their usefulness in feature selection for various inducers will be verified.

Acknowledgements The research works presented in the chapter were performed within the statutory project of the Department of Graphics, Computer Vision and Digital Systems (RAU-6, 2021), at the Silesian University of Technology, Gliwice, Poland.

References

1. J. Han, M. Kamber, J. Pei, *Data Mining: Concepts and Techniques* (Morgan Kaufmann, 2011)
2. M. Dash, H. Liu, Feature selection for classification. Intell. Data Anal. **1**, 131–156 (1997)
3. U. Stańczyk, Relative reduct-based estimation of relevance for stylometric features, in *Advances in Databases and Information Systems*. ed. by B. Catania, G. Guerrini, J. Pokorny, LNCS, vol. 8133 (Springer, Berlin, 2013), pp. 135–147
4. L. Yu, H. Liu, Efficient feature selection via analysis of relevance and redundancy. J. Mach. Learn. Res. **5**, 1205–1224 (2004)
5. J. Biesiada, W. Duch, A. Kachel, S. Pałucha, Feature ranking methods based on information entropy with Parzen windows, in *Proceedings of International Conference on Research in Electrotechnology and Applied Informatics, Katowice, Poland* (2005), pp. 109–119
6. I. Witten, E. Frank, M. Hall, *Data Mining. Practical Machine Learning Tools and Techniques*, 3rd edn. (Morgan Kaufmann, 2011)
7. Z. Pawlak, Rough sets and intelligent data analysis. Inf. Sci. **147**, 1–12 (2002)
8. Z. Pawlak, A. Skowron, Rough sets and boolean reasoning. Inf. Sci. **177**(1), 41–73 (2007)
9. U. Stańczyk, B. Zielosko, K. Żabiński, Application of greedy heuristics for feature characterisation and selection: a case study in stylometric domain, in *Proceedings of the International*

Joint Conference on Rough Sets, IJCRS 2018. Volume 11103 of Lecture Notes in Computer Science, ed. by H. Nguyen, Q. Ha, T. Li, Przybyla-Kasperek, M. (Springer, Quy Nhon, Vietnam, 2018), pp. 350–362
10. D. Holmes, Authorship attribution. Comput. Hum. **28**, 87–106 (1994). (April)
11. S. Argamon, K. Burns, S. Dubnov (eds.), *The Structure of Style: Algorithmic Approaches to Understanding Manner and Meaning* (Springer, Berlin, 2010)
12. H. Liu, H. Motoda, *Computational Methods of Feature Selection*. Data Mining and Knowledge Discovery Series (Chapman & Hall/Crc, 2007)
13. I. Guyon, S. Gunn, M. Nikravesh, L. Zadeh (eds.), *Feature Extraction: Foundations and Applications*. Volume 207 of Studies in Fuzziness and Soft Computing (Physica-Verlag, Springer, 2006)
14. E. Mansoori, Using statistical measures for feature ranking. Int. J. Pattern Recognit. Artifficial Intell. **27**(1), 1350003–14 (2013)
15. U. Stańczyk, Weighting attributes and decision rules through rankings and discretisation parameters, in *Machine Learning Paradigms: Theory and Application*. ed. by A.E. Hassanien (Springer International Publishing, Cham, 2019), pp. 25–43
16. U. Stańczyk, RELIEF-based selection of decision rules. Procedia Comput. Sci. **35**, 299–308 (2014)
17. B. Zielosko, M. Piliszczuk, Greedy algorithm for attribute reduction. Fundam. Inform. **85**(1–4), 549–561 (2008)
18. M. Reif, F. Shafait, Efficient feature size reduction via predictive forward selection. Pattern Recognit. **47**, 1664–1673 (2014)
19. Z. Pawlak, A. Skowron, Rudiments of rough sets. Inf. Sci. **177**(1), 3–27 (2007)
20. J.W. Grzymała-Busse, S.Y. Sedelow, W.A. Sedelow, Machine learning & knowledge acquisition, rough sets, and the english semantic code, in *Rough Sets and Data Mining: Analysis of Imprecise Data*. ed. by N. Cercone, T. Lin (Springer, Boston, 1997), pp. 91–107
21. X. Jia, L. Shang, B. Zhou, Y. Yao, Generalized attribute reduct in rough set theory. Knowl.-Based Syst. **91**, 204–218 (2016)
22. A. Janusz, D. Ślęzak, Rough set methods for attribute clustering and selection. Appl. Artif. Intell. **28**(3), 220–242 (2014)
23. U. Stańczyk,, B. Zielosko, Assessing quality of decision reducts, in *Knowledge-Based and Intelligent Information & Engineering Systems: Proceedings of the 24rd International Conference KES-2020, Verona, Italy, 16-18 September 2020*, ed. by M. Cristani, C. Toro, C. Zanni-Merk, R.J. Howlett, L.C. Jain. Volume 176 of Procedia Computer Science (Elsevier, 2020), pp. 3273–3282
24. B. Zielosko, U. Stańczyk, Reduct-based ranking of attributes, in *Knowledge-Based and Intelligent Information & Engineering Systems: Proceedings of the 24rd International Conference KES-2020, Verona, Italy, 16-18 September 2020*, ed. by M. Cristani, C. Toro, C. Zanni-Merk, R.J. Howlett, L.C. Jain. Volume 176 of Procedia Computer Science. (Elsevier, 2020), pp. 2576–2585
25. F. Mosteller, D. Wallace, Inference in an authorship problem. J. Am. Stat. Assoc. **58**(303), 275–309 (1963)
26. J. Rybicki, M. Eder, D. Hoover, Computational stylistics and text analysis, in *Doing Digital Humanities: Practice, Training, Research*, ed. by C. Crompton, R. Lane, R. Siemens, 1st edn. (Routledge, 2016), pp. 123–144
27. L. Pearl, M. Steyvers, Detecting authorship deception: a supervised machine learning approach using author writeprints. Lit. Linguist. Comput. **27**(2), 183–196 (2012)
28. M. Koppel, J. Schler, S. Argamon, Authorship attribution: what's easy and what's hard? J. Law Policy **21**(2), 317–331 (2013)
29. H. Baayen, H. van Haltern, F. Tweedie, Outside the cave of shadows: using syntactic annotation to enhance authorship attribution. Lit. Linguist. Comput. **11**(3), 121–132 (1996)
30. Y. Zhao, J. Zobel, Searching with style: authorship attribution in classic literature, in *Proceedings of the Thirtieth Australasian Conference on Computer Science - Volume 62*. ACSC '07, Darlinghurst, Australia, Australian Computer Society, Inc. (2007), pp. 59–68

31. M. Koppel, J. Schler, S. Argamon, Computational methods in authorship attribution. J. Am. Soc. Inf. Sci. Technol. **60**(1), 9–26 (2009)
32. E. Stamatatos, A survey of modern authorship attribution methods. J. Am. Soc. Inf. Sci. Technol. **60**(3), 538–556 (2009)
33. D. Khmelev, F. Tweedie, Using Markov chains for identification of writers. Lit. Linguist. Comput. **16**(4), 299–307 (2001)
34. S. García, J. Luengo, J.A. Sáez, V. López, F. Herrera, A survey of discretization techniques: taxonomy and empirical analysis in supervised learning. IEEE Trans. Knowl. Data Eng. **25**(4), 734–750 (2013)
35. H. Liu, F. Hussain, C. Tan, M. Dash, Discretization: an enabling technique. Data Min. Knowl. Discov. **6**(4), 393–423 (2002)
36. U. Stańczyk, B. Zielosko, G. Baron, Discretisation of conditions in decision rules induced for continuous data. PLOS ONE **15**(40), 1–33 (2020)
37. Y. Yang, G.I. Webb, X. Wu, Discretization methods, in *Data Mining and Knowledge Discovery Handbook*. ed. by O. Maimon, L. Rokach (Springer, US, Boston, MA, 2005), pp. 113–130
38. U. Fayyad, K. Irani, Multi-interval discretization of continuous valued attributes for classification learning, in *Proceedings of the 13th International Joint Conference on Artificial Intelligence*, vol. 2 (Morgan Kaufmann Publishers, 1993), pp. 1022–1027
39. U. Stańczyk, Evaluating importance for numbers of bins in discretised learning and test sets, in *Intelligent Decision Technologies 2017: Proceedings of the 9th KES International Conference on Intelligent Decision Technologies (KES-IDT 2017) – Part II*. Volume 72 of Smart Innovation, Systems and Technologies, ed. by I. Czarnowski, J.R. Howlett, C.L. Jain (Springer International Publishing, 2018), pp. 159–169
40. S.G. Weidman, J. O'Sullivan, The limits of distinctive words: re-evaluating literature's gender marker debate. Digit. Sch. Hum. **33**, 374–390 (2018)
41. U. Stańczyk, The class imbalance problem, in construction of training datasets for authorship attribution, in Man-Machine Interactions 4. ed. by A. Gruca, A. Brachman, S. Kozielski, T. Czachórski, AISC, vol. 391 (Springer, Berlin, 2016), pp. 535–547
42. M. Hall, E. Frank, G. Holmes, B. Pfahringer, P. Reutemann, I. Witten, The WEKA data mining software: an update. SIGKDD Explor. **11**(1), 10–18 (2009)
43. J. Bazan, M. Szczuka, The rough set exploration system, in *Transactions on Rough Sets III*, ed. by J.F. Peters, A. Skowron. Lecture Notes in Computer Science, vol. 3400 (Springer, Berlin, 2005), pp. 37–56
44. S. Theodoridis, K. Koutroumbas, *Pattern Recognit*, 4 edn. (Academic Press, 2008)
45. G. Baron, Analysis of multiple classifiers performance for discretized data in authorship attribution, in *Intelligent Decision Technologies 2017: Proceedings of the 9th KES International Conference on Intelligent Decision Technologies (KES-IDT 2017) – Part II*. Volume 73 of Smart Innovation, Systems and Technologies, ed. by I. Czarnowski, J.R. Howlett, C.L. Jain (Springer International Publishing, 2018), pp. 33–42
46. G. Baron, Influence of data discretization on efficiency of Bayesian Classifier for authorship attribution. Procedia Comput. Sci. **35**, 1112–1121 (2014); Knowledge-Based and Intelligent Information & Engineering Systems 18th Annual Conference, KES-2014 Gdynia, Poland, September 2014 Proceedings
47. J.R. Quinlan, *C4.5: Programs for Machine Learning* (Morgan Kaufmann Publishers Inc., San Francisco, CA, USA, 1993)
48. D.M. Farid, L. Zhang, C.M. Rahman, M. Hossain, R. Strachan, Hybrid decision tree and Naive Bayes classifiers for multi-class classification tasks. Expert Syst. Appl. **41**(4, Part 2), 1937–1946 (2014)
49. K. Stąpor, Evaluation of classifiers: current methods and future research directions, in *Proceedings of the Federated Conference on Computer Science and Information Systems (FedCSIS)*. Volume 13 of ACSIS (2017), pp. 37–40

Author Biography

Urszula Stańczyk received the M.Sc. degree in Computer Science, and Ph.D. degree (with honours) in technical sciences with specialisation in Informatics from the Silesian University of Technology (SUT), Gliwice, Poland. Her doctoral dissertation addressed application of elements from Boolean algebra to pre-processing of digital images. An Assistant Professor at the Institute of Informatics, SUT, till 2018, currently she is an Assistant Professor in the Department of Graphics, Computer Vision, and Digital Systems, SUT.

From 2003 to 2010 Editor-in-Chief of the "Activity Report" for the Institute of Informatics, Dr Stańczyk is a member of KES International (www.kesinternational.org), ADAA Group (http://adaa.polsl.pl/), MIR Labs (http://www.mirlabs.org/), International Rough Set Society (http://www.roughsets.org/), and a member of the Editorial Board of Intelligent Decision Technologies: An International Journal (http://www.iospress.nl/journal/intelligent-decision-technologies/). She is a member of Programme Committees for many scientific conferences, and one of key persons responsible for establishing a series of International Conferences on Man-Machine Interactions (ICMMI). She co-chaired Invited Sessions on Intelligent Data Analysis and Applications. Dr Stańczyk was an invited keynote speaker at KES 2020. She supports as reviewer many scientific journals included in JCR.

Her research interests include artificial intelligence, pattern recognition and classification, neural and rough processing, feature extraction, selection and reduction, induction of decision rules, rule quality measures and filtering, stylometric processing of texts, data mining.

She co-edited conference proceedings, and two multi-authored monographs on feature selection, authored and co-authored a two-volume monograph on synthesis and analysis of logic circuits, academic textbooks dedicated to arithmetic of digital systems, book chapters, conference papers, and journal articles focused on various applications of computational intelligence techniques to stylometry.

Chapter 4
Advances in Fuzzy Clustering Used in Indicator for Individuality

Mika Sato-Ilic

Abstract Clustering is one well-known data analysis used to summarize a large amount of data into a much small number of clusters based on the similarity between a pair of objects. For this purpose, it is self-evident that clustering is expected to be a powerful technique for dealing with the challenges of complex data. However, conventional clustering techniques still have problems with such data due to the strong noise and fluctuation of scalability. Fuzzy clustering is expected to overcome these problems since it has better performances such as robustness of the solution, tractability, less computation, and can obtain more accurate results for the real-world data. Recently, "new integration techniques of fuzzy clustering" to the conventional analysis have been developed for several kinds of complex data. In this chapter, we describe the advances of fuzzy clustering integrated to convex clustering for extracting differences of subjects based on data consisted of objects, variables, and subjects.

Keywords Fuzzy clustering · Convex clustering · Multidimensional scaling · A wearable activity recognition system

4.1 Introduction

Recently, the amount of data has become increasingly large and complex. This increased complexity of the data structure makes it more challenging to capture the latent vital structure of the data. For example, data which is consisted of objects with respect to variables through several subjects is a typical large and complex data. We call such data 3-way data. An example of the 3-way data is a data of times (objects), attributes (variables), and subjects. For example, when several persons (several subjects) wore sensor units on their bodies and did the same activity for a period of times. Then we can obtain data that consists of times (objects) in the period,

M. Sato-Ilic (✉)
Faculty of Engineering, Information and Systems, University of Tsukuba, Tennodai 1-1-1, Tsukuba 305-8573, Ibaraki, Japan
e-mail: mika@risk.tsukuba.ac.jp

M. Virvou et al. (eds.), *Advances in Selected Artificial Intelligence Areas*, Learning and Analytics in Intelligent Systems 24,
https://doi.org/10.1007/978-3-030-93052-3_4

measurement attributes (variables) of the sensors, and several subjects. In this case, when we consider creating a wearable activity recognition system by using this data, we tend to use the general tendency of all the subjects to the system. However, in this case, we cannot consider the individual difference of the subjects, and if we can include such individual difference in the system, then we can create a more suitable user-friendly system. Therefore, we need an indicator to obtain a score that indicates individuality rapidly. So, this chapter describes a methodology using an indicator based on a latent structure of 3-way data to obtain the individual difference [1]. The latent structure is obtained as each result of the dynamic fuzzy clustering [2] as each representative for each subject. In this case, the obtained clusters over the subjects are mathematically the same. Therefore, we can utilize these clusters as a common scale to measure the individual difference in the indicator. The indicator is proposed by using an idea of the objective function of convex clustering [3, 4]. By applying the dynamic fuzzy clustering result to the objective function of convex clustering, we can measure how much each representative of each subject fits the optimality of the convex clustering. Therefore, the score of the objective function for each subject can show the individual difference through the optimality of the convex clustering. The values of the indicator are mathematically comparable in the same criterion based on the objective function of convex clustering, so we can rapidly detect the individual difference of the subjects.

Moreover, if we consider the wearable activity recognition system, then the score of the indicator of individuality has to be robust for the difference of the captured latent structure of data which shows each representative of each subject. Based on this methodology, scores of the indicator based on different latent structures from the same data, for example, obtained result of the dynamic fuzzy clustering and obtained result of multidimensional scaling (MDS) [5–7], are also mathematically comparable. Therefore, we can investigate the robustness of individual scores by using different methods, such as dynamic fuzzy clustering or multidimensional scaling, in indicating differences among subjects. In numerical examples, we show a better performance of the proposed indicator of individuality by showing the robustness of individual scores in indicating differences among subjects. In addition, we show a better performance by discussing with comparisons of the result of the method that uses the indicator and the results of cases when we use the direct difference between a pair of distinct individual data matrixes corresponding to each subject of original data.

This chapter is organized as follows: In Sect. 4.2, fuzzy clustering methods for data with respect to variables [8] and dissimilarity among objects [9] are explained, and the dynamic fuzzy clustering method [2] is described. In Sect. 4.3, we describe convex clustering. In Sect. 4.4, the indicator for the individual difference [1] is described. In Sect. 4.5, numerical examples using the indicator are explained, and in Sect. 4.6, several conclusions and future directions of the work are described.

4.2 Fuzzy Clustering

Suppose X is a given data matrix consisting of n objects with respect to p variables as follows:

$$X = (x_{ia}), i = 1, \cdots, n, a = 1, \cdots, p. \tag{4.1}$$

The purpose of clustering is to classify the n objects shown in Eq. (4.1) into K clusters. The state of fuzzy clustering is represented by a partition matrix:

$$U = (u_{ik}), i = 1, \cdots, n, k = 1, \cdots, K$$

whose elements show the degree of belongingness of the objects to the clusters. In general, u_{ik} satisfies the following conditions:

$$u_{ik} \in [0, 1], \sum_{k=1}^{K} u_{ik} = 1. \tag{4.2}$$

The fuzzy c-means (FCM) method [8, 9] is one of the methods of fuzzy clustering which was originally formulated by using an idea of fuzzy logic [10, 11] and recently is well-known to provide better insight for capturing the latent structure of large amount and complex data [12, 13]. The purpose of this clustering method is to obtain solutions U and $\boldsymbol{v}_k = (v_{k1}, \cdots, v_{kp}), k = 1, \cdots, K$ which minimize the following weighted within-class sum of squares:

$$J(U, \boldsymbol{v}_1, \cdots, \boldsymbol{v}_K) = \sum_{i=1}^{n}\sum_{k=1}^{K} u_{ik}^m d^2(\boldsymbol{x}_i, \boldsymbol{v}_k), \tag{4.3}$$

where $\boldsymbol{v}_k = (v_{k1}, \cdots, v_{kp})$ denotes the values of the centroid of a cluster k, $\boldsymbol{x}_i = (x_{i1}, \cdots, x_{ip})$ is ith object, and $d^2(\boldsymbol{x}_i, \boldsymbol{v}_k)$ is the square Euclidean distance between $\boldsymbol{x}_i$ and $\boldsymbol{v}_k$. The exponent m which determines the degree of fuzziness of the clustering is chosen from $(1, \infty)$ in advance. By minimizing Eq. (4.3), we obtain the solutions U and $\boldsymbol{v}_1, \cdots, \boldsymbol{v}_K$ as follows:

$$u_{ik} = \left[\sum_{j=1}^{K}\left(\frac{d(\boldsymbol{x}_i, \boldsymbol{v}_k)}{d(\boldsymbol{x}_i, \boldsymbol{v}_j)}\right)^{\frac{2}{m-1}}\right]^{-1}, i = 1, \cdots, n, k = 1, \cdots, K. \tag{4.4}$$

$$\boldsymbol{v}_k = \frac{\sum_{i=1}^{n} u_{ik}^m \boldsymbol{x}_i}{\sum_{i=1}^{n} u_{ik}^m}, k = 1, \cdots, K.$$

Suppose Z_t be a given data matrix consisted of n objects with respect to p variables at a subject t called a 3-way data and shown as follows:

$$Z_t = \left(z_{ir}^{(t)}\right),\ i = 1, \cdots, n,\ r = 1, \cdots, p,\ t = 1, \cdots, T. \tag{4.5}$$

In order to obtain the same clusters over the T subjects, the following $nT \times p$ super matrix $\tilde{Z}$ is created.

$$\tilde{Z} = \begin{pmatrix} Z_1 \\ \vdots \\ Z_T \end{pmatrix} = (\tilde{z}_{jr}), j = 1, \cdots, nT, r = 1, \cdots, p. \tag{4.6}$$

The purpose of this fuzzy clustering is to classify the nT objects into K clusters. The state of the fuzzy clustering is represented by a partition matrix:

$$\tilde{U} = \begin{pmatrix} U_1 \\ \vdots \\ U_T \end{pmatrix} = (\tilde{u}_{jk}),\ U_t = \left(u_{ik}^{(t)}\right),\ j = 1, \cdots, nT,\ i = 1, \cdots, n,$$
$$k = 1, \cdots, K,\ t = 1, \cdots, T. \tag{4.7}$$

Where $\tilde{u}_{jk}$ is a degree of belongingness of an object j which is shown as $\tilde{z}_j = (\tilde{z}_{j1}, \cdots, \tilde{z}_{jp})$ to a fuzzy cluster k and $u_{ik}^{(t)}$ is a degree of belongingness of an object i to the same fuzzy cluster k at a subject t. From Eq. (4.7), it can be seen that the obtained K fuzzy clusters are the same over T subjects. In general, $\tilde{u}_{jk}$ satisfies the following conditions:

$$\tilde{u}_{jk} \in [0, 1], \sum_{k=1}^{K} \tilde{u}_{jk} = 1, j = 1, \cdots, nT. \tag{4.8}$$

Then the objective function of dynamic fuzzy clustering [2] is defined by using FCM as follows:

$$J\left(\tilde{U}, g_1, \cdots, g_K\right) = \sum_{j=1}^{nT} \sum_{k=1}^{K} \tilde{u}_{jk}^m d^2\left(\tilde{z}_j, g_k\right), \tag{4.9}$$

where $g_k = (g_{k1}, \cdots, g_{kp})$ denotes the values of the center of a cluster k, $d^2\left(\tilde{z}_j, g_k\right)$ is the squared Euclidean distance between $\tilde{z}_j$ and g_k. The exponent m that determines the degree of fuzziness of the clustering is chosen from $(1, \infty)$ in advance. By minimizing the objective function shown in Eq. (4.9) under the conditions shown in Eq. (4.8), we obtain the solutions $\tilde{U}, \boldsymbol{g}_1, \cdots, \boldsymbol{g}_K$.

By using the data shown in Eq. (4.1), we obtain the following squared Euclidean distance

$$d_{ij} = \sum_{a=1}^{p} (x_{ia} - x_{ja})^2, i, j = 1, \cdots, n. \tag{4.10}$$

In order to obtain a clustering result from $D = (d_{ij})$ shown in Eq. (4.10), we use a fuzzy clustering method named FANNY algorithm [14]. The objective function of the FANNY algorithm is defined as follows:

$$J(U) = \sum_{k=1}^{K} \left(\sum_{i=1}^{n} \sum_{j=1}^{n} u_{ik}^m u_{jk}^m d_{ij} / \left(2 \sum_{l=1}^{n} u_{lk}^m \right) \right). \tag{4.11}$$

By minimizing Eq. (4.11) under the conditions shown in Eq. (4.2), we obtain the solution U.

4.3 Convex Clustering

Convex clustering is a type of clustering method in which we obtain clustering results by solving a convex optimization problem. The idea is essentially based on the sparsity regularization of regression [15].

Suppose X be a given data matrix consisted of n objects and p variables as follows:

$$X = \begin{pmatrix} x_{11} & \cdots & x_{1p} \\ \vdots & \vdots & \vdots \\ x_{n1} & \cdots & x_{np} \end{pmatrix} = \begin{pmatrix} \boldsymbol{x}_1 \\ \vdots \\ \boldsymbol{x}_n \end{pmatrix}, \boldsymbol{x}_i = (x_{i1}, \cdots, x_{ip}), i = 1, \cdots, n. \tag{4.12}$$

Then suppose $\boldsymbol{q}_i$ be a centroid for the cluster containing $\boldsymbol{x}_i$ as follows:

$$Q = \begin{pmatrix} q_{11} & \cdots & q_{1p} \\ \vdots & \vdots & \vdots \\ q_{n1} & \cdots & q_{np} \end{pmatrix} = \begin{pmatrix} \boldsymbol{q}_1 \\ \vdots \\ \boldsymbol{q}_n \end{pmatrix}, \boldsymbol{q}_i = (q_{i1}, \cdots, q_{ip}), i = 1, \cdots, n. \tag{4.13}$$

The purpose of convex clustering is to obtain $\boldsymbol{q}_i$ which minimizes the following function:

$$F_\omega(Q) = \frac{1}{2}\sum_{i=1}^{n} d^2(\boldsymbol{x}_i, \boldsymbol{q}_i) + \omega \sum_{i<j} d^2(\boldsymbol{q}_i, \boldsymbol{q}_j). \tag{4.14}$$

In Eq. (4.14), $d^2(\boldsymbol{x}_i, \boldsymbol{q}_i)$ is the squared Euclidean distance between $\boldsymbol{x}_i$ and $\boldsymbol{q}_i$ and $d^2(\boldsymbol{q}_i, \boldsymbol{q}_j)$ is the squared Euclidean distance between $\boldsymbol{q}_i$ and $\boldsymbol{q}_j$. ω is a given parameter in which if $\omega = 0$, then we simply obtain a solution as $\boldsymbol{x}_i = \boldsymbol{q}_i, \forall i$, that is, we obtain n clusters in which only one object belongs to one cluster. If $\omega \to \infty$, then we obtain one cluster containing all objects. When we obtain the optimum solution, several values of the squared Euclidean distances between a pair of centroids are zeros. For example, if $d^2(\boldsymbol{q}_s, \boldsymbol{q}_l) = 0$, then $\boldsymbol{x}_s$ and $\boldsymbol{x}_l$ belong to the same cluster. Essentially this shows a merging process of the clustering.

4.4 Indicator of Individuality

Suppose we observe 3-way data shown in Eq. (4.5). Then based on the objective function of convex clustering shown in Eq. (4.14) when $\omega = 1$, the indicator of subject t based on a result of dynamic FCM as follows:

$$C_{FCM}(t) = \frac{1}{2}\sum_{i=1}^{n} d^2(\boldsymbol{z}_i^{(t)}, \boldsymbol{u}_i^{(t)}) + \sum_{i<j} d^2(\boldsymbol{u}_i^{(t)}, \boldsymbol{u}_j^{(t)}), t = 1, \cdots, T, \tag{4.15}$$

where $\boldsymbol{u}_i^{(t)} = (u_{i1}^{(t)}, \cdots, u_{ip}^{(t)})$ is a result of dynamic FCM for an object i at tth subject when $K = p$ shown in Eq. (4.7), when we apply data $\tilde{Z}$ shown in Eq. (4.6) to the dynamic fuzzy clustering method shown in Eq. (4.9). In Eq. (4.15), $U_t = \begin{pmatrix} \boldsymbol{u}_1^{(t)} \\ \vdots \\ \boldsymbol{u}_n^{(t)} \end{pmatrix}$ is representative of tth subject, and the value of $C_{FCM}(t)$ shows how much the representative of tth subject fits to the optimality of the convex clustering. And by comparisons among $\{C_{FCM}(1), \cdots, C_{FCM}(T)\}$, we can obtain the individual difference.

As another latent structure for the representative of tth subject, the indicator of subject t based on a result of Multidimensional scaling (MDS) is defined as follows:

$$C_{MDS}(t) = \frac{1}{2}\sum_{i=1}^{n} d^2(\boldsymbol{z}_i^{(t)}, \hat{\boldsymbol{x}}_i^{(t)}) + \sum_{i<j} d^2(\hat{\boldsymbol{x}}_i^{(t)}, \hat{\boldsymbol{x}}_j^{(t)}), t = 1, \cdots, T, \tag{4.16}$$

where $\boldsymbol{z}_i^{(t)} = (z_{i1}^{(t)}, \cdots, z_{ip}^{(t)})$ shown in Eq. (4.5) is a data of an object i at tth subject and $\hat{\boldsymbol{x}}_i^{(t)} = (\hat{x}_{i1}^{(t)}, \cdots, \hat{x}_{ip}^{(t)})$ is a result of MDS for an object i at tth subject when we apply data Z_t shown in Eq. (4.5) to MDS. The purpose of MDS is to obtain n points in a r ($r < n$) lower-dimensional space while holding approximately the same dissimilarity relationship among objects. Then we can reduce the number of dimensions for capturing efficient information from observed data by representing the data structure in a lower-dimensional spatial space. As a metric MDS, the following model has been proposed.

$$\tilde{d}_{ij} = \left\{ \sum_{\lambda=1}^{r} (\tilde{x}_{i\lambda} - \tilde{x}_{j\lambda})^2 \right\}^{\frac{1}{2}} + \varepsilon_{ij}, i, j = 1, \cdots, n. \tag{4.17}$$

In Eq. (4.17), $\tilde{d}_{ij}$ is a dissimilarity between objects i and j and $\tilde{x}_{i\lambda}$ is a point of an object i with respect to dimension λ in $r(r < n)$ dimensional configuration space. ε_{ij} is an error. When we observe data $X = (x_{ia})$ shown in Eq. (4.1), $\tilde{d}_{ij}$ is usually calculated by using Euclidean distance between objects i and j as follows:

$$\tilde{d}_{ij} = \left\{ \sum_{a=1}^{p} (x_{ia} - x_{ja})^2 \right\}^{\frac{1}{2}}, i, j = 1, \cdots, n. \tag{4.18}$$

That is, MDS finds r dimensional scaling (coordinate)$(\tilde{x}_{i1}, \cdots, \tilde{x}_{ir})$ and throws light on the structure of similarity relationship among the objects by representing the $\tilde{d}_{ij}$ shown in Eq. (4.18) as the distance between a point $(\tilde{x}_{i1}, \cdots, \tilde{x}_{ir})$ and a point $(\tilde{x}_{j1}, \cdots, \tilde{x}_{jr})$ in r dimensional space. Then $\tilde{X} = (\tilde{x}_{ia})$ can be estimated as follows:

$$\widehat{X} = \tilde{H}\tilde{\Delta}^{\frac{1}{2}}, \tag{4.19}$$

where $\widehat{X} = \left(\widehat{x}_{i\lambda}\right), i = 1, \cdots, n, \lambda = 1, \cdots, r$ shows values of coordinate of n objects in r dimensional space when $r < p < n$. $\tilde{H}$ is a matrix consisted of $\boldsymbol{h}_1, \cdots, \boldsymbol{h}_r, \boldsymbol{h}_\lambda = (h_{1\lambda}, \cdots, h_{n\lambda})', \lambda = 1, \cdots, r$ eigen vectors corresponding to $\delta_1, \cdots, \delta_r$ eigen values which satisfy $\delta_1 > \cdots > \delta_r$, by using eigenvalue decomposition of $P = -\frac{1}{2}JD^2J$, where $D^2 = (\tilde{d}_{ij}^2)$ and the matrix J is a symmetric matrix whose diagonal elements are $1 - 1/n$ and nondiagonal elements are $-1/n$. $\tilde{\Delta}^{\frac{1}{2}}$ is a diagonal matrix whose diagonal elements are values $\sqrt{\delta_1}, \cdots, \sqrt{\delta_r}$ [16]. In Eq. (4.16), $\widehat{\boldsymbol{x}}_i^{(t)} = (\widehat{x}_{i1}^{(t)}, \cdots, \widehat{x}_{ip}^{(t)})$ is a result of MDS for an object i at tth subject when we apply data Z_t shown in Eq. (4.5) to MDS represented in Eq. (4.19) when $r = p$.

In Eq. (4.16), $\widehat{X}^{(t)} = \begin{pmatrix} \widehat{\boldsymbol{x}}_1^{(t)} \\ \vdots \\ \widehat{\boldsymbol{x}}_n^{(t)} \end{pmatrix}$ is representative of tth subject, and the value of $C_{MDS}(t)$ shows how much the representative of tth subject fits to the optimality of the convex clustering. And by comparisons among $\{C_{MDS}(1), \cdots, C_{MDS}(T)\}$, we can obtain the individual difference.

Moreover, from Eqs. (4.15) and (4.16), we can compare the values of the indicator based on different latent structures which are $\widehat{X}^{(t)}$ and U_t from the same subject t. That is, we can compare values for each pair $(C_{MDS}(t), C_{FCM}(t)), t = 1, \cdots, T$. From this, we can investigate the robustness of individual scores by using different representatives in indicating differences among subjects.

4.5 Numerical Examples

We use a dataset of sensor data of daily and sports activities performed by 8 subjects with respect to 45 variables [17, 18]. The dataset consists of 19 regular daily and sports activities, and we select the activity of running on a treadmill at a speed of 8 km/h. For this activity, we use data that consisted of 7,500 times, 45 variables. For the variables, the data was observed for 45 (5 body positions × 9 kinds of sensors) variables which consist of 5 body-worn sensor units positioned on the torso (T), right arm (RA), left arm (LA), right leg (RL), left leg (LL), and each sensor unit has 9 kinds of information including x-axial accelerometers (xacc), y-axial accelerometers (yacc), z-axial accelerometers (zacc), x-axial gyroscopes (xgyro), y-axial gyroscopes (ygyro), z-axial gyroscopes (zgyro), x-axial magnetometers (xmag), y-axial magnetometers (ymag), z-axial magnetometers (zmag).

Figure 4.1 shows a result of values of the indicator based on dynamic FCM when $m = 2$ and MDS shown in Eqs. (4.15) and (4.16), respectively for the first 1,000 times which correspond to the first 40 s of data and whose number of times

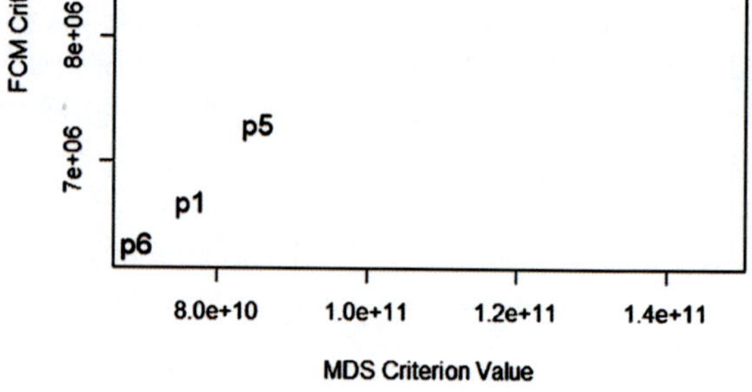

Fig. 4.1 Relationship between the results of indicator based on different latent structures (Times from 1 to 1000)

is 1000. In this figure, the ordinate shows the values of $\{C_{FCM}(1), \cdots, C_{FCM}(8)\}$ shown in Eq. (4.15), and the abscissa shows the values of $\{C_{MDS}(1), \cdots, C_{MDS}(8)\}$ shown in Eq. (4.16). The point "pt" shows a point of tth subject at the coordinate of $(C_{MDS}(t), C_{FCM}(t)), t = 1, \cdots, 8$. From this figure, it can be seen that the indicator based on different latent structures, which are the result of dynamic FCM at tth subject, U_t, and the result of MDS at the same subject, $\widehat{X}^{(t)}$, is an almost linear relationship in indicating difference among 8 subjects. From this, the indicator based on the objective function of convex clustering shows robustness for the difference of the kinds of representative, which are latent structures of data obtained from different kinds of methods such as dynamic FCM and MDS.

Figure 4.2a–m show figures of values of the coordinate of $(C_{MDS}(t), C_{FCM}(t)), t = 1, \cdots, 8$ over the several periods of times. From these figures, it can be seen that the values of the indicator based on different latent structures are almost a linear relationship in indicating difference among 8 subjects for all of the periods of times. Therefore, the indicator has robustness for the difference of the kinds of latent structures regardless of time periods.

Figure 4.3 shows the relationship between the difference of a pair of subjects of data and the difference of the same pair of the subjects of the results of dynamic FCM. We use the first 40 s of data and whose number of times is 1000. The difference of a pair of subjects t and t' of data, $\delta_{tt'}$, is calculated as follows:

$$\delta_{tt'} = \sum_{i=1}^{n} \sum_{r=1}^{p} (z_{ir}^{(t)} - z_{ir}^{(t')})^2, \quad t, t' = 1, \cdots, 8, \quad t \neq t', \tag{4.20}$$

and difference of the same pair of the subjects t and t' of the results of dynamic FCM, $\tilde{\delta}_{tt'}$, is calculated as follows:

$$\tilde{\delta}_{tt'} = \sum_{i=1}^{n} \sum_{k=1}^{p} (u_{ik}^{(t)} - u_{ik}^{(t')})^2, \quad t, t' = 1, \cdots, 8, \quad t \neq t'. \tag{4.21}$$

In the case of dynamic FCM, since we can obtain the same p clusters over the 8 subjects, we can compare mathematically the results of dynamic FCM of different subjects t and t'. Therefore, we can calculate $\tilde{\delta}_{tt'}$ such as Eq. (4.21). In Fig. 4.3, the abscissa shows the values of $\delta_{tt'}$ shown in Eq. (4.20), and the ordinate shows the values of $\tilde{\delta}_{tt'}$ shown in Eq. (4.21). The point "dtt'" is a point that shows a difference between subjects t and t' at the coordinate of $(\delta_{tt'}, \tilde{\delta}_{tt'}), t, t' = 1, \cdots, 8 (t \neq t')$. From this figure, it can be seen that there is no monotone relationship between the two differences. That is, the result of dynamic FCM captures the difference of subjects which differs from the direct difference of subjects of the original data.

However, through the indicator, the difference of subjects based on the result of dynamic FCM has a clear monotone relationship with the difference of subjects based on the result of MDS, which is shown in Fig. 4.1. So, the indicator can obtain the

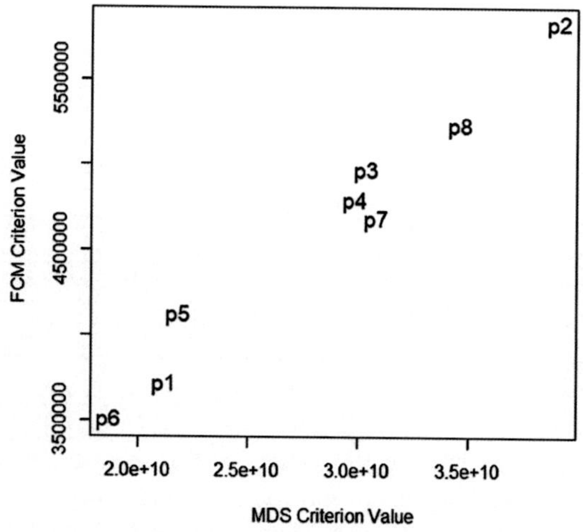

(a) Times from 1001 to 1500

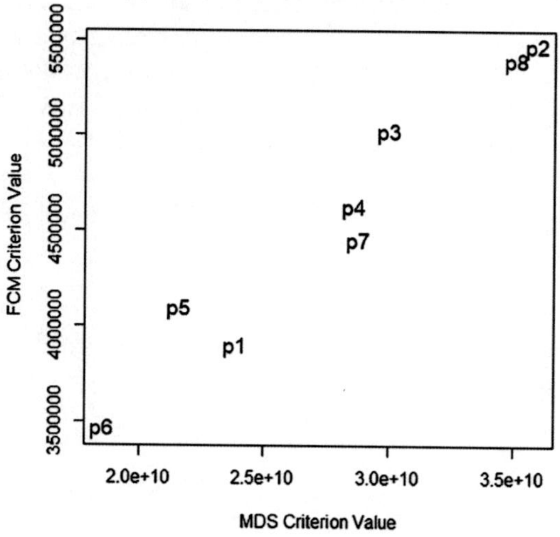

(b) Times from 1501 to 2000

Fig. 4.2 Relationship between the results of indicator based on different latent structures over the different periods of times

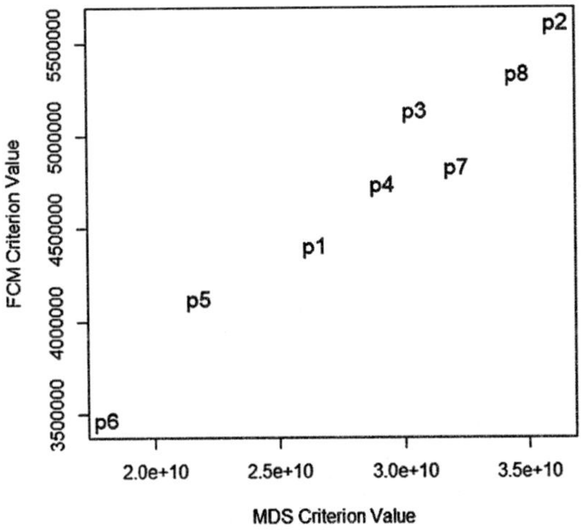

(c) Times from 2001 to 2500

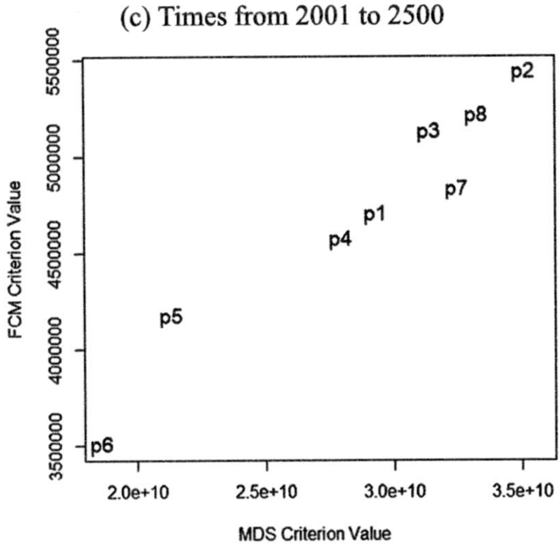

(d) Times from 2501 to 3000

Fig. 4.2 (continued)

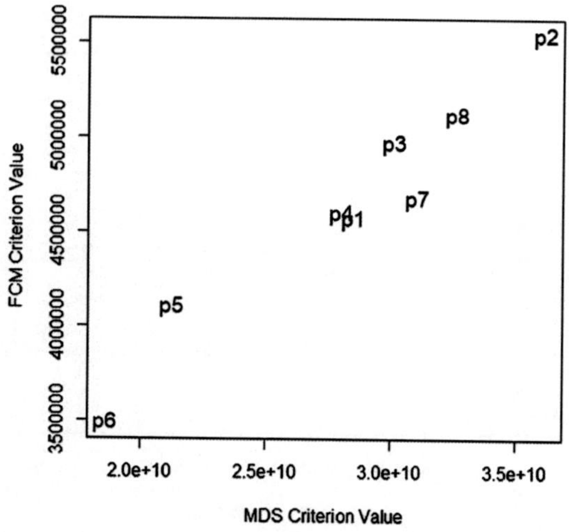

(e) Times from 3001 to 3500

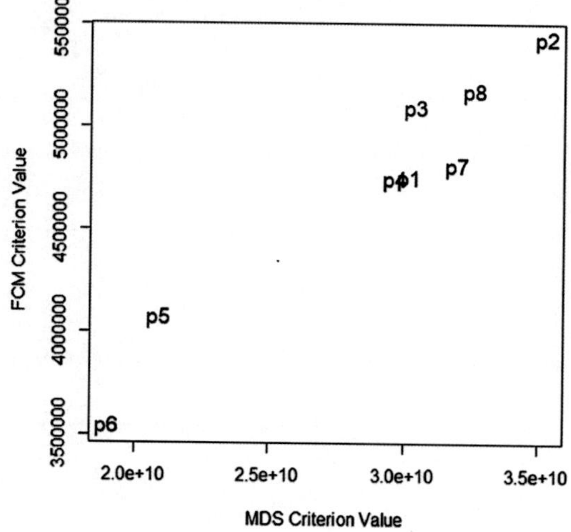

(f) Times from 3501 to 4000

Fig. 4.2 (continued)

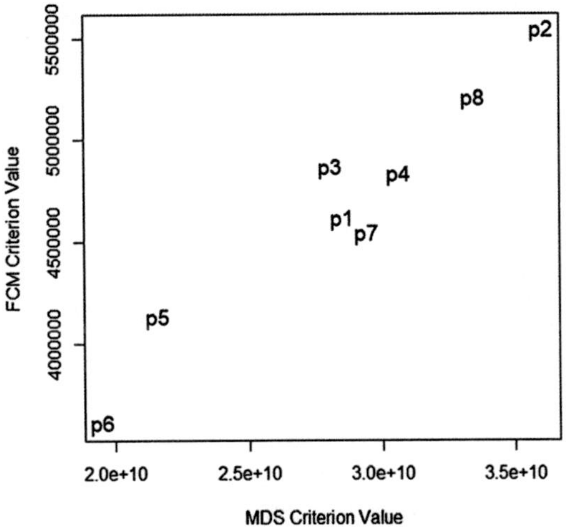

(g) Times from 4001 to 4500

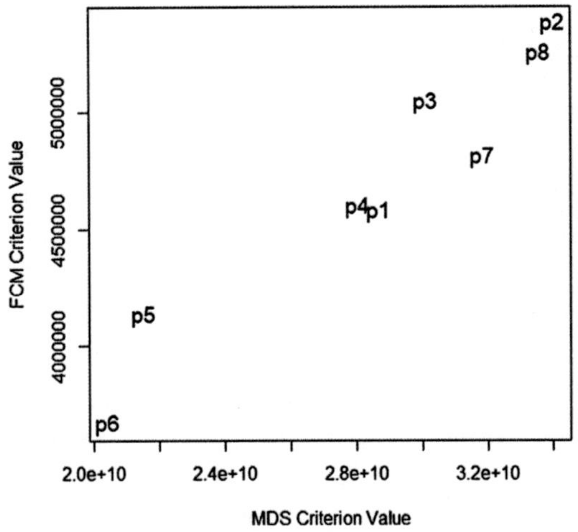

(h) Times from 4501 to 5000

Fig. 4.2 (continued)

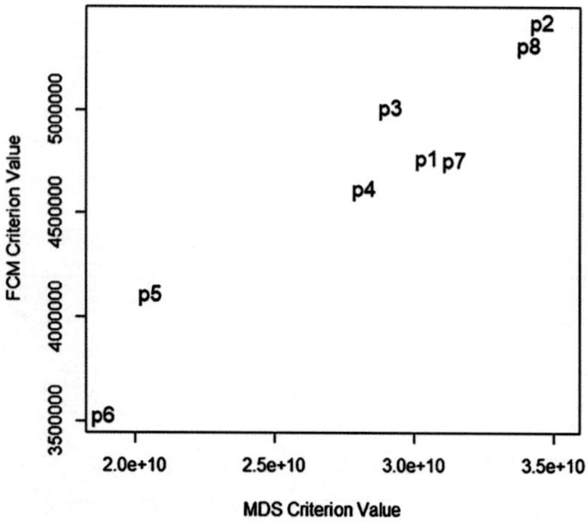

(i) Times from 5001 to 5500

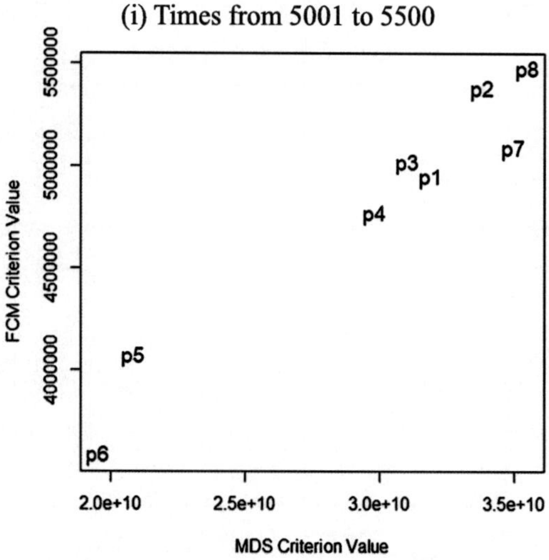

(j) Times from 5501 to 6000

Fig. 4.2 (continued)

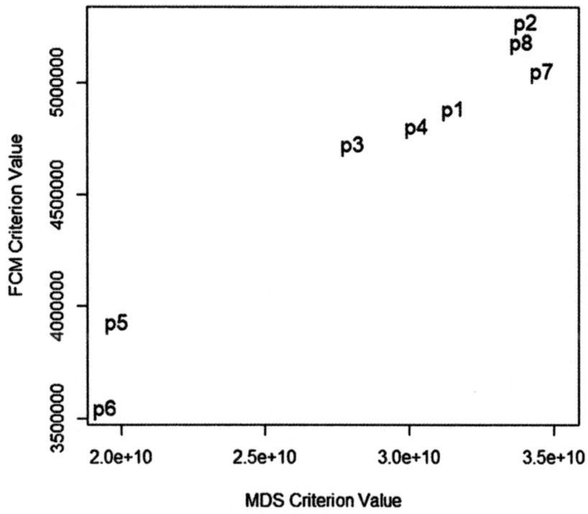

(k) Times from 6001 to 6500

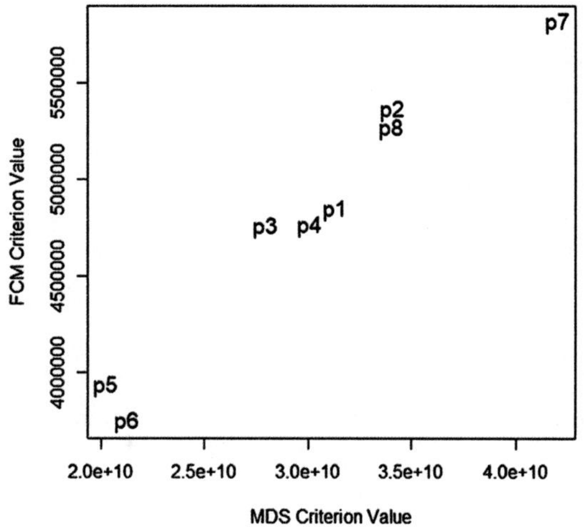

(l) Times from 6501 to 7000

Fig. 4.2 (continued)

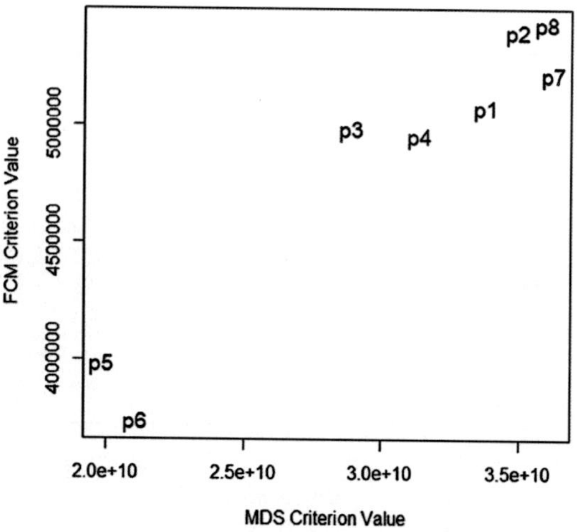

(m) Times from 7001 to 7500

Fig. 4.2 (continued)

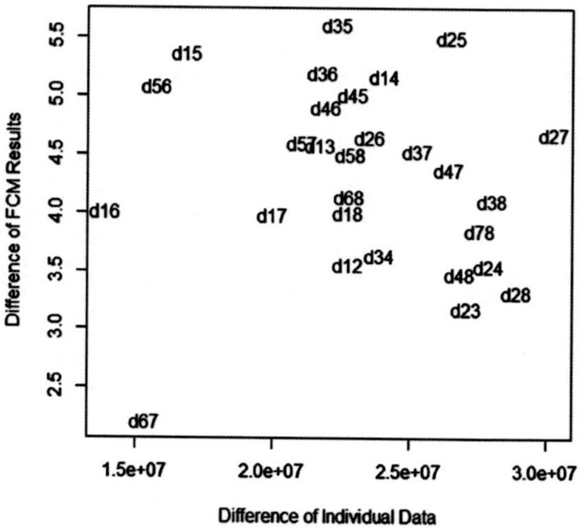

Fig. 4.3 Relationship between the difference of data and the difference of dynamic FCM results for each pair of subjects

difference of individuality of each subject, which has consistency for the captured latent structures and is different from simply calculating the difference from different data of different subjects.

In order to obtain clearer difference among 8 subjects from the directly calculated dissimilarity between a pair of subjects t and t' of data, $\delta_{tt'}$ shown in Eq. (4.20), we apply the dissimilarity $\delta_{tt'}$ to FANNY shown in Eq. (4.11) and obtain a result of fuzzy clustering when $m = 1.5$. We use the first 40 s of data and whose number of times is 1000. The result of FANNY is shown in Fig. 4.4. In this figure, the abscissa shows

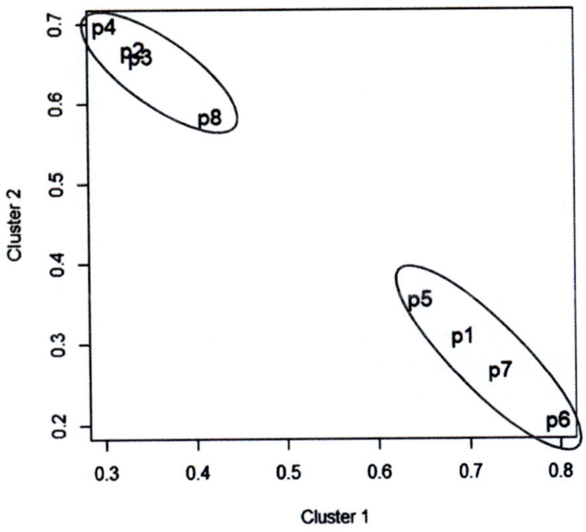

Fig. 4.4 Result of FANNY applied dissimilarity among subjects of original data

the values of degree of belongingness of objects to cluster 1, and the ordinate is the values of degree of belongingness of objects to cluster 2. The number of clusters is assumed to be 2. From this result, we can see two groups based on the similarity of subjects; one is a group of subjects 1, 5, 6, and 7, and another is a group of subjects 2, 3, 4, and 8. However, in Fig. 4.1, we cannot find such groups, and we can see different three groups such as a group of subjects 1, 5, and 6; a group of subjects 2 and 8; and a group of subjects 3, 4, and 7. From this, again, we can see that the proposed indicator captures different individuality subjects from the simply calculated difference among original data for subjects, and the captured individuality has consistency with the use of the differently captured latent structure of data such as a result of MDS.

4.6 Conclusions and Future Work

This chapter describes an indicator that shows the individuality of each subject by utilizing a fuzzy clustering result as a common scale over the subjects. The target data is 3-way data consisted of objects, variables, and subjects. The indicator is based on an idea of the objective function of convex clustering. The objective function is essentially used to obtain the clustering result, and the result is represented as unknown parameters in the objective function, which should be found in the algorithm.

However, in this method, conversely, we give the clustering result in the objective function as the latent structure of data which is a result of dynamic fuzzy clustering for each subject and obtain each value for each subject which can measure how much each latent structure of each subject fits the optimality of the convex clustering. Then, the value of the objective function for each subject can show individuality through the optimality of the convex clustering. Several numerical examples show a better performance of the indicator of individuality. In particular, we show that even if we use a different type of latent structure of data, such as a result of multidimensional scaling for each subject, through the optimality of the convex clustering in the indicator, the results of the indicator have consistency in the sense of difference of subjects. In addition, the numerical examples show that the indicator can capture the different features of subjects from the feature obtained when we directly calculate the difference of subjects based on original data.

In future studies, examinations of the indicator using more diverse latent structures captured by different types of clustering algorithms for adjusting larger amounts of data, including higher dimensional data, is necessary to know the generalization performance. In addition, we have to develop the adaptable tuning method for parameters of the convex clustering and fuzzy clustering methods.

References

1. M. Sato-Ilic, Indicator for individuality of subjects based on similarity of objects. Procedia Comput. Sci. Elsevier **185**, 193–202 (2021)
2. M. Sato-Ilic, Individual compositional cluster analysis. Procedia Comput. Sci. Elsevier **95**, 254–263 (2016)
3. F. Lindsten, H. Ohlsson, L. Ljung, *Just Relax and Come Clustering! A Convexification of k-means Clustering* (Linköpings universitet, Tech. rep., 2011)
4. E.C. Chi, K. Lange, Splitting methods for convex clustering. J. Comput. Graph. Stat. **24**, 994–1013 (2015)
5. J.C. Gower, Some distance properties of latent roots and vector methods used in multivariate analysis. Biometrika **53**, 325–338 (1966)
6. J.B. Kruskal, M. Wish, *Multidimensional Scaling* (Sage Publications, 1978)
7. W.S. Torgerson, *Theory and Methods of Scaling* (Wiley, New York, 1958)
8. J.C. Bezdek, *Pattern Recognition with Fuzzy Objective Function Algorithms* (Plenum, 1981)
9. J.C. Dunn, A fuzzy relative of the ISODATA process and its use in detecting compact well-separated clusters. J. Cybern. **3**(3), 32–57 (1973)
10. E.H. Ruspini, A new approach to clustering. Inform. Control **15**(1), 22–32 (1969)
11. L.A. Zadeh, Fuzzy sets. Inform. Control **8**, 338–353 (1965)
12. F. Klawonn, R. Kruse, R. Winkler, Fuzzy clustering: more than just fuzzification. Fuzzy Sets Syst. **281**, 272–279 (2015)
13. M.B. Ferraro, P. Giordani, Soft clustering. WIREs Comput. Stat. (2020)
14. L. Kaufman, P.J. Rousseeuw, *Finding Groups in Data: An Introduction to Cluster Analysis* (Wiley, 2005)
15. M. Yuan, Y. Lin, Model selection and estimation in regression with grouped variables. J. R. Stat. Soc. B **68**, 49–67 (2006)
16. G. Young, A.S. Householder, Discussion of a set of points in terms of their mutual distances. Psychometrika **3**, 19–22 (1938)
17. K. Altun, B. Barshan, Human activity recognition using inertial/magnetic sensor units, in *HBU 2010. LNCS 6219*, eds. by A.A. Salah, T. Gevers, N. Sebe, A. Vinciarelli (Springer, Berlin, Heidelberg, 2020), pp. 38–51
18. UCI Machine Learning Repository, http://archive.ics.uci.edu/ml/index.html

Mika Sato-Ilic holds the position of Professor in the Faculty of Engineering, Information and Systems at the University of Tsukuba. She is also a Special Counselor and former Vice President in the National Statistics Center of Japan, a committee member in the Financial Services Agency of Japan, and a member of the Science Council of Japan. She was an Invited Professor at the University of Paris in the Department of Information (Machine Learning, AI/Data Sciences) and a Research Contractor at the University of South Australia.

She is the founding Editor-in-Chief of the International Journal of Knowledge Engineering and Soft Data Paradigms, Associate Editor of the IEEE Transactions on Fuzzy Systems, Neurocomputing, Information Sciences, and on the editorial board of several journals. She is an ISI (International Statistical Institute) Elected Member and the Japan representative in a committee of women in ISI and was a Council of the International Association for Statistical Computing (a section of the International Statistical Institute), a Senior Member of the IEEE where she held several positions, including the Vice-Chair of the Fuzzy Systems Technical Committee of the IEEE Computational Intelligence Society, and worked on several IEEE committees.

Her academic output includes four books, 14 book chapters, and over 140 journal and conference papers. Her research interests include the development of methods for data mining, multi-dimensional data analysis, multi-mode multi-way data theory, pattern classification, and computational intelligence techniques, for which she has received 24 academic awards.

Chapter 5
Pushing the Limits Against the No Free Lunch Theorem: Towards Building General-Purpose (GenP) Classification Systems

Alessandra Lumini, Loris Nanni, and Sheryl Brahnam

Abstract In this chapter, we provide an overview of the tools our research group is exploiting to build general-purpose (GenP) classification systems. Although the "no free lunch" (NFL) theorem claims, in effect, that generating a universal classifier is impossible, the goals of GenP systems are more modest in requiring little to no parameter tuning for performing competitively across a range of tasks within a domain or with specific data types, such as images, that span across several fields. The tools outlined here for building GenP systems include methods for building ensembles, matrix representations of data treated as images, deep learning approaches, data augmentation, and classification within dissimilarity spaces. Each of these tools is explained in detail and illustrated with a few examples taken from our work building GenP systems, which spans nearly fifteen years. We note both our successes and some of our limitations. This chapter ends by pointing out some developments in quantum computing and quantum-inspired algorithms that may allow researchers to push the limits hypothesized by the NFL theorem even further.

Acronyms

ACT	Activation Layer
CBD	Compact Binary Descriptor
CLASS	Classification Layer
CLBP	Complete LBP

A. Lumini
University of Bologna, Campus di Cesena, Via Macchiavelli, 47521 Cesena, FC, Italy

L. Nanni
Dipartimento Di Ingegneria Dell'Informazione, University of Padova (Padua), Padova, Italy

S. Brahnam (✉)
Department of Information Systems and Cybersecurity, Missouri State University, Springfield, MO, USA
e-mail: SBrahnam@MissouriState.edu

M. Virvou et al. (eds.), *Advances in Selected Artificial Intelligence Areas*, Learning and Analytics in Intelligent Systems 24,
https://doi.org/10.1007/978-3-030-93052-3_5

CNN	Convolutional Neural Network
CONV	Convolutional Layer
DCT	Discrete Cosine Transform
DT	Decision Trees
FC	Fully-Connected Layer
GenP	General Purpose (classifier)
GOLD	Gaussians of Local Descriptors
GWT	Gabor Wavelet Transform
HASC	Heterogeneous Auto-Similarities of Characteristics
IDE	Input Decimated Ensemble
INC	Inception Module
LBP	Local Binary Pattern
LDA	Linear Discriminant Analysis
LPQ	Local Phase Quantization
LTP	Local Ternary Pattern
ML	Machine Learning
MRELBP	Median Robust Extended LBP
NFL	No Free Lunch (theorem)
PCA	Principal Component Analysis
PCAN	Principal Component Analysis Network
PDV	Pixel Difference Vectors (generated in CBD)
POOL	Pooling Layer
QC	Quantum Computation
QI	Quantum Inspired (algorithms)
RES	Residual Layer
RF	Rotation Forest
RICLBP	Rotation Invariant Co-occurrence LBP
RLBP	Rotated LBP
RS	Random Subspace
STFT	Short-Term Fourier Transform
SVM	Support Vector Machine
TL	Transfer Learning

5.1 Introduction

Within the last couple of decades, the trend in machine learning (ML) has shifted away from focusing on building optimal classification systems for very specific and well-defined problems to seeking solutions that generalize across many domains and over a broad set of recognition tasks. The dream, the ultimate pursuit in ML, is to create an algorithm that works optimally straight out of the box on any problem whatsoever. Let us call such an ideal *the universal classifier*. The first question that arises is whether accomplishing such a goal is even possible.

The answer is no, at least not according to Wolpert's [1] famous "No Free Lunch" (NFL) theorem, which states, in effect, that developing a stand-alone classifier that works well across every possible problem is nothing but a pipe dream. The NFL theorem claims that all optimization algorithms perform equally well when their performances are averaged across all possible problems. What this means is that there is no single best optimization algorithm—a pessimistic view, to say the least. Be that as it may, there are some solid criticisms and challenges facing the NFL theorem. A significant argument against the NFL hypothesis is that it only applies when the target function is selected from a uniform distribution of all possible functions. When it comes to practical problems, one algorithm may do better overall than another [2]. For this reason, some suggest that NFL holds little relevance to ML research. Let us assume some optimism and suppose that certain algorithms do indeed work better on specific practical problems: does that provide us any hope that some algorithms will generalize better than others across a broader range of practical problems?

No one knows. Our research goal is to determine how far the limits can be pushed to generate more generalizable approaches. Thus, our goal is not to pursue the holy grail of developing the universal classifier but rather the more modest pursuit of building what we call General-Purpose (GenP) classifier systems. We fully recognize that GenP systems will be more limited in their generalizability than the ideal held up by the universal classifier.

Our research objective then is to produce GenP systems that possess some of the characteristics of a universal classifier, namely:

- Simplicity and convenience: a GenP system should require little to no parameter tuning for its stated range of tasks;
- Competitive performance against (if not better than) less flexible systems that have been optimized for particular problems and specific datasets;
- Generalizability across domains (e.g., biomedical and biometric imagery) within a particular classification problem (image classification) and across many data sets representative of the problem [3].

Our research team has been working on GenP systems for more than a decade. This chapter briefly describes the basic building blocks, the set of tools we use to push against the NFL theorem. These building blocks include the following:

1. Multiclassifier/ensemble systems (Sect. 5.2);
2. Matrix representations of vector data that are then treated as images (Sect. 5.3);
3. Deep learning approaches, which include building heterogeneous ensembles of deep learners and training more traditional learners on combinations of handcrafted and deep features (Sect. 5.4);
4. Data augmentation techniques (Sect. 5.5);
5. Classification within dissimilarity spaces as well as within feature spaces (Sect. 5.6).

Each of these tools for developing GenP systems builds upon each other and will be described below, along with some examples taken from our work. We will note some of our successes and some of the limitations we have had to confront. We will end this chapter optimistically by pointing out new developments in quantum computing and quantum-inspired algorithms that offer more hope for overcoming the limitations hypothesized by the NFL theorem.

5.2 Multiclassifier/Ensemble Methods

Multiclassifier/ensemble systems aggregate the decisions of a group of classifiers to produce better classification decisions and form the bedrock upon which all our GenP systems are erected. The terms *multiclassifier* and *ensemble* are sometimes distinguished in the literature, as will be defined below, but more often than not, they are used interchangeably, as is the case here. The power of multiclassifier systems depends on the ability of the classifiers that make up the ensemble to meet two fundamental conditions: (1) they cannot be random classifiers, i.e., they must produce an accuracy above 50%, and (2) they must be independent of each other [4], i.e., some diversity must be introjected into them. If all the classifiers in the system produce nearly the same results, nothing new will be introduced into the ensemble, and thus, little if any improvement in recognition or classification will be obtained. In this context, diversity is considered intuitively as the difference among individual members in an ensemble and is measured based on each member's classification results, i.e., two classifiers are considered diverse if they make different errors given new samples. Introjection of diversity is a sine qua non for a successful multiclassifier system.

5.2.1 *Canonical Model of Single Classifier Learning*

A review of the canonical model of single classifier learning is necessary to understand the full power of multiclassifiers. As illustrated in Fig. 5.1, a single classifier system can be divided into three levels [5]:

1. The data level, where some raw data is at a minimum divided into a training and test set;

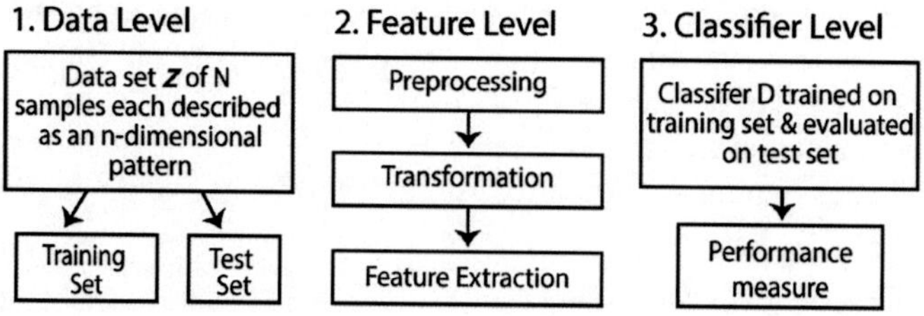

Fig. 5.1 Canonical model: single classifier training and evaluation in a feature space

2. The feature level, where the data is preprocessed and transformed, after which relevant features (sometimes called *descriptors*) are extracted;
3. The classifier level, where the parameters of a learning algorithm are fine-tuned until it optimally learns from the training set of data to assign the appropriate labels. The model is then evaluated on unknown samples in the test set using a performance metric.

The objective in training a traditional classifier is to learn to assign class labels to numerical representations (commonly referred to as *features*) extracted from raw data describing objects found in the real world. These features are attributes considered relevant to a particular classification problem. For most simple classifiers, the numerical representations must be in the form of a one-dimensional input vector of n numerical values, where each feature (f) is a value on one of the axes in the n-dimensional feature space. Thus, each vector fed into a classifier normally defines a specific point within a given *feature space.*

If the extracted features are genuinely relevant for the classification problem at hand, then objects belonging to the same class will share features and thus cluster together. Classification problems are divided into two major types depending on whether the division of objects into classes is known or unknown. If the set of c classes is unknown, then the task is for the classifiers to discover the clusters, the number typically defined in advance by the researchers. This type of learning is called *unsupervised learning.* If class labels are provided in the data set, the task of the classifier is to learn the labels. This process is called *supervised learning.*

Given the above formalizations, in supervised learning, a classifier can be defined as *D(f)*, where *D(f)* is any function that maps a sample's features within a feature space F to the appropriate label in the set of classes. The job of a classifier system in supervised learning is often a matter of partitioning a feature space into c decision regions so that when it is given an unknown feature vector, it can assign to it the correct label.

Let us mention here a popular supervised classifier we often use when building ensembles, especially on the data and feature levels, and that is the Support Vector Machine (SVM) [6]. SVMs are binary statistical learners that generate a hypothesis space of linear functions within some high dimensional feature space F [7]. The objective of an SVM is to find a hyperplane in F that accurately classifies the data points. As there are many potential qualifying hyperplanes, the task is to find the one that maximizes the margin, or distance between the data points of both classes, while simultaneously dividing both classes. Kernel functions project features onto a higher-dimensional feature space in those cases where a linear boundary between the two classes cannot be found. Several strategies are available as well that enable SVMs to handle more than two classes and multilabel problems [8].

A data set of samples is defined as a collection $\mathbf{Z} = \{z_1, \ldots, z_N\}, z_j \in \Re^n$ of input vectors. With supervised learning, the elements of $\mathbf{Z}$ will also include the class labels for each z_j, denoted as $l(z_j) \in \Omega, \text{j} = 1, \ldots, N$. Each dataset $\mathbf{Z}$ is divided into at least two subsets, a training set for classifier parameter learning/tuning and a test set for evaluating performance. A problem arises when training the classifier using the same samples in the training and test set: *overtraining*, which happens when a classifier learns to discriminate perfectly all samples in **Z** but fails to classify new samples correctly that are not included in **Z**. To overcome this problem, some datasets specify the training and test set along with a separate validation set for adjusting classifier parameters. If training, test, and validation sets are not specified in the data set, different methods can be employed to generate these sets. Two of the most common methods for dividing **Z** include cross-validation and bootstrapping. In cross-validation, **Z** is divided into *K* subsets (with *K* ideally a factor of *N*). One subset is selected as the test set, with the remaining *K*-1 subsets form the training set. This process is repeated *K* times, each time excluding a different subset. Classifier performance is computed as an average of all *K* performances. In the case where $K = N$, the method is called *leave-one-out*. Bootstrapping generates new datasets artificially from the original dataset. Given **Z**, then a new data set **Z*** is a random sampling with replacement of the set **Z**. Classifier performance is calculated as the average performance of the classifiers trained on the new datasets produced by bootstrapping.

There are many measures of classifier performance. One is the *counting estimator*, which counts misclassifications as follows: Error(D)= $\frac{N_{error}}{N_T}$, where T is the training set, N_T the number of samples in T, and N_{error} the number of samples in T that were misclassified by D. Another common measure of classifier performance is accuracy, which is calculated as 1−Error(D).

Thus far, we have mostly described the data and classifier levels of the canonical model. As mentioned above, objects belonging to the same class should share some critical attributes or features. Most input vectors contain a large amount of information that is extraneous to the classification problem. In face recognition, for example, the vast majority of pixels define the background and irrelevant aspects of people; only a small portion of pixels relate directly to the face. Many samples in large data sets also contain erroneous information that is contradictory or simply missing. Preprocessing methods are needed to clean up the data and reduce noise, for example, by predicting missing values or, in the case of image recognition, by sharpening images and extracting backgrounds.

Reducing the number of features is especially critical for contending with the *curse of dimensionality*: the larger the size of the input vector, the slower the training process and the more errors that are introduced. Transforming raw input vectors (z_j) sets the stage for extracting the most relevant information. Transformation can take many forms: (1) normalization, where inputs are scaled within a small range (e.g., −1.0 to 1.0), (2), smoothing techniques (binning, clustering, and regression), and (3) aggregation, where summarizing operators compress data (daily figures collapsed into weekly averages). A transform maps the *n* values of the raw input vectors into another set of values that retains the same amount of energy and information as the former set. The objective behind the transform is to decorrelate values and redistribute the energy into a smaller number of components so that the most relevant features can be extracted. Transformation can occur within the spatial or frequency domain. Standard transforms in the spatial domain include such statistical methods as principal component analysis (PCA) and linear discriminant analysis (LDA); in the frequency domain, the discrete Fourier transform (DFT), Gabor wavelet transform (GWT), and discrete cosine transform (DCT) are typical.

Many types of features can be extracted either from the raw input vectors or from transformed values. In the classical model, most features are handcrafted; that is, they are designed by researchers to handle specific issues like variations in scale and illumination with image data. The type of descriptor selected by the researcher depends on the ML problem and the type of classifier. In computer vision, the most common descriptors extract information from the texture of images. Table 5.1 provides a list and a brief description of some of the most powerful handcrafted texture descriptors, many based on Local Binary Patterns (LBPs) (for a comprehensive discussion of LBP and its variants, see [9]).

Table 5.1 Some popular handcrafted descriptors

Acronym	Description	Refs.
LBP	The canonical Local Binary Pattern (LBP) is computed at each pixel location by considering the values of a small circular neighborhood (with radius *R* pixels) around the value of a central pixel, which is used as a threshold for mapping the surrounding values to a 1 (if of higher value that the threshold) or 0 (if of lower value). The descriptor is the concatenated histogram of these binary numbers	Ojala et al. [10]
LTP	A variant of LBP that has inspired many other variants is the Local Ternary Pattern (LTP). This version of LBP takes advantage of a three-valued $(1, 0, -1)$ vs binary encoding that thresholds around zero	Tan and Triggs[11]
LPQ	Local Phase Quantization (LPQ) is a blur invariant form of LBP that utilizes the local phase information extracted from the 2D short-term Fourier transform (STFT) computed over a rectangular neighborhood at each pixel position of the image	Ojansivu and Heikkila [12]
RICLBP	Rotation Invariant Co-occurrence LBP combines the concept of co-occurrence among LBPs with rotation equivalence classes	Nosaka et al. [13]
RLBP	Rotated LBP calculates the descriptor with respect to a local reference	Mehta and Egiazarian [14]
CLBP	Complete LBP encodes a local region using a central pixel	Guo et al. [15]
MRELBP	Median Robust Extended LBP compares regional image medians rather than raw image intensities	Liu et al. [16]
HASC	Heterogeneous Auto-Similarities of Characteristics models linear and non-linear feature dependencies	San Biagio et al. [17]
GOLD	Gaussians of Local Descriptors is a flexible local feature representation that leverages parametric probability density functions	Guo et al. [18]

5.2.2 *Methods for Building Multiclassifiers*

Multiclassifier systems are a collection of classifiers that predict class labels from previously unseen samples in the test set by aggregating the predictions of the individual classifiers. As illustrated in Fig. 5.2, multiclassifier systems can be constructed on each of the three levels defining the canonical model or on various combinations of each level. For instance, classifiers can be trained on different divisions of the data (Z_i) combined, possibly, with augmented data artificially generated from the training set (see Sect. 5.4) or on different preprocessing methods (P_i), transformations (T_i), feature sets (F_i) and various combinations thereof. The whole training set can also be trained on the same feature set as long as the classifiers are not identical.

Classifiers can be either homogeneous (uniform) or heterogeneous (diverse). The terms *multiclassifier* and *ensemble* originally referred to these two distinctions, but today both terms are used interchangeably. An example of a uniform collection of classifiers would be a set of SVMs, each supplied with a different kernel function or trained on different training sets. SVM combined with neural network (NNs) trained on the same data set would be an example of a heterogeneous collection. Hybrid systems that combine uniform and diverse classifiers are also possible. Given a set of K classifiers $\boldsymbol{D} = \{D_1, \ldots, D_k\}$, the results of $\boldsymbol{D}$ can either be combined (fused) using some decision rule (sum rule, product rule, max rule, etc.) or selected for its competency. For a discussion of fusion and selection methods, see [19].

Constructing multiclassifier systems on the data level typically involves three steps applied to the training set T: (1) K new training sets are created from T, (2) a uniform or diverse classifier D_{i} is trained on each new training set T_i, and (3) the decisions of the K classifiers are combined using a decision rule.

The new data sets are constructed by perturbing the patterns in the training set. This process can be accomplished in one go or iteratively. A few common methods for constructing new training sets include:

- *Bagging* [20], where the new training sets $T_1, \ldots, T_\text{K}$ are simply subsets of T;
- *Class Switching* [21], where the new training sets are randomly generated by changing the labels of some subset of training samples;

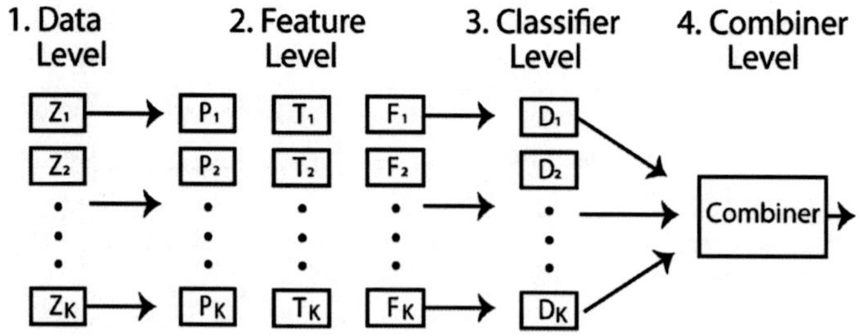

Fig. 5.2 Model of multiclassifier systems

- *Arcing* [22], where misclassified patterns form the basis for calculating patterns to be included in the next new training set generated;
- *Decorate* [23], where the training sets are iteratively constructed by adding artificially generated data constructed according to some distribution and then labeled to maximally differ from the current ensemble prediction; a new classifier is also added to the ensemble at each training iteration;
- *Clusterization* [24], where new training sets are generated based on the membership of the training patterns in the clusters as calculated by Fuzzy C-Means (or another clustering method); different feature sets are selected for each cluster with the goal of maximizing the discrimination between clusters;
- *Boosting/AdaBoost* [25], where each sample in the training set is assigned a weight that increases at each iteration for those that prove most difficult to classify;

On the feature level, new training sets can be generated by changing a given feature in some manner (by perturbing it or using different preprocessing methods and transforms) or by building ensembles on subsets of features or on sets of different handcrafted features. Some popular perturbation methods include:

- *Random Subspace* (RS) [26], where new training sets, $T_1,\ldots, T_K$, are randomly selected subsets of the feature set; RS trains K classifiers on each of these modified sets;
- *Input Decimated Ensemble* (IDE) [27], where PCA, applied to the training patterns in class I, is used to generate the new training set T_i. Because the size of IDE is bounded by the number of classes, so too is the number classifiers in the ensemble. This limitation can be avoided by partitioning the training patterns into clusters and applying PCA on the training patterns within each cluster;
- *Cluster-based Pattern Discrimination* [28], where new training sets are created by partitioning the classes into clusters and selecting different features for each cluster.

Finally, there are hybrid methods that combine different perturbation methods. Some of the most popular mixed methods are based on decision trees (DTs):

- *Rotation Forest* (RF) [29], an ensemble of DTs, where the new training sets are created by applying several PCA projections on subsets of the training patterns;
- *Random Forest* [30], where a random selection of features splits a given node in a bagging ensemble of Decision Trees (DT);
- *RotBoost* [31], where an ensemble of DTs is created by combining RF and Adaboost;

Our research team continues to build GenP systems from ensembles. At the start, our GenP ensembles were generated using hybrid methods. In [32], for example, an ensemble based on combining RS and an editing approach to reduce outliers was evaluated on many data sets representing diverse problems (different medical tasks and image classification problems, a vowel dataset, a credit data set, etc.). The best classifier system was developed empirically by testing different combinations of approaches. This GenP ensemble, whose parameters were held constant, demonstrated promising results across nearly all the image data sets and consistently performed better compared to a single SVM whose parameters had been optimally tuned on each dataset.

The GenP ensemble developed in [32] was refined in [33], where the focus of our team was on testing different classifiers and their combinations across twenty-five diverse datasets. The most interesting result obtained from our experiments lent weight to the NFL hypothesis (no single approach was discovered that outperformed all the other classifier systems in all the tested datasets). Be that as it may, the best GenP ensemble (based on the simple sum rule) outperformed each stand-alone classifier without any ad hoc hyperparameters tuning on the datasets. In other words, the GenP ensemble experimentally derived in [33] would likely perform better out-of-the-box on a similar but different classification problem than would any of the single classifiers tested in this work if trained explicitly on a data set representing the new task.

In yet another study [34], we conducted a survey study in the field of biometrics. In this work, various sources and architectures related to the combination of different biometric systems (both unimodal and multimodal) were discussed, and methods for combining biometric systems at different levels of fusion were reviewed. Experiments showed that "mixed" ensembles, based on the concatenation of scores used to train a second-level classifier, outperformed all other approaches—but at the expense of a significant increase in computational cost. Moreover, in a work that examined fusion approaches in fingerprint recognition [35], we showed that ensembles provide significant advantages over single classifier systems not only in terms of accuracy but also in terms of robustness against the different sensors used in the acquisition step.

5.3 Matrix Representation of the Feature Vector

A common transformation initially applied to raw sensor data is to reshape the original high dimensional matrix values into a 1D input vector. The main reasons for turning 2D sensor data into 1D input vectors are (1) to ready the data for transforms like PCA and (2) to prepare an input vector acceptable to classifiers requiring vector inputs. Reshaping matrix data has consequences, however, that are not always desirable. First, converting matrices to vectors adds an unnecessary step that increases computational complexity. Leaving matrix data in its original form has been shown to significantly reduce computation time when applying some classic feature transforms [36]. Second, many feature extractors, like LBP and Gabor filters [37], extract descriptors directly from matrices, and 2D versions of common 1D feature transforms are now available. For example, in [38], the authors present a powerful 2D PCA transform. What is of significance here is that there can be a loss of information in transforming 2D matrices into vectors.

Interestingly, there is also an advantage in doing the reverse, in transforming 1D feature vectors into matrices, a transformation that is especially useful when building

GenP systems. Several early studies have shown how matrix reshaping of 1D vectors can increase classifier diversity (good for generating ensembles) and generalizability [36, 39].

In [40], we developed a GenP system by randomly rearranging 1D vectors into a set of fifty matrices from which LTP descriptors were extracted and trained on separate SVMs. By extracting features from these randomly rearranged matrices, our system was able to investigate the correlation among sets of features within arbitrarily generated neighborhoods. We were able to show that the generated matrices extracted additional information from the input vectors.

In [41], we went a step further and demonstrated that applying many different methods for reshaping vectors into matrices increased the generalizability of multi-classifier systems. Finally, in a paper that explored building GenP ensembles on the data and feature levels in the medical domain [33], we extracted texture descriptors from many different matrix representations of 1D protein representations that we treated as images and perturbed for increasing even further the diversity of a set of classifiers. This approach was shown to increase the system's ability to classify proteins across many data sets representing different protein classification problems.

5.4 GenP Systems Based on Deep Learners

Most research into different ML domains identifies a division between systems designed before the advent of deep learning and those proposed afterward, a division that recognizes the revolutionary power of deep learners. Deep learners are neural networks with hidden layers composed of interconnected neurons with learnable weights, biases, and activation functions. First proposed by Hinton in 2006 [42], the distinguishing characteristic of deep learners is the cascade of a large number of specialized hidden layers organized in a hierarchical structure within the neural network architecture.

An robust class of deep learners especially suited to matrix and image classification is the Convolutional Neural Network (CNN). CNNs are composed of layers whose neurons are arranged in three dimensions (height, width, and depth) such that every layer transforms a 3D input volume into a 3D output of neuron activations. As illustrated in Fig. 5.3, there are five specialized processing classes of neurons in CNN: *conv*olutional (CONV), *act*ivation (ACT), *pool*ing (POOL), followed by a final stage that includes a Fully-Connected (FC) and a *class*ification (CLASS) layer (for instance, Softmax in Fig. 5.3, which is often used in multiclass classification as it takes a vector of scores and computes a probability distribution).

The CONV layer is the kernel of a CNN and is computationally very expensive. CONV computes the outputs of neurons connected to local regions by means of a convolution operation on the input. The extent of the connectivity of the local regions is a hyperparameter (the *receptive field*) and a parameter sharing strategy that controls the number of parameters. The parameters of CONV layers are shared sets of weights (the kernels) that have relatively small receptive fields.

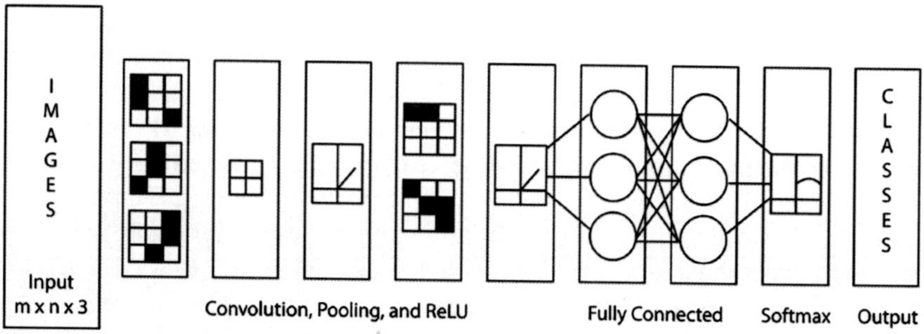

Fig. 5.3 General CNN architecture

POOL layers apply non-linear downsampling operations, such as max pooling, the most popular pooling operator (it partitions the input into non-overlapping rectangles and outputs the maximum for each one). The main purpose of the POOL layers is to reduce the spatial size of the representation mainly by eliminating parameters, which controls overfitting and the computational complexity of the network. POOL layers are typically inserted between CONV layers.

ACT layers apply different activation functions of which there are many: the non-saturating ReLU (Rectified Linear Unit) function and the sigmoid function being two of the most common.

FC layers, as the name suggests, have neurons that are fully connected to all the activations in the previous layer. FC follows CONV and POOL layers and often makes up the last hidden layers.

There are many different CNN architectures that combine these different layers along with the addition of other specialized ones. Five of the most common are AlexNet [43], GoogleNet [44], VGGNet [45], ResNet [46], and DenseNet: [47]. AlexNet is composed of five CONV layers followed by three FC layers, with some max-POOL layers inserted in the middle. A ReLU is applied to each CONV along with an FC for faster training. GoogleNet introduced a subnetwork called the inception module (INC) made up of parallel convolutional filters with concatenated outputs. One advantage of INC is that it significantly reduces the number of parameters. It is composed of twenty-seven layers in total. VGGNet is a very deep CNN that has sixteen CONV/FC layers, with a POOL layer inserted every two or three CONV layer instead of after every CONV layer as is the case with AlexNet. The CONV layers are homogeneous with very small kernels. ResNet is nearly twenty times deeper than AlexNet and eight times deeper that VGGNet. Although ResNet is deeper than VGGNet, the model is smaller and easier to optimize. The main advantage of ResNet is the introduction of residual (RES) layers with special skip connections and batch normalization. Furthermore, FC layers at the end of the ResNet are replaced by global average pooling. VGGNet has three FC layers; ResNet only has one which outputs the class probabilities. DenseNet extends ResNet by connecting each layer to all the other layers.

CNNs can help create GenP ensembles in four main ways:

1. Pretrained CNNs can be used as a complementary deep feature extractor, and these deep learned features can be combined with popular handcrafted features, such as LBP;
2. Ensembles of pretrained CNNs can be generated;
3. CNNs composed of different architectures can be trained on the data and combined;
4. CNN architectures can form the backbones on a set of Siamese networks (this will be discussed in Sect. 5.6)

The remainder of this section will explore the first three methods for generating GenP systems from CNNs.

5.4.1 *Deep Learned Features*

We will begin our discussion of CNNs for GenP systems by examining pretrained CNNs as feature extractors. As illustrated in Fig. 5.4, the outputs of different pretrained CNN layers can be harvested and fed into simpler classifiers. These features are often called *learned features* because, in training, a CNN network is discovering new representations of the data layer by layer. These representations have points of similarity with handcrafted features. For example, layers trained on images that lie closest to the input tend to detect edges and textural information [48] that resembles Gabor filters and color blobs. These features are followed in later layers by the discovery of image patches and contours, with each subsequent layer representing higher-level features [49]. As with handcrafted features, learned features are highly generalizable and largely independent of the data source the CNN is trained on. For this reason, classical classifiers can successfully be trained on learned feature sets.

The source of learned features is a pretrained CNN D_S that has been trained on a large data set Z_s. The generalization power of CNNs is amplified when trained on enormously large datasets, such as ImageNet [50], a database organized around WordNet's hierarchy of nouns. Hundreds of images depicting these nouns have been amassed in ImageNet, which now has over 14 million images depicting over 20,000 categories. Matrix data taken from a new data set of images Z_T can be fed into CNN models pretrained on ImageNet or on another massive image data set. Learned features are then collected from a particular layer of the D_S and fed into a traditional classifier.

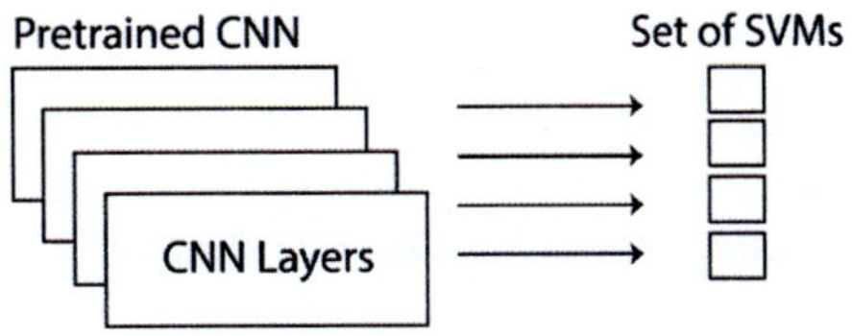

Fig. 5.4 Learned features extracted from different layers of a pretrained CNN and trained on a set of SVMs

Ensembles are easily generated by combining (1) learned and handcrafted features and (2) learned features taken from several layers in pretrained models (e.g., AlexNet and ResNet trained on ImageNet). More and more studies have been investigating the value of extracting features from both the lower CNN levels [51, 52] and the top layers [53, 54]. For an investigation of features extracted layer by layer, see [55].

One of our early efforts building deep GenP ensembles on the feature level is a paper entitled "Handcrafted versus non-handcrafted features for computer vision classification" [56]. The objective of this paper was to produce a GenP image classifier that combined generalizability with maximum classification power. The system experimentally derived in this work combined and evaluated eleven handcrafted features (many of which are listed in Table 5.1) with features extracted from ten pretrained CNNs. Different sets were tested on 18 image datasets representing very different image classification problems. The focus was on exploiting features extracted from the deep layers of the pretrained deep learners.

Three approaches were taken for generating the learned features:

1. Deep transfer learning features based on CNN (specifically, different versions of VGGNet [45]); sets of features were taken from different layers of these CNNs trained on ImageNet. PCA and DCT were applied to reduce the number of features extracted from the deep layers as not applying some sort of reduction method could result in the curse of dimensionality;
2. Deep features extracted from a Principal Component Analysis Network (PCAN) [49], which is a simple deep neural network (implemented in three stages) that relies on cascading PCA to learn multistage filter banks.
3. Compact binary descriptor (CBD) [57], which is a binary descriptor that is learned by extracting Pixel Difference Vectors (PDVs) from local patches and computing the difference between each pixel and its neighbors. The PDVs are projected in an unsupervised way onto low-dimensional binary vectors.

All learned and handcrafted vectors were each processed by a separate SVM, and combinations were evaluated across 18 data sets representing different classification problems with the aim of generating a high-performing GenP that worked well across the data sets without parameter tuning. Handcrafted features were compared with learned features, and experiments demonstrated that both types extract different kinds of information from the input images. Thus, we showed that the fusion of handcrafted features with learned features significantly outperformed SVMs trained separately on the different feature sets.

5.4.2 Transfer Learning

Not only can a pretrained CNN be used as a feature extractor, but it can also learn a new problem by fine-tuning the weights of the network for a new classification problem, a process called transfer learning (TL). TL is particularly valuable when the number of samples in a data set is small [48]. CNN accuracy improves as a function of data size; thus, training CNNs on large data sets is preferable.

As illustrated in Fig. 5.4, given a pretrained CNN model D_S that has learned to classify a large data set Z_s (e.g., ImageNet), the layers of D_S are adapted in TL to learn a new target task defined by another data set Z_T. D_S is the source predictive function $D_s(\bullet)$, and the adaptive model D_T is the predictive function for the new task defined by Z_T. In TL, the early CONV layers of D_S are fixed (see Fig. 5.5) and connected to one or more new layers. The adapted CNN model D_T is then either partially or completely retrained on Z_T. Sometimes it is necessary to retrain or fine-tune a given D_T multiple times [58] to find the best hyperparameters thereby significantly increasing training time (see [59] for a survey of TL and a detailed discussion of some TL challenges).

Our investigations in applying TL as a GenP building tool are mainly focused on studying the feasibility of designing ensembles based on the fusion of CNNs that differ in their training (fine-tuning) or in their architectures. In the task of Plankton

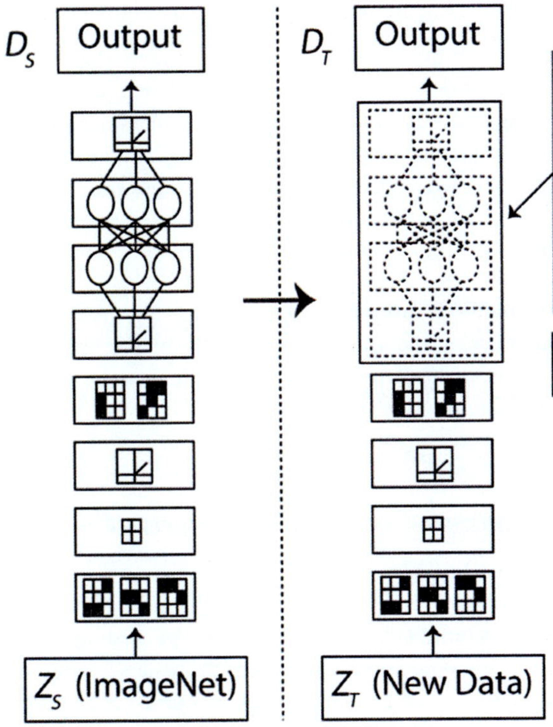

Fig. 5.5 Transferred learning. A CNN network D_S (left) pretrained on a large data set Z_s is altered (right) to learn a task represented by a new (typically smaller) data set Z_T by freezing the first layers of D_S and adding a new network (NEW) to the end of D_S. This new adapted network D_T(right) is then either fully or partially retrained on Z_T

classification [35], for example, three TL approaches for fine-tuning pretrained models were evaluated:

1. One-round tuning: which is the standard approach for fine tuning pretrained networks. In one-round tuning the network is initialized with the pretrained weights (obtained by training the given deep learner on ImageNet) and then retrained on the training set of the target problem.
2. Two-rounds tuning: in this case, a first round of fine-tuning is performed on a dataset similar to the target, and a second round is performed on the actual training set of the target problem. The idea behind this method is to teach the network to recognize the domain patterns (underwater patterns in Plankton classification, which have little in common with the images included in the ImageNet database); the second round adjusts the classification weights based on the specific target problem.
3. Incremental tuning: which takes advantage of a crucial setting in the training phase, the number of iterations, or epochs, used in training. The rationale for this method is to introduce variability and, therefore, some diversity into the collection of deep learners, making them suitable for building ensembles.

Experimental results comparing the three methods showed that the standard approach of retraining a network (the one-round strategy) was good enough for the target problems. There are possibly two reasons for this outcome: either the data sets chosen for the two-round method were not similar enough to the target problem to make much of a difference or, more likely, because the dimension of the training set of the target problem was sufficient for performing training. The ensemble based on the incremental version only slightly improved the performance and only in some cases.

The fusion of the three methods, however, succeeded in significantly improving performance in the Plankton classification tasks. The use of preliminary training created classifiers that were more diverse than those obtained by training deep learners with only one-round tuning.

5.4.3 Multiclassifier System Composed of Different CNN Architectures

The third way to use CNNs to build GenP systems takes into account the fusion of different CNN architectures and parameters. Because of the instability of the training process [4], ensembles of CNNs and other neural networks perform better than their stand-alone counterparts. The depth of the deep learner changes the character of the network. On the one hand, shallow CNNs, for example, fail to capture subtle differences in the data set because these shallow networks are too generalized; on the other hand, deep CNNs fail to capture general similarities because deeper networks are highly sensitive to subtle differences. This diversity in the type of information picked up by deep and shallow CNNs makes them perfect candidates for generating ensembles on the classifier level.

In addition to combing different CNN architectures, ensembles can also be built with different activation functions. Combining different activation functions is an effective way to produce robust classifier systems [60]. In the early days, the most common activation functions were the sigmoid and the hyperbolic tangent, neither of which worked very well with deep learners due to the vanishing gradient problem. It was quickly discovered that nonlinearities, such as Rectified Linear Units (ReLU) [61], work better with deep neural networks. ReLU is the identity function for positive inputs and zero with negative inputs [62]. Many variations of ReLU have been proposed: Leaky ReLU [63], which has a hyperparameter $\alpha > 0$ applied to the negative inputs to ensure the gradient is never zero, Exponential Linear Units (ELU) [64], which always produces a positive gradient, and Scaled Exponential Linear Unit (SELU) [65], a version of ELU multiplied by the constant $\lambda > 1$ to maintain the mean and variance in the input features, to name a few. Some ReLU variants have learnable parameters, such as Parametric ReLU (PReLU) [66], which provides Leaky ReLU with a learnable parameter on the negative slope, and Adaptive Piecewise Linear Unit (APLU) [67], which learns piecewise slopes and points of nondifferentiability for each neuron using gradient descent.

Our research group has conducted extensive studies comparing different CNN architectures in several domains [35, 68] based on the idea of perturbing CNN architectures by randomly changing the activation functions as described above. In [69], for instance, a method for CNN model design was proposed that was based on changing all the activation layers of the best-performing CNN models by stochastic layer replacement. In this method, each activation layer of a CNN was replaced with a different activation function stochastically drawn from a given set that included more than ten "static" and "dynamic" activation functions. This process introduced diversity among the models, preparing them for ensemble generation. The fusion of Stochastic ResNet50-like models was shown to outperform each stand-alone approach compared to ensembles composed of standard ResNet50. Extensive evaluation carried out on a wide array of benchmark data sets for image classification and segmentation and on a variety of biomedical datasets demonstrated that the proposed idea is very effective for building a high-performing ensemble of CNNs.

5.5 Data Augmentation

As already noted, it is best to train deep learners, such as CNNs, on large image data sets as this improves performance and avoids overfitting [70], but large numbers of samples are not always possible to collect, especially when involving human subjects [71]. Another way around this problem aside from transfer learning is to augment small data sets by generating new samples from the original set. Common methods of data augmentation for images include reflection, translation, and rotation [43, 72–74], as well as changing the contrast, saturation, and brightness [43, 73, 74] levels of the original images. It is also possible to apply what is called PCA jittering, which multiplies a certain number of the most important components by a small number [43, 73]. So effective is data augmentation that fast augmentation libraries, such as Albumentations [75], have been developed to provide scientists and engineers with easy-to-use wraparounds for the most common methods and other augmentation libraries.

The problem of data availability is particularly acute in bioinformatics, where collecting a large image dataset for training deep learning models is generally too expensive and time-consuming. In [76], our research team proposed the application of different data augmentation strategies to perturb patterns in images in the training set. The resulting ensemble was the fusion of CNNs built using different batch sizes, different learning rates, and different data augmentation methods. In this work, we proposed a general method that worked efficiently on many bioimage classification problems.

5.6 Dissimilarity Spaces

As described above, a major focus in our work building GenP systems has been on extracting different feature sets, whether handcrafted or learned, from samples and then training sets of SVMs to discriminate samples within the given feature spaces. Recently, our work has shifted its focus to another classification strategy that is beginning to attract much research interest: training patterns within one or several (dis)similarity spaces. Distinguishing objects by comparing their similarities and dissimilarities is fundamental to how human beings learn to classify objects [77]. Thus, training classifiers within (dis)similarity spaces is warranted when the discriminative task involves classifying things with discernible patterns, objects whose distinguishing features, for example, implicate shapes and texture [78]. As is the case in human learning, classifying objects based on (dis)similarity requires the classifier to estimate an unknown sample's class label based on a pairwise analysis of its (dis)similarities with all the other samples within the training set, a process which in ML has traditionally involved the applications of distance measures such as shape matching distance [79], tangent distance [8], and the earth mover's distance (EMD) [80]. The sample space can be any set as long as the (dis)similarity function employed is defined for all possible pairs of samples [81].

Not all learning based on (dis)similarity is a function of a distance measure, however. Rather (dis)similarity can involve a range of functions that not only measure distances between samples within a space but also build other spaces. Although often not distinguished in the literature, methods based on dissimilarity and similarity offer different perspectives. As noted in [82], the type of data and the problem itself determine which viewpoint is most appropriate. As our work thus far has been based on building GenP systems within dissimilarity spaces, our discussion here will be on approaches based on dissimilarities.

According to the taxonomy introduced in [82], dissimilarity classification is of two types: those concerned with dissimilarity vectors [83] and those focused on building dissimilarity spaces [84]. Strategies based on dissimilarity vectors treat a multiclass problem as a binary one by calculating pairwise distances between feature vectors extracted from the pairs. If the two samples under consideration have the same label, the result is positive; otherwise, it is negative. The binary objective of the classifier is to discern whether the vectors of pairs of samples belong to the same or a different class. Studies that have investigated dissimilarity vectors have revolved around both traditional learners [83] and deep learners [85, 86]. Feature vectors selected for examination have included LBP and its many variants as well as many other handcrafted features (see [87]). Both handcrafted and learned descriptors have also been investigated [86].

Strategies based on dissimilarity spaces generate classifiers from feature vector spaces. A dissimilarity space is different from the classical feature space described in Sect. 6.2.1. In a feature space, the feature vector represents a sample measured across all features. In a dissimilarity space derived from feature vectors, a vector is the distance between pairs of samples. Some studies based on dissimilarity spaces include [88], where prototype selection was used to generate classifiers and the dissimilarity representations were treated as a vector space, and [89], where a dissimilarity space was based on deep convolutional features.

A number of papers, including some of our own, have exploited Siamese neural networks (SNN) [90] composed of identical CNNs (see Fig. 5.6) to produce distance models. An SNN contains two identical subnetworks whose parameter alterations during training are mirrored. SNNs, rather than learning to classify vector inputs are designed to find (dis)similarities of samples by comparing feature vectors (see [91] for an overview).

The study of dissimilarity spaces is a major step forward in the direction of building GenP classifiers since many classification problems cannot be dealt with in a vector space. Our research has been aimed at designing a general system based on an ensemble designed by perturbing the feature (dissimilarity) space.

The classification system we proposed in [92], for example, combined classifiers trained on different dissimilarity spaces generated by a large set of Siamese Neural

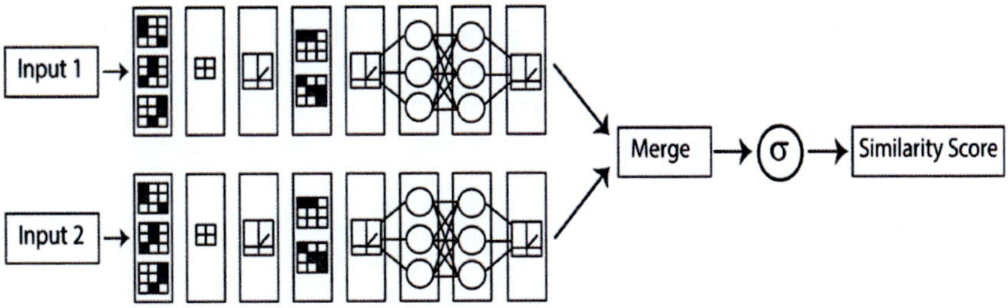

Fig. 5.6 Siamese Neural Network (SNN) with of two identical CNNs

Networks (SNNs). A set of centroids from the patterns in the training data was calculated with supervised k-means clustering, and the resulting centroids were employed to generate the dissimilarity space via the Siamese networks. Vector space descriptors were extracted by projecting patterns onto the similarity spaces. SVMs then classified an image by its dissimilarity vector. The generalizability of this strategy in image classification was strongly demonstrated across data sets representing very different types of images in two distinct domains: medical images and animal audio vocalization represented as spectrograms.

5.7 Conclusion

In our work over the last fifteen years, we have shown that it is possible to generate GenP systems, as defined in the introduction, based on ensembles that perform well across an array of databases within a given domain or even across domains if the data is of a similar type, such as images, which share many primitive elements, or matrix data, which may or may not have much in common. Although generally the NFL theorem appears to hold when these GenP ensembles are asked to perform across databases representing widely different problems and data types, we have nonetheless progressively pushed against the limits of the NFL theorem using the tools outlined in this chapter: ensembles, matrix representations of data treated as images, deep learning approaches, data augmentation, and classification within dissimilarity spaces. The reader may now wonder what new developments in ML lie on the horizon that can be exploited to push the limits even further.

We believe that thirty years from now the period from 2019 to 2021 will be remembered not so much for the COVID19 pandemic but rather for some momentous developments in quantum computing: Google's claim of quantum supremacy [93] and simulation of a chemical reaction with their quantum computer [94], their unveiling of a Quantum AI campus in Santa Barbara with a quantum data center and research labs, the teleportation of 3-dimensional quantum states by Chinese and Austrian scientists [95], and the race to offer cloud access to quantum computing and programming tools by the tech giants Microsoft, Intel, Amazon, and Google.

A major bottleneck in harnessing deep learning for GenP ensembles is the lack of computing power. The computational power of quantum computation (QC) will offer more of a solution to this challenge. Already quantum processors are computing in a Hilbert space of $2^{53} \approx 9 \times 10^{15}$, which far exceeds our fastest supercomputers [96]. To simulate 42 qubits on Intel's quantum simulator is said to require five trillion transistors [97]. In the field of ML, the expectation of QC is an exponential increase in handling dimensionality compared with classical computing [98]. As an example, currently, an artificial neuron can only process its inputs in N dimensions, but a quantum neuron can process its inputs in 2^N dimensions [99].

As shown in [100], research in quantum ML is increasing exponentially as well. Much of this research is inspired and guided by the field's current understanding of ML. Of particular interest to our research group is a wave of classical algorithms reconceptualized to build quantum ensembles [101–103]. In [102] and [101], for instance, frameworks have been developed that exploit superposition to store sets of parameters for generating ensembles of quantum classifiers that compute in parallel. In [103], a quantum version of Random Forest is proposed that quadratically increases the speed of prediction of the classical Random Forest. Similarly, many quantum versions of current classifiers have been described, such as quantum SVMs [98, 104].

Perhaps more consequential for GenP development in the near future are quantum-inspired (QI) algorithms intended for classical computing rather than for quantum computing. Some QI algorithms proposed in the last couple of years include a QI algorithm for linear regression [105], QI SVMs [106], QI linear genetic programming [107], QI deep belief networks [108], QI similarity measures [109], and a new QI binary classifier [110]. We hope to investigate whether these QI methods will enhance diversity in building GenP ensembles.

Finally, according to some researchers, experiments and the new models being developed in QC are challenging the extended Church-Turing thesis formulated by Bernstein and Vazirani [111] that states that the Turing machine can simulate any reasonable model of computation, including quantum models. Whether the extended Church-Turing thesis holds is debated [96]. It is hard to say today how future developments in QC will challenge the NFL theorem, but we are hopeful that new quantum algorithms will. Though it is doubtful a universal classifier is possible, we believe that future GenP's will continue to approach such an ideal.

References

1. D.H. Wolpert, The supervised learning no-free-lunch theorems, in *6th Online World Conference on Soft Computing in Industrial Applications* (2001), pp. 25–42
2. M. Delgado et al., Do we need hundreds of classifiers to solve real world classification problems? J. Mach. Learn. Res. **15**, 3133–3181 (2014)
3. L. Nanni, S. Ghidoni, S. Brahnam, Ensemble of convolutional neural networks for bioimage classification. Appl. Comput. Inf. **17**(1), 19–35 (2021)
4. L.K. Hansen, P. Salamon, Neural network ensembles. IEEE Trans. Pattern Anal. Mach. Intell. **12**, 993–1001 (1990)
5. D. Lu, Q. Weng, A survey of image classification methods and techniques for improving classification performance. Int. J. Remote Sens. **28**, 823–870 (2007)
6. V. Vapnik, The support vector method, in *Artificial Neural Networks ICANN'97*. (Springer, Lecture Notes in Computer Science, 1997), pp. 261–271
7. N. Cristianini, J. Shawe-Taylor, *An Introduction to Support Vector Machines and Other Kernel-Based Learning Methods* (Cambridge University Press, Cambridge, UK, 2000)
8. R.O. Duda, P.E. Hart, D.G. Stork, *Pattern Classification*, 2nd edn. (Wiley, New York, 2000)
9. S. Brahnam, et al., (eds)., *Local Binary Patterns: New Variants and Application*. (Springer, Berlin, 2014)
10. T. Ojala, M. Pietikainen, T. Maeenpaa, Multiresolution gray-scale and rotation invariant texture classification with local binary patterns. IEEE Trans. Pattern Anal. Mach. Intell. **24**(7), 971–987 (2002)
11. X. Tan, B. Triggs, *Enhanced local texture feature sets for face recognition under difficult lighting conditions*. Anal. Model. Faces Gestures **LNCS 4778**, 168–182 (2007)

12. V. Ojansivu, J. Heikkila, Blur insensitive texture classification using local phase quantization, in *ICISP* (2008), pp. 236–243
13. R. Nosaka, C.H. Suryanto, K. Fukui, Rotation invariant co-occurrence among adjacent LBPs, in *ACCV Workshops* (2012), pp. 15–25
14. R. Mehta, K. Egiazarian, Dominant rotated local binary patterns (drlbp) for texture classification. Pattern Recogn. Lett. **71**(1), 16–22 (2015)
15. Z. Guo, L. Zhang, D. Zhang, A completed modeling of local binary pattern operator for texture classification. IEEE Trans. Image Process. **19**(6), 1657–1663 (2010)
16. L. Liu, et al., Median robust extended local binary pattern for texture classification. IEEE Trans. Image Process. In press
17. M. San Biagio et al., Heterogeneous auto-similarities of characteristics (hasc): Exploiting relational information for classification, in *IEEE Computer Vision (ICCV13)*. (Sydney, Australia, 2013), pp. 809–816
18. Y. Guo, G. Zhao, M. Pietikainen, Discriminative features for texture description. Pattern Recogn. Lett. **45**, 3834–3843 (2012)
19. L. Nanni, S. Brahnam, A. Lumini, Classifier ensemble methods, in *Wiley Encyclopedia of Electrical and Electronics Engineering*, ed by J. Webster (Wiley, New York, 2015), pp. 1–12
20. L. Breiman, Bagging predictors. Mach. Learn. **24**(2), 123–140 (1996)
21. G. Martínez-Muñoz, A. Suárez, Switching class labels to generate classification ensembles. Pattern Recogn. **38**(10), 1483–1494 (2005)
22. G. Bologna, R.D. Appel, A comparison study on protein fold recognition. in *The 9th International Conference on Neural Information Processing* (Singapore, 2020)
23. P. Melville, R.J. Mooney, Creating diversity in ensembles using artificial, information fusion. Spec. Issue Divers. Multiclassifier Syst. **6**(1), 99–111 (2005)
24. L. Nanni, A. Lumini, FuzzyBagging: a novel ensemble of classifiers. Pattern Recogn. **39**(3), 488–490 (2006)
25. Y. Freund, R.E. Schapire, A decision-theoretic generalization of on-line learning and an application to boosting. J. Comput. Syst. Sci. **55**(1), 119–139 (1997)
26. T.K. Ho, The random subspace method for constructing decision forests. IEEE Trans. Pattern Anal. Mach. Intell. **20**(8), 832–844 (1998)
27. K. Tumer, N.C. Oza, Input decimated ensembles. Pattern Anal Appl **6**, 65–77 (2003)
28. L. Nanni, Cluster-based pattern discrimination: a novel technique for feature selection. Pattern Recogn. Lett. **27**(6), 682–687 (2006)
29. J.J. Rodriguez, L.I. Kuncheva, C.J. Alonso, Rotation forest: a new classifier ensemble method. IEEE Trans. Pattern Anal. Mach. Intell. **28**(10), 1619–1630 (2006)
30. L. Breiman, Random forest. Mach. Learn. **45**(1), 5–32 (2001)
31. C.-X. Zhang, J.-S. Zhang, RotBoost: a technique for combining Rotation forest and AdaBoost. Pattern Recogn. Lett. **29**(10), 1524–1536 (2008)
32. L. Nanni, S. Brahnam, A. Lumini, Double committee adaBoost. J. King Saud Univ. **25**(1), 29–37 (2013)
33. L. Nanni, et al., Toward a general-purpose heterogeneous ensemble for pattern classification. Comput. Intell. Neurosci. **Article ID 909123**, 1–10 (2015)
34. A. Lumini, L. Nanni, Overview of the combination of biometric matchers. Inf. Fusion **33**, 71–85 (2017)
35. A. Lumini, L. Nanni, Deep learning and transfer learning features for plankton classification. Ecol. Inf. **51**, 33–43 (2019)
36. Z. Wang, et al., Pattern representation in feature extraction and classification-matrix versus vector. IEEE Trans. Neural Netw. **19**(758–769) (2008)
37. R. Eustice et al., UWIT: Underwater image toolbox for optical image processing and mosaicking in MATLAB, in *International Symposium on Underwater Technology*. (Tokyo, Japan, 2002), pp. 141–145
38. J. Yang et al., Two-dimension pca: a new approach to appearance-based face representation and recognition. IEEE Trans. Pattern Anal. Mach. Intell. **26**(1), 131–137 (2004)

39. Z. Wang, S.C. Chen, Matrix-pattern-oriented least squares support vector classifier with AdaBoost. Pattern Recogn. Lett. **29**, 745–753 (2008)
40. L. Nanni, Texture descriptors for generic pattern classification problems. Expert Syst. Appl. **38**(8), 9340–9345 (2011)
41. L. Nanni, S. Brahnam, A. Lumini, Matrix representation in pattern classification. Exp. Syst. Appl. **39.3**, 3031–3036 (2012)
42. G. Hinton, S. Osindero, Y.-W. Teh, A fast learning algorithm for deep belief nets. Neural Comput. **18**(7), 1527–1554 (2006)
43. A. Krizhevsky, I. Sutskever, G.E. Hinton, ImageNet Classification with Deep Convolutional Neural Networks, in *Advances In Neural Information Processing Systems*. ed. by F. Pereira et al. (Curran Associates Inc., Red Hook, NY, 2012), pp. 1097–1105
44. C. Szegedy, et al., Going deeper with convolutions, in *IEEE Computer Society Conference on Computer Vision and Pattern Recognition* (2015), pp. 1–9
45. K. Simonyan, A. Zisserman, *Very Deep Convolutional Networks for Large-Scale Image Recognition* (Cornell University, 2014). arXiv:1409.1556v6
46. K. He et al., Deep residual learning for image recognition, in *2016 IEEE Conference on Computer Vision and Pattern Recognition (CVPR)*. (IEEE, Las Vegas, NV, 2016), pp. 770–778
47. G. Huang, et al., Densely connected convolutional networks. CVPR **1**(2), 3 (2017)
48. J. Yosinski, et al., *How Transferable are Features in Deep Neural Networks?* (Cornell University, 2014) arXiv:1411.1792.
49. T.-H. Chan et al., Pcanet: a simple deep learning baseline for image classification? IEEE Trans. Image Process. **24**(12), 5017–5032 (2015)
50. J. Deng, et al. ImageNet: a large-scale hierarchical image database. in *CVPR* (2009)
51. B. Athiwaratkun, K. Kang, *Feature Representation in Convolutional Neural Networks* (2015). arXiv:1507.02313.
52. B. Yang, et al., Convolutional channel features, in *IEEE International Conference on Computer Vision (ICCV)* (2015)
53. C. Barat, C. Ducottet, String representations and distances in deep convolutional neural networks for image classification. Pattern Recogn. Bioinf. **54**(June), 104–115 (2016)
54. A.S. Razavian, et al., CNN features off-the-shelf: an astounding baseline for recognition. CoRR (2014). arXiv:1403.6382
55. R.H.M. Condori, O.M. Bruno, Analysis of activation maps through global pooling measurements for texture classification. Inf. Sci. **555**, 260–279 (2021)
56. L. Nanni, S. Ghidoni, S. Brahnam, Handcrafted versus non-handcrafted features for computer vision classification. Pattern Recogn. **71**, 158–172 (2017)
57. J. Lu, et al., Learning compact binary face descriptor for face recognition. IEEE Trans. Pattern Anal. Mach. Intell. (2015)
58. H. Li, et al., *Rethinking the Hyperparameters for Fine-Tuning* (2020). arXiv:2002.11770
59. R. Ribani, M. Marengoni, A survey of transfer learning for convolutional neural networks. in *2019 32nd SIBGRAPI Conference on Graphics, Patterns and Images Tutorials (SIBGRAPI-T)* (2019), pp. 47–57
60. G. Maguolo, L. Nanni, S. Ghidoni, Ensemble of convolutional neural networks trained with different activation functions. Exp. Syst. Appl. **166**, 114048 (2021)
61. X. Glorot, A. Bordes, Y. Bengio, Deep sparse rectifier neural networks. in *AISTATS* (2011)
62. V. Nair, G.E. Hinton, Rectified linear units improve restricted boltzmann machines, in *27th International Conference on Machine Learning*. (Haifa, Israel, 2010), pp. 1–8
63. A.L. Maas, Rectifier nonlinearities improve neural network acoustic models (2013)
64. D.-A. Clevert, T. Unterthiner, S. Hochreiter, Fast and accurate deep network learning by exponential linear units (ELUs). CoRR (2015). arXiv:1511.07289
65. G. Klambauer, et al., *Self-normalizing neural networks*, in *31st Conference on Neural Information Processing Systems (NIPS 2017)* (Long Beach, CA, 2017)
66. K. He et al., Delving deep into rectifiers: surpassing human-level performance on imagenet classification. IEEE Int. Conf. Comput. Vis. (ICCV) **2015**, 1026–1034 (2015)

67. F. Agostinelli, et al., Learning activation functions to improve deep neural networks. CoRR (2014).arXiv:1412.6830
68. A. Lumini, et al., Image orientation detection by ensembles of Stochastic CNNs. Mach. Learn. Appl. **6**, 100090 (2021)
69. L. Nanni, et al., Stochastic selection of activation layers for convolutional neural networks. Sensors (Basel, Switzerland) **20** (2020)
70. M. Hutter, *Learning Curve Theory* (2021). arXiv:2102.04074
71. B. Sahiner et al., Deep learning in medical imaging and radiation therapy. Med. Phys. **46**(1), e1–e36 (2019)
72. O. Ronneberger, P. Fischer, T. Brox, U-Net: convolutional networks for biomedical image segmentation, in *MICCAI 2015 LNCS*. ed. by N. Navab et al. (Springer, Cham, 2015), pp. 234–241
73. J. Shijie, et al., Research on data augmentation for image classification based on convolution neural networks, in *Chinese Automation Congress (CAC) 2017* (Jinan, 2017), pp. 4165–4170
74. A. Dosovitskiy et al., Discriminative unsupervised feature Learning with exemplar convolutional neural networks. IEEE Trans. Pattern Anal. Mach. Intell. **38**(9), 1734–1747 (2016)
75. A. Buslaev, et al., *Albumentations: Fast and Flexible Image Augmentations* (2018). arXiv: 1809.06839
76. L. Nanni, S. Brahnam, G. Maguolo, Data augmentation for building an ensemble of convolutional neural networks, in *Smart Innovation, Systems and Technologies*. ed. by Y.-W. Chen et al. (Springer Nature, Singapore, 2019), pp. 61–70
77. A. Tversky, Features of similarity. Psychol. Rev. **84**(2), 327–352 (1977)
78. E. Pękalska, R.P. Duin, *The Dissimilarity Representation for Pattern Recognition - Foundations and Applications* (World Scientific, Singapore, 2005)
79. S. Belongie, J. Malik, J. Puzicha, Shape matching and object recongtiion using shape contexts. IEEE Trans. Pattern Anal. Mach. Intell. **24**(24), 509–522 (2002)
80. Y. Rubner, C. Tomasi, L.J. Guibas, The earth mover's distance as a metric for image retrieval. Int. J. Comput. Vision **40**, 99–121 (2000)
81. Y. Chen et al., Similarity-based classification: concepts and algorithms. J. Mach. Learn. Res. **10**, 747–776 (2009)
82. Y.M.G. Costa et al., The dissimilarity approach: a review. Artif. Intell. Rev. **53**, 2783–2808 (2019)
83. S. Cha, S. Srihari, Writer identification: statistical analysis and dichotomizer, in *SSPR/SPR* (2000)
84. E. Pękalska, R.P. Duin, Dissimilarity representations allow for building good classifiers. Pattern Recognit. Lett. **23**, 943–956 (2002)
85. R.H.D. Zottesso et al., Bird species identification using spectrogram and dissimilarity approach. Ecol. Inf. **48**, 187–197 (2018)
86. V.L.F. Souza, A. Oliveira, R. Sabourin, A writer-independent approach for offline signature verification using deep convolutional neural networks features. in *2018 7th Brazilian Conference on Intelligent Systems (BRACIS)* (2018), pp. 212–217
87. J.G. Martins et al., Forest species recognition based on dynamic classifier selection and dissimilarity feature vector representation. Mach. Vis. Appl. **26**, 279–293 (2015)
88. E. Pękalska, R.P. Duin, P. Paclík, Prototype selection for dissimilarity-based classifiers. Pattern Recogn. **39**, 189–208 (2006)
89. M. Hernández-Durán, Y.P. Calaña, H.M. Vazquez, Low-resolution face recognition with deep convolutional features in the dissimilarity space. in *IWAIPR* (2018)
90. J. Bromley, et al. Signature verification using a "Siamese" time delay neural network. Int. J. Pattern Recognit. Artif. Intell. (1993)
91. D. Chicco, Siamese neural networks: an overview, in *Artificial Neural Networks. Methods in Molecular Biology*, ed. by H. Cartwright (Springer Protocols, Humana, New York, NY, 2020), pp. 73–94

92. L. Nanni et al., Experiments of image classification using dissimilarity spaces built with siamese networks. Sensors **21**(1573), 2–18 (2021)
93. E. Gibney, Hello quantum world! Google publishes landmark quantum supremacy claim. Nature **574**, 461–462 (2019)
94. F. Arute et al., Hartree-Fock on a superconducting qubit quantum computer. Science **369**, 1084–1089 (2020)
95. Y.-H. Luo, et al., Quantum teleportation in high dimensions. Phys. Rev. Lett. **123**(7), 070505 (2019)
96. F. Arute et al., Quantum supremacy using a programmable superconducting processor. Nature **574**(7779), 505–510 (2019)
97. L. Greenemeier, How close are we—really—to building a quantum computer. Sci. Am. (2018)
98. V. Havlícek et al., Supervised learning with quantum-enhanced feature spaces. Nature **567**, 209–212 (2019)
99. F. Tacchino, et al., An artificial neuron implemented on an actual quantum processor. NPJ Quantum Inf. **5**, 1–8 (2018)
100. G. Acampora, Quantum machine intelligence. Quantum Mach. Intell. **1**(1), 1–3 (2019)
101. M. Schuld, F. Petruccione, *Quantum ensembles of quantum classifiers.* Sci. Rep. **8** (2018)
102. A. Abbas, M. Schuld, F. Petruccione, On quantum ensembles of quantum classifiers. Quantum Mach. Intell. **2**, 1–8 (2020)
103. K. Khadiev, L. Safina. *The Quantum Version of Random Forest Model for Binary Classification Problem* (2021)
104. D. Willsch, et al., Support vector machines on the D-Wave quantum annealer. Comput. Phys. Commun. **248**, 107006 (2020)
105. A.A. Gily'en, Z. Song, E. Tang, *An Improved Quantum-Inspired Algorithm for Linear Regression* (2020). arXiv:2009.07268
106. C. Ding, T. Bao, H.-L. Huang, Quantum-inspired support vector machine. IEEE Trans. Neural Netw. Learn. Syst. (2021)
107. D.M. Dias, M. Pacheco, Describing quantum-inspired linear genetic programming from symbolic regression problems. IEEE Congr. Evol. Comput. **2012**, 1–8 (2012)
108. W. Deng et al., An improved quantum-inspired differential evolution algorithm for deep belief network. IEEE Trans. Instrum. Meas. **69**, 7319–7327 (2020)
109. L. Bai et al., A quantum-inspired similarity measure for the analysis of complete weighted graphs. IEEE Trans. Cybern. **50**, 1264–1277 (2020)
110. P. Tiwari, M. Melucci, Towards a quantum-inspired binary classifier. IEEE Access **7**, 42354–42372 (2019)
111. E. Bernstein, U. Vazirani, Quantum complexity theory. SIAM J. Comput. **26**, 1411–1473 (1997)

Alessandra Lumini received her Master's Degree cum laude from the University of Bologna in 1996, and in 2001 she received her Ph.D. degree in Computer Science for her work on "Image Databases". She is currently an Associate Professor at DISI, University of Bologna. She is a member of the Biometric Systems Lab and of the Smart City Lab and she is interested in biometric systems, pattern recognition, machine learning, image databases, and bioinformatics. She is coauthor of more than 200 research papers. Her Google H-index is 45.

Sheryl Brahnam received her PhD. in computer science at The Graduate Center of the City University of New York in 2002, where she won a science fellowship from 1997 to 1999. Currently, she is a professor in the Department of Information Systems and Cybersecurity at Missouri State University. There she held the Daisy Portenier Loukes Research Professorship from 2015 and 2018. Her research focus is on machine learning and the social aspects of computing, for which she has received attention in the media. She has been invited to deliver several keynote addresses and has been an associate editor of several journals as well as an editor of several books. She has over 175 publications, and her Google H-index is 34.

Chapter 6
Bayesian Networks: Theory and Philosophy

Dawn E. Holmes

Abstract This chapter explores the theory of Bayesian networks with particular reference to Maximum Entropy Formalism. A discussion of objective Bayesianism is given together with some brief remarks on applications.

Keywords Bayesian networks · Maximum entropy · Objective Bayesianism

6.1 Introduction

The literature on Bayesian networks is vast and a general overview is unlikely to be useful. This chapter, therefore, concentrates on the use of the maximum entropy formalism to find the prior distribution required by a Bayesian network before updating algorithms can be applied. Some of the important results of this methodology are brought together in this chapter and a full example is given showing the method in detail. Philosophical issues, including objective Bayesianism, are discussed. A very brief introduction to the many real-world applications is given, as well as a similarly brief contrast with artificial neural networks (ANN's).

6.2 Bayesian Networks

6.2.1 Bayesian Networks Background

Judea Pearl, the grandfather of Bayesian networks, published the seminal work *'Probabilistic Reasoning in Intelligent Systems'* [1] in 1988 resulting from research

D. E. Holmes (✉)
Department of Statistics and Applied Probability, University of California, Santa Barbara, CA 93106, USA
e-mail: holmes@pstat.ucsb.edu

M. Virvou et al. (eds.), *Advances in Selected Artificial Intelligence Areas*, Learning and Analytics in Intelligent Systems 24,
https://doi.org/10.1007/978-3-030-93052-3_6

conducted over the previous ten years. It remains the standard textbook and reference on the subject and is now deservedly regarded as a classic. Pearl coined the term 'Bayesian network' in 1985 but had introduced the fundamental idea of using Bayesian probabilities in intelligent systems in an earlier 1982 paper titled *'Reverend Bayes on Inference Engines: A Distributed Hierarchical Approach'* [2]. Since then, research into Bayesian networks has proliferated. Unlike earlier intelligent or expert system, which were not based on probability theory, Bayesian networks provide a sound theoretical framework for reasoning under uncertainty, using an underlying graphical structure and the probability calculus.

However, in situations where insufficient data is available, there will be too many conditional probabilities for an expert to confidently estimate and, on a practical level, although experts can provide the causal knowledge from which the Bayesian network may be constructed, their knowledge may not be in a form that enables them to assign the associated probabilities.

A method for providing minimally prejudiced estimates of the missing conditional probabilities is obviously attractive. The maximum entropy formalism has been shown to provide the required method. If we are to apply this methodology to Bayesian networks, we must show that the estimates for missing information can be found efficiently, and this has been done previously by Holmes, Rhodes and Garside for certain classes of Bayesian network [3]. One of the main difficulties in this work has been in meeting the independence requirements imposed by a Bayesian network. Holmes [4] has shown that, for singly connected Bayesian networks, only a subset of the independencies implied by d-separation are preserved by the maximum entropy formalism and thus the remaining independencies must be explicitly included as constraints.

6.2.2 Bayesian Networks Defined

Let V be a finite set of vertices and B a set of directed edges between vertices with no feedback loops, the vertices together with the directed edges form a directed acyclic graph (DAG). Formally, a Bayesian network is defined as follows. Let:

(i) **V** be a finite set of vertices
(ii) **B** be a set of directed edges between vertices with no feedback loops. The vertices together with the directed edges form a directed acyclic graph $\boldsymbol{G} = \langle \boldsymbol{V}, \boldsymbol{B} \rangle$
(iii) a set of events be depicted by the vertices of **G** and hence also represented by **V**, each event having a finite set of mutually exclusive outcomes.
(iv) E_i be a variable with possible outcomes e_i of the event, where $i, j = 1, \ldots, n_i$
(v) P be a probability distribution over the combination of events. Hence **P** consists of all possible $P\left(\bigcap_{i \in V} E_i\right)$

Let C be the following set of constraints:

(2i) the elements of **P** sum to unity.

(2ii) for each event i with set of parents M_i there are associated conditional probabilities $P\left(E_i| \bigcap_{j\in M_i} E_j\right)$ for each possible outcome that can be assigned to E_i and E_j

(2iii) the independence relationships implied by d-separation in the directed acyclic graph.

Then $\mathbf{N} = \langle \mathbf{G}, \mathbf{P}, \mathbf{C}\rangle$ is a Bayesian network if **P** satisfies **C**. The causal information in (ii) is of the form $P(\mathbf{E}_{c_1}|\mathbf{M}_{c_1})$.

Consider the Bayesian network shown Fig. 6.1. The nodes $a_0, a_1, b_1, ..., b_q, c_1, ..., c_t \in$ V. Triangles, labelled with bold capital letters, represent all subtrees below a node. **T** depicts the entire graph except the subtree with root node a_1. The general state S of the tree is given by $\bigcap_{v\in V} E_v$. A state is

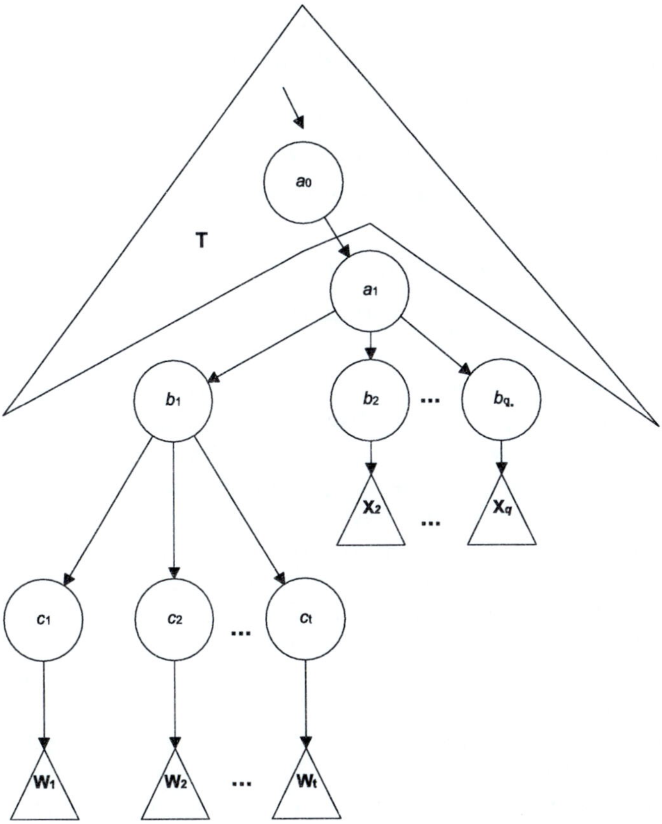

Fig. 6.1 A multiway Bayesian network **N** with a small part shown in detail

determined by assigning e_v^j to E_v. The probability of a state is assumed to be non-zero.

Let T denote the general state of $\mathbf{T}$.

$\boldsymbol{X}_2 ... \boldsymbol{X}_q$ and $\boldsymbol{W}_1 ... \boldsymbol{W}_r$ denote the descendants of $b_1, ..., b_q$ and $c_1, ..., c_t$ respectively.

$\boldsymbol{X}_2 ... \boldsymbol{X}_q$ and $\boldsymbol{W}_1 ... \boldsymbol{W}_r$ denote the general state of $\boldsymbol{X}_2 ... \boldsymbol{X}_q$ and $\boldsymbol{W}_1 ... \boldsymbol{W}_r$ respectively.

Then the probability of a general state of $\mathbf{N}$ is given by: $P(S) = P\left(\boldsymbol{T} E_{a_0} E_{b_1} ... E_{b_q} E_{c_1} ... E_{c_t} \boldsymbol{W}_1 ... \boldsymbol{W}_r \boldsymbol{X}_2 ... \boldsymbol{X}_q\right)$.

This equation is expanded using the chain rule and all the variables instantiated with arbitrary values. Event E_{a_1} is assigned $e^i_{a_1}$ and so on. The instantiation of $\mathbf{T}$ is denoted by $\boldsymbol{T}^u$, $\boldsymbol{X}_k$ by $\boldsymbol{X}_k^{x_k}$ and similarly for $\boldsymbol{W}_1 ... \boldsymbol{W}_r$. If we denote this state by $S_{b_1^j a_1^i}$ then:

$$P\left(S_{b_1^j a_1^i}\right) = P\left(\boldsymbol{T}^u e^i_{a_1} e^{j_1}_{b_1} ... e^{j_q}_{b_q} e^{f_1}_{c_1} ... e^{f_q}_{c_q} \boldsymbol{X}_x^{x_2} ... \boldsymbol{X}_q^{x_q} \boldsymbol{W}_1^{w_1} ... \boldsymbol{W}_r^{w_r}\right)$$

Again, we can expand this expression using the chain rule. Another state $S_{b_1^m a_1^i}$ is considered with E_{b_1} instantiated with $e^m_{b_1}$, An analogous expression to that for $P\left(S_{b_1^j a_1^i}\right)$ is determined and expanded using the chain rule. Noting that for the tree depicted in Fig. 6.1, the causal information required is in the form $P\left(e^j_b | e^i_a\right) = \beta(b_1, a_1)$, except for the root node, division of these two expressions results in:

$$\frac{P\left(S_{b_1^j a_1^i}\right)}{P\left(S_{b_1^m a_1^i}\right)} = \frac{\beta(b_1^j, a_1^i) \prod_{k=1}^t \beta\left(c_k^{j_k}, b_1^j\right)}{\beta(b_1^m, a_1^i) \prod_{k=1}^t \beta\left(c_k^{j_k}, b_1^m\right)}$$

These two particular states were chosen because many terms cancel and this expression has been used in showing that the maximum entropy approach returns the same result as the above Bayesian approach, under the same constraints.

Before considering the maximum entropy formalism, it is crucial to note that d-separation, a fundamental concept in Bayesian network theory, is required to identify the conditional dependencies in Bayesian network theory. For a given causal network, d-separation categorizes all constraints as either dependencies or independencies. Independence will be discussed in more detail later in this chapter, but for now it suffices to say that using the chain rule and d-separation, it is possible to determine the joint probability distribution of the system under consideration, given a fully specified network.

Although Cooper [5] has shown that for multi-connected networks the problem of determining the probability distribution is *NP*-hard, there are now many methods for updating probabilities in completely determined singly connected Bayesian networks. Methods also exist for transforming fully specified multiply connected graphs into singly connected networks so that information propagation can still be performed. See [6, 7] for example.

Pearl's method of propagation through local computation in a single-connected network was implemented in early expert systems, such as MUNIN. However, one of the major drawbacks of using Bayesian networks is that complete information, in the form of marginal and conditional probabilities must be specified before the usual updating algorithms are applied. Following on from the work of Rhodes and Garside, Holmes and Rhodes [8] have shown that when all or some probabilities are missing in a multivalued multiway BN, it is possible to determine unbiased estimates using maximum entropy and thus preserve the integrity of the network. This original technique has been further developed and details will be given in the next section.

6.3 Maximizing Entropy for Missing Information

6.3.1 Maximum Entropy Formalism

Traditionally, the concept of entropy has its roots in theoretical physics and is embodied in the 2nd Law of Thermodynamics. Shannon was the first to propose entropy as a basis for information theory by Shannon and Weaver [9]. In 1996, Jaynes [10] proved that the Maximum Entropy Formalism provides a minimally prejudiced distribution, i.e. a distribution in accordance with the Principle of Insufficient Reasoning. A key feature of the Principle of Maximum Entropy is that it conforms to the current state of knowledge and reflects that ignorance may be reduced or increased when additional information becomes available. As such, it provides an ideal mechanism for dealing with Bayesian networks, where knowledge is uncertain and changeable.

The techniques developed by Rhodes and Garside [11] for estimating missing information in two special cases of Bayesian networks, namely 2-valued multiway trees and 2-valued binary inverted trees were the starting point for the current authors' work. Holmes and Rhodes showed that when all or some of this information is missing in a multi-valued multiway tree-like Bayesian network, it is possible to determine unbiased estimates using maximum entropy. The techniques thus developed depend on the property that the probability of a state of a fully specified Bayesian network, found using the standard Bayesian techniques, can be equated to the maximum entropy solution. For a *generalized* Bayesian network, that is one in which the graphical structure is not a special case and in which some, all or none of the essential information is missing, Holmes shown that missing information can be estimated using the maximum entropy formalism (MaxEnt) alone, thus divorcing these results from their dependence on Bayesian techniques. The next section gives an example of this work.

6.3.2 Maximum Entropy Method

In this section, the set of partial differential equations generated by applying the maximum entropy formalism to a Bayesian network with missing priors is considered. Specifically, **a generalized Bayesian network with maximum entropy is discussed. In order the simply the communication of the PDE results**, terms *contributing* to explicit expressions from partial differentiation are given, with no loss of generality.

Let K be a set of multivalued events representing a knowledge domain of multi-valued events a_i.

$\boldsymbol{E}_V$ is a variable associated with each event. Then the general state of the Bayesian network $N = \langle \boldsymbol{G}, \boldsymbol{P}, \boldsymbol{C} \rangle$ is given by $\bigcap_{v \in V} \boldsymbol{E}_v$. Assuming the probability of each state is non-zero, individual states are given by assigning some e_v^j to each $\boldsymbol{E}_V$. The number of states in the Bayesian network is given by $N_S = \prod\limits_{i \in V} n_i$ where n_i is the number of values in the ith event. States are numbered $1 \ldots N_S$ and denoted by S_i where $i = 1 \ldots N_S$. The probability of a state is denoted by $P(S_i)$.

The minimally prejudiced probability distribution **P** is found by maximizing

$$H = -\sum_{i=1}^{N_S} P(S_i) ln P(S_i) \tag{6.1}$$

in such a way that the constraints implied by d-separation, given as marginal or conditional probabilities, hold.

A set of sufficient constraints **C** is given where each constraint $C_j \in \boldsymbol{C}$. Each constraint is assigned a Lagrange multiplier λ_j where j represents the subscripts corresponding to the events on the associated edge of the tree. For example, a typical edge $\langle a_1, b_1 \rangle$ in the Bayesian network shown in Fig. 6.1, with a_1 the parent of b_1 where event a_1 has p outcomes and event b_1 has m outcomes has associated Lagrange multipliers $\lambda\left(b_1^1, a_1^1\right), \ldots, \lambda\left(b_1^m, a_1^p\right)$. The constraints, under the assumption of d-separation, associated with this arbitrary edge are given by:

$P\left(e_b^j | e_a^i\right) = \beta(b_1, a_1)$ where $i = 1, \ldots, N_S$ and $j = 1, \ldots, m$

together with the normalization constraint:

$\sum\limits_{i=1}^{N_S} P(S_i) = 1$ with associated Lagrange multiplier λ_0.

Using the Lagrange multiplier technique Jaynes showed that the general expression for the probability of a state given a set of linear constraints is:

$$P(S_i) = \prod_{j=0}^{N_C} \exp\left(-\lambda_j \sigma_{i,j}\right) i = 1, \ldots, N_S \tag{6.2}$$

where $\sigma_{i,j}$ is the coefficient of the i th state in the j th constraint and N_C is the number of constraints, excluding the sum to unity.

The method continues by transforming constraints into expressions containing the sums of probabilities of states. For example, the family of constraints on the arbitrary edge $\langle a_1, b_1 \rangle$ given by $P\left(e_b^j | e_a^i\right) = \beta(b_1, a_1)$ are written:

$$\left(1 - \beta\left(b_1^j, a_1^i\right)\right) \sum_{x \in X} P(S_x) - \beta\left(b_1^j, a_1^i\right) \sum_{y \in Y} P(S_y) = 0$$

$$\text{where } X = \left\{ x | \sum_x P(S_x) = P\left(e_{a_1}^i e_{b_1}^j\right) \right\} \text{ and } Y = \left\{ y | \sum_y P(S_y) = \sum_{\substack{k=1 \\ k \neq j}}^{k=m} P\left(e_{a_1}^i e_{b_1}^k\right) \right\} \tag{6.3}$$

Further manipulation, by substitution, of (2) and (3) results in the following expression:

$$\begin{aligned} &\left(1 - \beta\left(b_1^j, a_1^i\right)\right) \sum_{x \in X} \prod_{C_j \in C} \exp\left(\left(-\lambda_{b_1^j, a_1^i}\right) \frac{\partial C_j}{\partial P(S_x)}\right) \\ &-\beta\left(b_1^i, a_1^i\right) \sum_{y \in Y} \prod_{C_j \in C} \exp\left(\left(-\lambda_{b_1^i, a_1^i}\right) \frac{\partial C_j}{\partial P(S_y)}\right) = 0 \end{aligned} \tag{6.4}$$

Let the probability of a state with events a_1 and b_1 instantiated with their ith and jth outcomes respectively be denoted by $P\left(S_{b_1^j, a_1^i}\right)$. Then from (6.5), (6.7) and (6.8) $P\left(S_{b_1^j, a_1^i}\right)$ contains the expressions:

$$\exp\left(\left(-\lambda\left(b_1^j, a_1^i\right)\right)\left(1 - \beta\left(b_1^j, a_1^i\right)\right)\right) \tag{6.5}$$

when $x \in X$ and

$$\exp\left(-\lambda\left(b_1^j, a_1^i\right)\right) \prod_{k=1}^{k=m-1} \exp\left(\left(-\lambda\left(b_1^k, a_1^i\right)\right)\left(-\beta\left(b_1^k, a_1^i\right)\right)\right) \tag{6.6}$$

when $y \in Y$.

Using (6.5) and (6.6) and further algebraic manipulation equation, Eq. (6.4) gives an expression for an arbitrary Lagrange multiplier:

$$\exp\left(-\lambda\left(b_1^j, a_1^i\right)\right) = \frac{\beta\left(b_1^j, a_1^i\right) \sum_{y \in Y} \prod_{Cj \in C - C\left(b_1^j, a_1^i\right)} \exp\left(\left(-\lambda b_1^j, a_1^i\right) \frac{\partial C_j}{\partial P(S_y)}\right)}{\left(1 - \beta\left(b_1^j, a_1^i\right)\right) \sum_{x \in X} \prod_{Cj \in C_{C - C_{\left(b_1^j, a_1^i\right)}}} \exp\left(\left(-\lambda b_1^j, a_1^i\right) \frac{\partial C_j}{\partial P(S_x)}\right)} \tag{6.7}$$

An iterative algorithm using the above expression allows us to update the Lagrange multipliers. However, as shown in Holmes [12], it is also possible to solve for the Lagrange multipliers algebraically and thus an expression for the probability of each state can be found.

6.3.3 Solving for the Lagrange Multipliers

In this section, a detailed example is given, showing how to solve for the Lagrange Multipliers derived in Sect. 3.2. Assume a Bayesian network has the graphical structure of a binary tree, with three nodes, A, B and C, where A is the root node with B and C as A's children as shown in Fig. 6.2.

Events *a*, *b* and *c*, defined by the variables E_a, E_b and E_c, have mutually exclusive and exhaustive outcomes, as follows $E_a = \left\{e_a^1, e_a^2, e_a^3\right\}, E_b = \left\{e_b^1, e_b^2, e_b^3\right\}, E_c = \left\{e_c^1, e_c^2, e_c^3\right\}$.

Consider the information required for a completely specified prior distribution. The sum to unity requirement will be designated as constraint zero. The remaining constraints are given as the conditional probabilities associated with each outcome. Without loss of generality, the special case of the root node was not considered in the above development. However, since the information concerning the root node is required specifically as marginal probabilities, we introduce a simple definitional device for root node constraints.

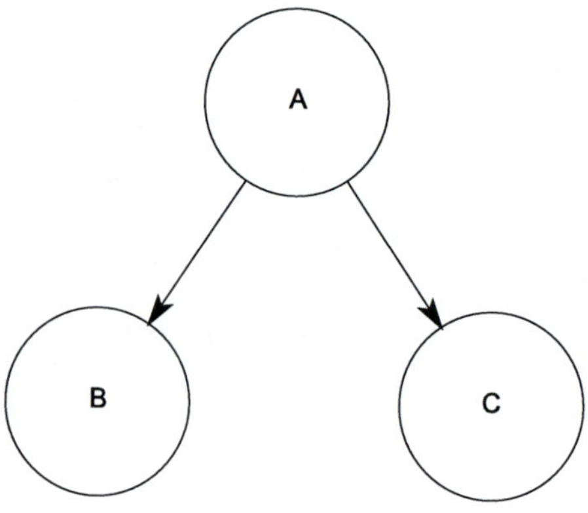

Fig. 6.2 A simple Bayesian network

$$P\left(e_a^i\right) = \alpha(a_i)\, i = 1, 2, 3$$

Clearly, since the outcomes are mutually exclusive and exhaustive $\alpha(a_3) = 1 - \alpha(a_2) - \alpha(a_1)$ and so we need only conder cases where i = 1,2. For child nodes, the number of conditional probability constraints can also be reduced. For example, from Eq. (6.1) we have:

$P\left(e_c^3|e_a^1\right) = \beta(c_3 a_1)$, $P\left(e_c^2|e_a^1\right) = \beta(c_2 a_1)$ and $P\left(e_c^1|e_a^1\right) = \beta(c_1 a_1)$. Hence $\beta(c_3 a_1) = 1 - \beta(c_2 a_1) - \beta(c_1 a_1)$ making $\beta(c_3, a_1)$ redundant. Similarly, the following constraints are also redundant:

$$\beta(c_3 a_2), \beta(c_3 a_3), \beta(b_3 a_1), \beta(b_3 a_2) \text{ and } \beta(b_3 a_3)$$

The complete list of the 15 required constraints is the following:

The sum to unity, or normalization constant

$$P\left(e_a^i\right) = \alpha(a_i)\, i = 1, 2 \tag{6.8}$$

$$P\left(e_b^j|e_a^i\right) = \beta\left(b_j a_i\right) i = 1, 2, 3 \text{ and } j = 1, 2 \tag{6.9}$$

$$P\left(e_c^j|e_a^i\right) = \beta\left(c_j a_i\right) i = 1, 2, 3 \text{ and } j = 1, 2 \tag{6.10}$$

The Bayesian network we are working with can be in any of 27 states, which we will number 0–26. A table showing all these states is given below. States will be referred to by their subscript as we go through the example.

Next, the knowledge given as marginal and conditional probabilities is expressed as a family of linear constraints. Using the notation described earlier, the sum to unity is expressed in terms of the probability of the states given in the state table.$\sum_{i=0}^{26} P(S_i) = 1$.

From Table 6.1, the two constraint equations $P\left(e_a^i\right) = \alpha(a_i)\, i = 1, 2$ associated with the root node, and given in Eq. (6.8), are written:

$$(1 - \alpha(a_1))\left(\sum_{i=0}^{8} P(S_i)\right) - \alpha(a_1)\left(\sum_{i=9}^{17} P(S_i) + \sum_{i=18}^{26} P(S_i)\right) = 0 \tag{6.11}$$

Table 6.1 The state table

$S_0 : e_a^1 e_b^1 e_c^1$	$S_1 : e_a^1 e_b^1 e_c^2$	$S_2 : e_a^1 e_b^1 e_c^3$	$S_3 : e_a^1 e_b^2 e_c^1$	$S_4 : e_a^1 e_b^2 e_c^2$	$S_5 : e_a^1 e_b^2 e_c^3$	$S_6 : e_a^1 e_b^3 e_c^1$
$S_7 : e_a^1 e_b^3 e_c^2$	$S_8 : e_a^1 e_b^3 e_c^3$	$S_9 : e_a^2 e_b^1 e_c^1$	$S_{10} : e_a^2 e_b^1 e_c^2$	$S_{11} : e_a^2 e_b^1 e_c^3$	$S_{12} : e_a^2 e_b^2 e_c^1$	$S_{13} : e_a^2 e_b^2 e_c^2$
$S_{14} : e_a^2 e_b^2 e_c^3$	$S_{15} : e_a^2 e_b^3 e_c^1$	$S_{16} : e_a^2 e_b^3 e_c^2$	$S_{17} : e_a^2 e_b^3 e_c^1$	$S_{18} : e_a^3 e_b^1 e_c^1$	$S_{19} : e_a^3 e_b^1 e_c^2$	$S_{20} : e_a^3 e_b^1 e_c^3$
$S_{21} : e_a^3 e_b^2 e_c^1$	$S_{22} : e_a^3 e_b^2 e_c^2$	$S_{23} : e_a^3 e_b^2 e_c^3$	$S_{24} : e_a^3 e_b^3 e_c^1$	$S_{25} : e_a^3 e_b^3 e_c^2$	$S_{26} : e_a^3 e_b^3 e_c^3$	

$$(1-\alpha(a_2))\left(\sum_{i=9}^{17} P(S_i)\right) - \alpha(a_2)\left(\sum_{i=0}^{8} P(S_i) + \sum_{i=18}^{26} P(S_i)\right) = 0 \tag{6.12}$$

Let us look at the conditionals, for example $\beta(b_1a_1) = P\left(E_b^1 | E_a^1\right)$. By the definition of conditional probability $\beta(b_1a_1)P\left(E_a^1\right) = P\left(E_a^1 E_b^1\right)$. Applying the chain rule, we get: $\beta(b_1a_1)\left(P\left(E_a^1 E_b^1\right) + P\left(E_a^1 E_b^2\right) + P\left(E_a^1 E_b^3\right)\right) = P\left(E_a^1 E_b^1\right)$.

Subsequently, factorization gives:

$$(1-\beta(b_1a_1))\left(P\left(E_a^1 E_b^1\right)\right) - \beta(b_1a_1)\left(P\left(E_a^1 E_b^2\right) + P\left(E_a^1 E_b^3\right)\right) = 0 \tag{6.13}$$

Looking at Table 6.1, we can see that the states contributing to $\beta(b_1a_1)$ are 0,1, and 2 from $P\left(e_b^1 e_a^1\right)$, states 3, 4 and 5 from $P\left(e_b^1 e_a^2\right)$ and states 6, 7, 8 from $P\left(e_a^1 e_b^3\right)$.

We can now write Eq. (6.13) in terms of probabilities of states:

$$(1-\beta(b_1a_1)) \sum_{i=0,1,2} P(S_i) - \beta(b_1a_1)\left(\sum_{i=3,4,5} P(S_i) + \sum_{i=6,7,8} P(S_i)\right) = 0 \tag{6.14}$$

Each of the constraint equations is written in a form similar to (6.14). Each constraint is assigned a Lagrange multiplier λ_j $j = 0, \ldots, 14$ and the results summarized in the Table 6.2. The columns labelled λ refer to the Lagrange multipliers by subscript. λ_0 is the Lagrange multiplier assigned with the normalization constant contributes to all states. λ_1 and λ_2 are the Lagrange multiplier assigned to $P\left(e_a^i\right) = \alpha(a_i)$ $i = 1, 2$ respectively, λ_3 is assigned to $\beta(b_1a_1)$ and so on.

Table 6.2 Constraint/State table

λ				λ			
0	1	1	1		1	1	1
1	$1-\alpha(a_1)$ 0–8	$-\alpha(a_1)$ 9—17	$-\alpha(a_1)$ 18 -26	8	$-\beta(b_2a_2)$ 9,10,11	$1-\beta(b_2a_2)$ 12,13,14	$-\beta(b_2a_2)$ 15,16,17
2	$-\alpha(a_2)$ 0–8	$1-\alpha(a_2)$ 9—17	$-\alpha(a_2)$ 18 -26	9	$1-\beta(c_1a_2)$ 9,12,15	$-\beta(c_1a_2)$ 10,13,16	$-\beta(c_1a_2)$ 11,14,17
3	$1-\beta(b_1a_1)$ 0,1,2	$-\beta(b_1a_1)$ 3,4,5	$-\beta(b_1a_1)$ 6,7,8	10	$-\beta(c_2a_2)$ 9,12,15	$1-\beta(c_2a_2)$ 10,13,16	$-\beta(c_2a_2)$ 11,14,17
4	$-\beta(b_2a_1)$ 0,1,2	$1-\beta(b_2a_1)$ 3,4,5	$-\beta(b_2a_1)$ 6,7,8	11	$1-\beta(b_1a_3)$ 18,19,20	$-\beta(b_1a_3)$ 21,22,23	$-\beta(b_1a_3)$ 24,25,26
5	$1-\beta(c_1a_1)$ 0,3,6	$-\beta(c_1a_1)$ 1,4,7	$-\beta(c_1a_1)$ 2,5,8	12	$-\beta(b_2a_3)$ 18,19,20	$1-\beta(b_2a_3)$ 21,22,23	$-\beta(b_2a_3)$ 24,25,26
6	$-\beta(c_2a_1)$ 0,3,6	$1-\beta(c_2a_1)$ 1,4,7	$-\beta(c_2a_1)$ 2,5,8	13	$1-\beta(c_1a_3)$ 18,21,24	$-\beta(c_1a_3)$ 19,22,25	$-\beta(c_1a_3)$ 20,23,26
7	$1-\beta(b_1a_2)$ 9,10,11	$-\beta(b_1a_2)$ 12,13,14	$-\beta(b_1a_2)$ 15,16,17	14	$-\beta(c_2a_3)$ 18,21,24	$1-\beta(c_2a_3)$ 19,22,25	$-\beta(c_2a_3)$ 20,23,26

Looking next at the probability of a state, we see that $P(S_0)$ has contributions $(1-\alpha(a_1))$ from (6.11), $-\alpha(a_2)$ from (6.12) and $(1-\beta(b_1a_1))$ from (6.13) as well as contributions, not derived here, from other constraint equations.

From Table 6.2, the constraint/state table, we can see which Lagrange multipliers contribute to a given state, which will enable us to determine the Lagrange multipliers in terms of the probability of a state. We will consider state zero as an example of a typical state. From Table 6.2, we see that:

$$\begin{aligned} P(S_0) = &\exp(-\lambda_0)\exp(-\lambda_1(1-\alpha(a_1)))\exp(-\lambda_2(-\alpha(a_2))) \\ &\exp(-\lambda_3(1-\beta(b_1a_1)))\exp(-\lambda_4(-\beta(b_2a_1))) \\ &\exp(-\lambda_5(1-\beta(c_1a_1)))\exp(-\lambda_6(-\beta(c_2a_1))) \end{aligned} \tag{6.15}$$

Similarly,

$$\begin{aligned} P(S_2) = &\exp(-\lambda_0)\exp(-\lambda_1(1-\alpha(a_1)))\exp(-\lambda_2(-\alpha(a_2))) \\ &\exp(-\lambda_3(1-\beta(b_1a_1)))\exp(-\lambda_4(-\beta(b_2a_1))) \\ &\exp(-\lambda_5(-\beta(c_1a_1)))\exp(-\lambda_6(-\beta(c_2a_1))) \end{aligned} \tag{6.16}$$

Next, we note that the probability of a state can be found in terms of marginal and conditional probabilities. For state 0 we have:

$$\begin{aligned} P(S_0) = &P\left(e_a^1e_b^1e_c^1\right) = P\left(e_a^1\right)P\left(e_b^1|e_a^1\right)P\left(e_c^1|e_a^1\right) \\ &= \alpha(a_1)\beta(b_1a_1)\beta(c_1a_1) \end{aligned} \tag{6.17}$$

Similarly, for state 2:

$$P(S_2) = \alpha(a_1)\beta(b_1a_1)\beta(c_3a_1) \tag{6.18}$$

The above method is applied to each state. We then proceed by determining a series of quotients to find expressions for each Lagrange multiplier. These expressions are then equated with the corresponding Bayesian expression for the probability of a state. By this means, using Eqs. (6.15), (6.16), (6.17) and (6.18) we find an expression for $\exp(-\lambda_5)$.

Thus:

$$\frac{P(S_0)}{P(S_2)} = \exp(-\lambda_5) = \frac{\beta(c_1a_1)}{\beta(c_3a_1)}. \tag{6.19}$$

Quotients isolating a given Lagrange multiplier are not unique and the choice among the possibilities is arbitrary. Similar expressions can be found for each Lagrange multiplier except for the special case of those associated with the root node, which we consider next.

We find an expression for $\exp(-\lambda_0)$. Again using the quotient method, we start by finding $\frac{P(S_0)}{P(S_{18})}$ using the maximum entropy method

$$\frac{P(S_0)}{P(S_{18})} = \frac{\begin{array}{c}\exp(-\lambda_1)\exp(-\lambda_3(1-\beta(b_1a_1)))\exp(-\lambda_4(-\beta(b_2a_1)))\\ \exp(-\lambda_5(1-\beta(c_1a_1)))\exp(-\lambda_6(-\beta(c_2a_1)))\end{array}}{\begin{array}{c}\exp(-\lambda_{11}(1-\beta(b_1a_3)))\exp(-\lambda_{12}(1-\beta(b_2a_3)))\\ \exp(-\lambda_{13}(1-\beta(c_1a_3)))\exp(-\lambda_{14}(-\beta(c_2a_3)))\end{array}} \quad (6.20)$$

Equating (6.20) with the Bayesian expression results in the following:

$$\exp(-\lambda_1) = \frac{\alpha(a_1)\Delta(ca_3)\mathrm{B}(ba_3)}{\alpha(a_3)\Delta(ca_1)\mathrm{B}(ba_1)}$$

where $\mathrm{B}(ba_3) = \beta(b_3a_3)^{-\beta(b_3a_3)}\beta(b_2a_3)^{-\beta(b_2a_3)}\beta(b_1a_3)^{-\beta(b_1a_3)}$.

and $\Delta(ca_3) = \beta(c_3a_3)^{-\beta(c_3a_3)}\beta(c_2a_3)^{-\beta(c_2a_3)}\beta(c_1a_3)^{-\beta(c_1a_3)}$.

$\mathrm{B}(ba_1)$ and $\Delta(ca_1)$ are defined similarly.

Similar expressions can be found for $\exp(-\lambda_1)$ and $\exp(-\lambda_2)$.

6.3.4 Independence

As remarked in Sect. 2.2, *d*-separation is a fundamental concept in Bayesian network theory. Verma and Pearl [13] proved the following theorem, which determines all the conditional independencies implied by a DAG in a causal network.

Theorem (Verma and Pearl) *Given* $N = \langle \mathbf{G}, \boldsymbol{P}, \boldsymbol{C}\rangle$ *and* $\mathbf{G} = \langle \boldsymbol{V}, \boldsymbol{B}\rangle$. *Let* $X, Y, Z \subseteq \boldsymbol{V}$ *such that X and Y are d-separated by Z in* **G**, *then X and Y are conditionsally independent given Z.*

It is necessary to prove that the maximum entropy model preserves the conditional independencies implied by *d*-separation in the Bayesian model. Garside [14] proved that for a 2-valued binary tree it is not necessary to explicitly include these independencies as constraint equations in the maximum entropy model. The technique is to show that the maximum entropy model derived using only the causal constraints, retains the independence relationships embodied in the Bayesian model. The reader can find details in Holmes [15] where independence proofs for the multiway, multi-valued trees and inverted trees are presented. All independence proofs therefore must not use techniques outside of MaxEnt.

6.3.5 Overview

In this section, we have shown that maximum entropy can be used to find minimally prejudiced estimates of missing information in a class of Bayesian networks. For this class, the independence constraints implied by *d*-separation are preserved by the maximum entropy formalism and do not need to be explicitly stated. For tree-like Bayesian networks, as discussed in the example, these estimates can be found using iterative algorithms that run in linear time. Generally, in tree and inverted-tree structures, the set of independence constraints used for the maximum entropy model, have been proved elsewhere to be sufficient.

The original motivation for this work arose from the need to overcome the main objection to Bayesian networks at the time, which was that no theoretically sound method existed for determining the required prior probabilities. In using the maximum entropy formalism, a prior distribution based on all the information currently available and updating as new information becomes available is relatively simple. In practical terms, the initial development of the Bayesian network using these methods can incorporate both expert knowledge, derived from knowledge elicitation techniques, as well as existing database sources.

6.4 Philosophical Considerations

6.4.1 Thomas Bayes and the Principle of Insufficient Reason

Reverend Thomas Bayes is famous for Bayes' Theorem, but it was not until after his death in April 1761 that his work in probability was discovered and published. The work of this eighteenth century statistician, has had a profound effect on the fields of probability, statistics and artificial intelligence. From his single research paper on probability, almost lost, an extremely rich environment for data analysis emerged. Without the foundation of the work of Thomas Bayes, the Enigma code would not have been broken when it was, Sally Clark would have been convicted of a crime she did not commit, and robots would not know where they are!

The Principle of Insufficient Reason, first formulated by Bernoulli, states that if an event has many possible outcomes and there is insufficient reason for doing otherwise, equal values must be assigned to the probability of each outcome. This is a reasonable common-sense principle, but problems arise when trying to apply it to determining a probability distribution. For example, Rhodes [16] has pointed out that, in a physical single six-sided die rolling experiment, an unbiased die results in an expected value of 3.5. However, many possible probability distributions exist with this expected value and without prior knowledge, the Principle of Insufficient Reason provides no means for choosing between them. The Principle of Maximum Entropy, which can be viewed as a generalization of the Principle of Insufficient Reason, provides a way forward, in that it allows the probability distribution for any combination of partial knowledge, and hence partial ignorance, to be determined. However, although Bayesian networks are based on Bayes' Theorem for probabilistic inference, practitioners are generally comfortable with using frequentist probabilities exclusively to determine a prior distribution and so, in this sense, Bayesian networks are not Bayesian.

6.4.2 Objective Bayesianism

In the work of De Finetti [17] for example, we see a theory of subjective probability proposed as a response to the problems inherent in the earlier logical theory of probability. In the subjective interpretation, probabilities are interpreted as an individual's rational degrees of belief and consequently repeated experimental trials are not required to determine probabilities. Tribus [18] and other prestigious researchers have shown that the axioms of probability are a necessary consequence of properties a measure of rational degrees of belief should possess. These are determined based on experience. For example, early medical expert systems relied on doctors and knowledge engineers working together to elicit probabilities, with varying degrees of success. Of course, under such scenarios there are many possible prior distributions, and the use of domain experts was crucial if other experts were to take such systems seriously.

Jaynes developed a theory, based on Shannon and Weaver's information model, wherein probabilities should satisfy the constraints placed on them by the system, and from the many prior distributions satisfying this condition, the one maximizing entropy should be chosen. Jayne's interpretation, known as the maximum entropy formalism, is the basis for objective Bayesianism.

The optimal determination of the prior distribution is therefore crucial to the legitimacy of any Bayesian network and, as we have seen in this chapter, the maximum entropy formalism provides the necessary means for achieving this level of performance.

Holmes [19] argued that objective Bayesian networks, defined as those whose prior distribution is based on all and only the available information, have properties and strengths that those using frequentist probabilities along, do not possess. On this view, a prior distribution should include both subjective and frequentist probabilities concerning a given domain, as available, themselves to be based on the complete information available. One of many, Jaynes supported the objective Bayesian interpretation of probability.

6.4.3 Bayesian Networks Versus Artificial Neural Networks

Many papers have been written comparing the results of Bayesian networks with those of artificial neural networks (ANN's) in practical situations. For example these two machine learning classification methods have been contrasted for systems as diverse as those for predicting surface roughness in high-speed machining Correa.et al. [20] where the authors concluded that BN's had advantages over ANNs for their application, and the diagnosis of headache types Trojan Fenerich et al.[21], where the authors concluded that the results *'...suggest that BNs give better accuracy when comparing to ANNs for this problem'*.

There are several issues that ANN's do not adequately address, simply because for current practical considerations they are not important. For example, the difficulty in ensuring that a global minimum has been found is no longer considered important. With ever-larger ANN's, performance is barely affected by arriving at a sufficiently close local minimum. A problem that is reminiscent of the early work on prior distribution determination in BN's is that of parameter initialization.

As so often happens, the applications of a new idea have out-stripped theoretical considerations. There is clearly a lot of foundational work to be done on ANN's. The original idea of modelling the functioning of the human brain when we do not know how *that* works proved to be contentious and research has moved on, while still maintaining some of the language. E.g. 'neuron' in an ANN no longer attempts to model the neuron in a human brain.

There are many choices to be made when constructing a neural network and these have generally been decided on an ad hoc basis, analogous to the situation for determining prior probabilities in the early days of Bayesian network theory. That ANN's work is not disputed but how and why they work remains a matter of interest. A new book, *'The Principles of Deep Learning Theory An Effective Theory Approach to Understanding Neural Networks'* [22] by Daniel A. Roberts and Sho Yaida based on research in collaboration with Boris Hanin, due to be published in 2022, is expected to provide a significant contribution to the foundations of ANN's.

6.5 Bayesian Networks in Practice

Bayesian networks are used extensively in a wide range of applications. In this section, a flavor of a few of the main applications is given. Increasingly, specialist knowledge outside the realm of the underlying computer science and statistics is necessary in order to understand the research literature on BN's. this has led to a wealth of papers reviewing BN's in relatively narrow areas, a few of which are cited here.

In their paper *'Bayesian networks in healthcare: Distribution by medical condition'*, Scott McLachlan et al., [23], point out that, although BN's are used extensively in medical research there is scant evidence to suggest that they are being used to their fullest extent in practical healthcare.

Cyber security is a key area where Bayesian network methodology is particularly useful. The absence of historical data lends itself readily to the user of the techniques described in this chapter. The graphical structure also means that all available knowledge concerning a particular situation can be utilized. Using minimally prejudiced estimates, it is possible to detect patterns and offer predictions are where the next attack is likely to occur. A review is given by Chockalingam S et al. in their paper *'Bayesian Network Models in Cyber Security: A Systematic Review'* [24].

Healthcare, Medical Bioinformatics, Cyber Security and Big Data are some of the main areas of Bayesian network application. There are many excellent papers in specialized areas, and interested reader is encouraged to search accordingly.

References

1. J. Pearl, Probabilistic reasoning in intelligent systems, in *Networks of Plausible Inference* (Morgan Kaufmann Publishers, San Francisco, 1988)
2. J. Pearl, Reverend Bayes on inference engines: a distributed hierarchical approach, in *Proceedings, AAAI National Conference on AI* (Pittsburgh, PA, 1982), pp. 133–136
3. D.E. Holmes, P.C. Rhodes, G.R. Garside, Efficient computation of marginal probabilities in multivalued Bayesian inverted multiway trees given incomplete information. Int. J. Intell. Syst. **14**(6), 535–558 (1998)
4. D.E. Holmes, Independence in multivalued multiway causal trees, in *Proceedings of the 19th Conference on Maximum Entropy and Bayesian Methods* (Boise, Idaho. USA)
5. G.F. Cooper, The computational complexities of probabilistic inference using Bayesian belief networks. Artif. Intell. **42**, 393–405 (1990)
6. R. Cowell, P. Dawid, S. Lauritzen, D. Spiegelhalter, *Probabilistic Networks and Expert Systems* (Springer, Berlin, 1999)
7. K.P. Murphy, Y. Weiss, M.I. Jordan, Loopy belief propagation for approximate inference: an empirical study, in *Proceedings of the 15th Conference on Uncertainty in Artificial Intelligence* (San Francisco, CA, 1999)
8. D.E. Holmes, P.C. Rhodes, Reasoning with incomplete information in a multivalued multiway causal tree using the maximum entropy formalism. Int. J. Intell. Syst. 13(9), 841–859 (1998)
9. C. Shannon, *E and Weaver W* (University of Illinois Press, The Mathematical Theory of Communication, 1948)
10. Jaynes, Prior Probabilities. IEE Trans. Syst. Sci. Cybern. SSC-4, 227–241 (1968)
11. P.C. Rhodes, G.R. Garside, The use of maximum entropy as a methodology for probabilistic reasoning.Knowl Based Syst 8(5), 249–258 (1995)
12. D.E. Holmes, Efficient Estimation of Missing Information in Multivalued Singly Connected Networks Using Maximum Entropy, in *Maximum Entropy and Bayesian Methods*. ed. by W. von der Linden, V. Dose, R. Fischer, R. Preuss (Kluwer Academic, Dordrecht, 1999), pp. 289–300
13. T. Verma, J. Pearl, Causal networks: semantics and expressiveness, in *Proceedings of the 4th AAAI Workshop on Uncertainty in Artificial Intelligence* (1988)
14. G.R. Garside, *Conditional Independence Between Siblings in a causal Binary Tree with two-values Events* (Research Report, Department of Computing, University of Bradford, UK, 1997)
15. D.E. Holmes, Independence in multivalued multiway causal trees. in *Proceedings of the 19th Conference on Maximum Entropy and Bayesian Methods* (2000)
16. G.R. Rhodes, *Decision Support Systems: Theory and Practice* (Alfred Waller, 1993)
17. B. De Fineeti, *Theory of Probability* (Wiley, New York, 1974)
18. M. Tribus, *Rational Descriptions, Decisions and Designs* (Pergamon, 1969)
19. D.E. Holmes, Why making objective Bayesian networks objectively Bayesian makes sense, in *Causality and the Sciences*, eds by Illari Russo and Williamson (2011), pp. 583–600
20. M. Correa, Comparison of Bayesian networks and artificial neural networks for quality detection in a machining process. Exp. Syst. Appl. 36(3), Part 2, 7270–7279 (2009)
21. A. Trojan Fenerich, M.T. Arns Steiner, J.C. Nievola, K. Borges Mendes, D.P. Tsutsumi, B. Samways dos Santos, Diagnosis of headaches types using artificial neural networks and bayesian networks. IEEE Latin Am Trans **18**(01), 59–66 (2020). https://doi.org/10.1109/TLA.2020.9049462.
22. D.A. Roberts, S. Yaida, Based on research in collaboration with Boris Hanin, The Principles of Deep Learning Theory An Effective Theory Approach to Understanding Neural Networks due to be published by CUP in 2022
23. S. McLachlan, K. Dube, G.A. Hitman, N.E. Fenton, E. Kyrimi, Bayesian networks in healthcare: distribution by medical condition. Artif. Intell. Med. **107** (2020)
24. S. Chockalingam, W. Pieters, A. Teixeira, P. van Gelder, Bayesian network models in cyber security: a systematic review, in *Secure IT Systems. Lecture Notes in Computer Science*, vol. 10674, eds. by H. Lipmaa, A. Mitrokotsa, R. Matulevičius (Springer, Berlin, 2017)

Dawn E. Holmes BA (Hons), MA, M.Sc., Ph.D., is a Senior Member IEEE, and serves as a faculty member in the Department of Statistics and Applied Probability, University of California, Santa Barbara, USA. Dr. Holmes has research interests in Bayesian Networks, Maximum Entropy and Machine Learning and has published extensively in these areas. As a recipient of the University of California Academic Senate Distinguished Teaching Award, she continues to enjoy teaching at all levels. She is an Associate Editor for the International Journal of Knowledge-Based and Intelligent Engineering Systems and serves on the international program committees of several conferences, as well as reviewing for many journals.

Part II
Advances in Artificial Intelligence Applications

Chapter 7
Artificial Intelligence in Biometrics: Uncovering Intricacies of Human Body and Mind

Marina Gavrilova, Iryna Luchak, Tanuja Sudhakar, and Sanjida Nasreen Tumpa

Abstract Human identity recognition is one of the key mechanisms of ensuring proper information access to individuals. It forms the basis of many government, social, consumer, financial and recreational activities in the society. Biometrics are also increasingly used in a cybersecurity context to mitigate vulnerabilities, estimate potential risks and to ensure protection against an unauthorized access. This chapter will provide a reader with a comprehensive overview of the current progress and state-of-the-art approaches to solving problems of biometric security through artificial intelligence methods, from classical machine learning paradigms to novel deep learning architectures. In addition, the chapter will discuss new types of biometric problems: cancelable biometric systems, multi-modal systems, and social behavioral systems. The chapter will conclude with open problems and applications to classical identity problems as well as outline the emerging paradigms.

Keywords Artificial intelligence · Deep learning · Biometric security · Social behavioral biometrics · Cancelable biometrics · Cloud computing · Privacy · Aesthetic · Cybersecurity · Information fusion

M. Gavrilova (✉) · I. Luchak · T. Sudhakar · S. N. Tumpa
Biometric Technologies Laboratory, University of Calgary, Calgary, Canada
e-mail: mgavrilo@ucalgary.ca
URL: https://ucalgary.ca/labs/biometric-technologies/home

I. Luchak
e-mail: iryna.luchak@ucalgary.ca

T. Sudhakar
e-mail: tanuja.sudhakar1@ucalgary.ca

S. N. Tumpa
e-mail: sanjidanasreen.tumpa@ucalgary.ca

M. Virvou et al. (eds.), *Advances in Selected Artificial Intelligence Areas*, Learning and Analytics in Intelligent Systems 24,
https://doi.org/10.1007/978-3-030-93052-3_7

7.1 Introduction

We witness a new epoch where biometric data explosion leads to redefining our understanding of human identity. In this new world, human physical appearance, behavioral expressions, social interactions, online activities, emotions, and human psychology are highly entwined. The epoch of digital human identity, which started with the popularization of inexpensive data capturing sensors in the early 2020s, and evolved into the macrocosm of biometric big data, systems, and methodologies, has its foundation on two orthogonal principles. The first foundation is in the combination of physical and digital worlds, that encompasses human physical characteristics with their behavior in digital domains. The second one is the emergence of advanced cognitive systems and artificial intelligence methods, including classical statistical data analytics, information fusion techniques, traditional machine learning and sophisticated deep learning architectures. Those foundations propelled the domain of digital human identity and biometric research to new heights, that were impossible to envision even a decade ago.

Through the lens of a biometric researcher, our deeply interconnected society provides a rich ground for exploration into the nature and complexity of a human being, encompassing a multi-dimensional avenue of traits. Our personal and public lives are becoming more and more entwined: our professional affiliations might affect our leisurely activities or hobbies, our interests in art might find expression through our work relationships, and our online activities might lead to revealing our hidden emotions and psychological traits. Different facets of life become even more prominent through our cyberworld profiles and activities, to which we collectively refer to as online digital identities. Not surprising, such areas as artificial intelligence, big data analytics, trustworthy decision-making, information fusion, pattern recognition, biometrics and security, rely on a gamut of data collected from online domains.

Biometric system can be defined as a pattern recognition system, that can find uniquely identifying features among different individuals [1]. In a typical biometric security system, features are extracted from physiological and behavioral traits as well as from social and soft data in a specific environment. All features are stored in the training database as biometric profiles of enrolled individuals. During an authentication phase, physiological, behavioral, and social features can be extracted depending upon the availability of data; then the test profile will be created using available features. Finally, identification or verification decision is made based on the matching score of the test and training profiles [1].

The original research in biometric pattern recognition had its roots in statistics, data synthesis and classical machine learning methods [2]. Fundamental computational geometry structures, such as Voronoi diagrams and Delaunay triangulations, were used both to find patterns in data and to perform template matching [3]. They were successfully utilized in fingerprint identification, facial expression synthesis, iris recognition, face, and gait analysis, and so on. In 2009, the book summarizing geometry-based approach to computational intelligence in the biometric domains appeared in the Springer-Verlag series, Studies in Computational Intelligence [3].

This research laid a solid foundation for intelligent processing of biometric systems, and proposed ideas of combining information fusion with biometric processing, which was thoroughly explored in the 2012 book "Multimodal Biometrics and Intelligent Image Processing for Security System" [4]. The work also predicted a shift towards automated machine learning and cognitive systems within the information security domain, fueled by the emerging deep learning paradigms [5].

The past decade witnessed a significant interest in research based on users of social networks (Twitter, Flickr, etc.), focusing on their preferences and interactions from the spheres of their interests or affiliations [6]. Substantial number of current human-computer interaction studies include computational aesthetics - understanding someone's artistic preferences and interests with the goal of identifying their gender or their identity [7]. In the big data analytics domain, web browsing history and social network activity patterns of millions of users are being collected and analyzed on a daily basis and being used not only for commercial purposes but also for disaster planning, drug discovery and medical diagnostics [8]. On the other hand, in security research, artimetrics (artificial biometrics) domain and social behavioral biometric (SBB) emerged as new research realms [9]. They study the multitude of features exhibited by users in their online interactions in virtual worlds [10] as well as popular social networks [11], respectively. Unique social behavioural patterns can be exploited as a modality of biometric trait and fused at feature, match score, rank, or decision levels. One of the intriguing phenomena is that social behavioural biometrics can be extracted by observing the known behavioural biometrics (e.g. expression, interactions, gestures, voice, activities, etc.) of individuals in a specific social setting over a period of time. For instance, the idiosyncratic way of person's start to a speech can be revealed by analyzing voice data acquired from regular meetings and can act as social behavioral biometric during authentication [8]. The multi-modal system research within the context of social behavioral traits explores the major advantage of the information fusion among different biometric types: the fusion of classical physiological and behavioral data with social data and soft biometrics. This results in reliable decision-making from the biometric system, resilient to the distortion or low-quality data of some physiological or behavioral biometrics [12].

Artificial intelligence, specifically machine learning and deep learning methods, allow to learn complex patterns extracted from multitude of data (security, social, public, behavioral, aesthetic, etc.) [9]. In addition, social behavior and contextual information, taken into the consideration during decision-making process, can improve the accuracy of existing multimodal biometric decision-making systems [4]. For instance, it is possible to fuse such social information with existing physiological or behavioral biometrics as part of a multimodal system to make a confident decision on the person's identity [11, 13]. In such a system, auxiliary features that include social behavioral features (time and location of communication, topics, lexicographical profiles), contextual information (profession, interest, community) and soft biometrics (person audio or video aesthetic preference, age, gender, emotion) can be obtained by analyzing everyday social activities of a person in a social environment. The social environment can be either physical (e.g. meeting room, family gathering, office), synchronous online (i.e. Zoom or Teams meeting), asynchronous online (Twitter,

Flickr, Facebook, etc.), or a virtual world communication (Second Life, Altspace etc.) [6].

All traditional and emerging directions of biometric and cybersecurity research were greatly enhanced by the advancements made in the domains of cognitive intelligence and deep learning [5, 9]. The first working prototype of chaotic neural network multi-modal biometric system was created in the Biometric Technologies Laboratory at the University of Calgary, Canada in 2011 [14, 15]. Since that time, a gamut of Deep Neural Network architectures has been developed and prominently made its way into biometric security domain. Their applications can be found in video surveillance [16], risk analysis and prediction [17], privacy-sensitive applications [18], real-time emergency response systems [19], physio-rehabilitation medical practice [20], online recommender systems [21], online social networks [22] and other domains [8].

This book chapter reviews major developments in machine learning methods for biometric research in stand-alone, multi-modal and cloud-based security systems. It starts with the introduction to the biometric security research, proceeds with defining unimodal and multi-modal architectures, surveys both traditional and emerging biometric domains, and introduces the concepts of biometrics on the cloud. Deep learning methodology is discussed in the context of biometric security systems. The chapter then proceeds by focusing on two major directions of research: social behavioral systems and cancelable multi-modal systems using deep learning. It lists key application domains of importance to industries and general public, and outlines future directions of vibrant research on the intersection of machine learning and biometrics.

7.2 Background and Literature Review

Biometric technology has emerged as an immensely popular and sought after form of user authentication, that can successfully replace Personal Identification Numbers (PINs), passwords, and tokens [23]. A 2019 report by Verizon showed that passwords account for 81% of all data breaches [24]. Thus, many organizations are turning to more reliable forms of biometric authentication as a proof of identity. Biometric systems are found in a wide spectrum of applications ranging from consumer devices such as mobile phones to law enforcement, surveillance, immigration, nationwide identity schemes, physical and logical access control [25]. Moreover, they are used in multiple consumer domains, such as emergency response, medicine, physio-rehabilitation, emotion prediction, human-computer interaction, robotics, recommender systems and others [8]. This section will cover an overview of biometric systems: biometric system architecture, unimodal and multi-modal biometric systems, physiological, behavioural, and soft biometric traits. It also will provide an introduction to social behavioural biometrics, aesthetic biometrics, and will explore the application of deep learning to biometric technology.

7.2.1 Biometric Systems Overview

Biometric systems are classified according to their purpose: biometric verification systems and biometric identification systems [1].

The goal of a biometric verification system is to verify a user's authenticity already enrolled in the system using a one-to-one matching scheme. It is commonly used for physical access control, logical access control, user verification in consumer devices, border control and immigration. The biometric identification system based on one-to-many matching scheme. Biometric identification is used by government and forensic agencies to identify unknown persons. Some examples of biometric identification systems are surveillance systems, systems to identify traffic violators, and criminal identification in forensic investigation [26]. The four basic components that constitute a biometric system (see Fig. 7.1) are: (a) Data capture module, (b) Processing module, (c) Data storage module, and (d) Matching module. User enrollment is the initial step for both verification and identification, during which user's profile is created using specific biometric and personal information of a user.

In the data capture module, biometric data is acquired from the subject using biometric sensors. The data acquisition process may be time consuming until biometric data of sufficient data quality is obtained from the subject [27]. Data quality may affect greatly biometric system recognition performance, and in case data cannot be retaken, the biometric system design can be modified to compensate for this deficiency [12].

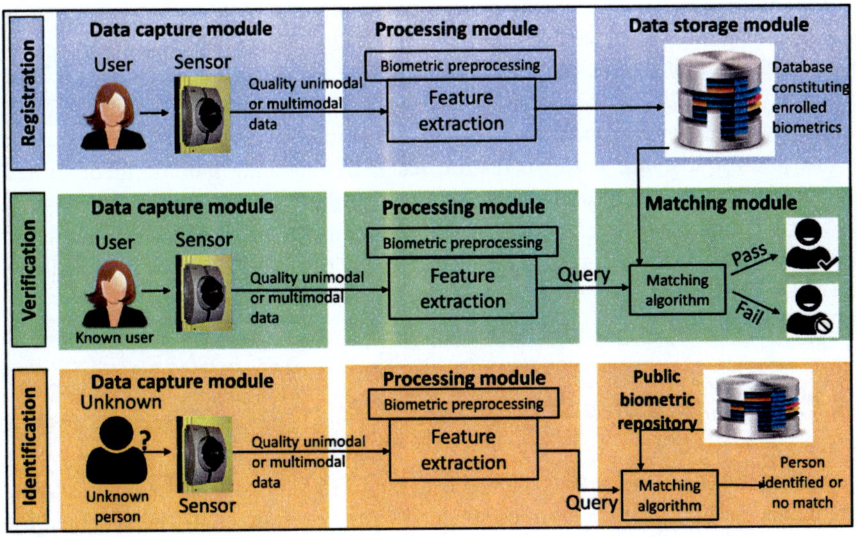

Fig. 7.1 A representation of biometric registration, biometric verification, and biometric identification workflow

The biometric processing module performs biometric pre-processing and biometric feature extraction. During the pre-processing stage, biometrics are enhanced and segmented to separate background noise [28]. Examples are detecting a face from a cluttered image and segmentation of eyelids, eyelashes in iris images. After segmentation, quality enhancement is usually performed to further reduce distortion. Examples are enhancement algorithms like image sharpening, histogram equalization, or contrast modifications to minimize illumination differences by the camera [29] [30]. The main task of the processing module is biometric feature extraction. Feature extraction refers to the generation of a compressed but definitive representation of the biometric trait. For example, position and orientation of fingerprint minutia points, striations in the iris, and fingervein vascular pattern are unique feature representations of a user. Extracting precise biometric features is essential to the uniqueness and discriminatory property of a template, the greater is the authentication accuracy. Precise feature extraction helps to reduce false positives (when a biometric system authenticates a wrong user) and false negatives (when a biometric system fails to recognize a genuine user). There are various feature extraction techniques used for different modalities based on geometric processing, specifically Voronoi Diagrams and Delaunay Triangulations based methods [31], Circular Contourlet Transform (CCT) [32], Discrete Fourier Transforms (DFT) [33], and Histogram of Oriented Gradients (HOG) [34].

Enrolled biometric templates are then stored in a central repository. Certain personal information such as name, Personal Identification Number (PIN), gender, or address may also be stored. It is very important for the data storage module to be secure from malicious software or intruders who can compromise user's biometrics.

The fourth module is the matching module. The input biometric query is compared to stored templates in the database to establish an individual's identity. It may involve one-to-one matching for biometric verification or one-to-many matching process for biometric identification. Matching is performed by generation of match scores, similarity scores, or confidence scores [35]. Dissimilarity scores can also be used to measure the distance between feature sets. The matching module incorporates the decision-making module, which provides a ranking of identities and generates a success/failure of authentication. Current research is focused on establishment of user identity not only accurately, but with a high degree of confidence in decision [6].

7.2.2 *Classification and Properties of Biometric Traits*

According to [36], a biometric trait needs to conform to certain pre-requisites to be established as a reliable trait. An established biometric trait should be accepted by the society and should be convenient to provide. Second, it should be a universal trait found in the majority of the population. Third, it should possess discriminability or uniqueness that clearly separates one person's biometric trait from another. Fourth, it should be permanent, that means it should not fade or vary over time. Fifth, it should

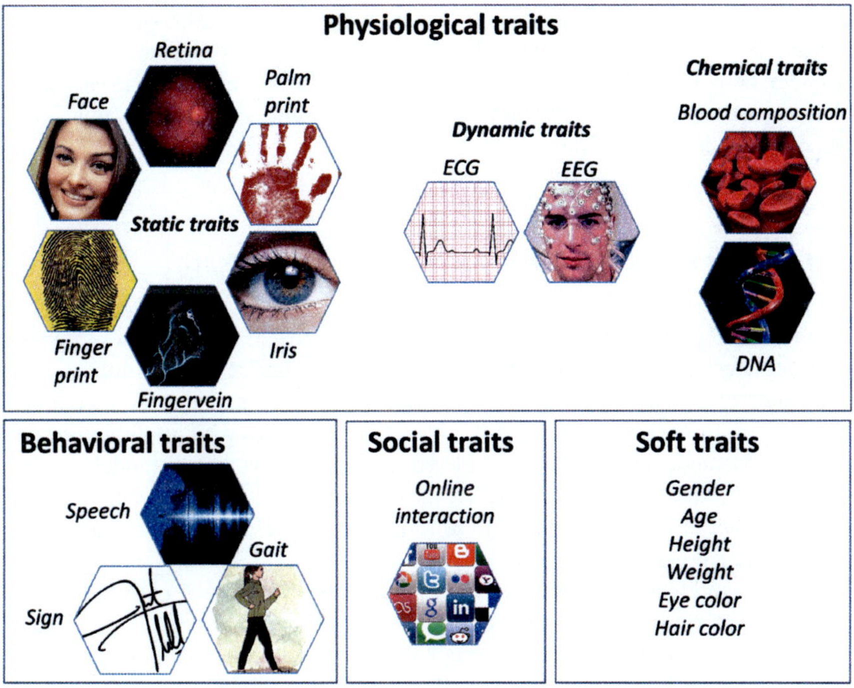

Fig. 7.2 A broad classification of various types of traits into physiological, behavioural, social, and soft biometric traits

be complex to recreate or forge, thereby making circumvention and non-repudiation (denial of biometric) difficult [36].

Biometric classification is illustrated in Fig. 7.2 and includes physiological, behavioral, social, and soft traits. Physiological traits can be further classified into (a) Static traits, (b) Dynamic traits, and (c) Chemical traits. Static traits are permanent body traits that do not change significantly over time, such as iris, fingervein, fingerprint, palmprint, retina, and face. Dynamic traits display a pattern over time such as electrocardiogram (ECG or EKG—a graph that measures electrical activity of the heart), and electroencephalogram (EEG—a graph that measures electrical activity of the brain). A third type of physiological traits is chemical-based traits, such as DNA sequences forming genes and blood composition. Such chemical traits are predominantly used in forensic investigation for biometric identification in a crime scene. The second very well known type is behaviour-based biometric traits. Traits associated with human behavior, such as walking style (gait), signature, and typing patterns are included in this category. Applications of behavioral traits can be found in emotion recognition [37], medical diagnosis of diseases such as arthritis, Parkinson's [17], and patient rehabilitation [17]. A new category of biometric recognition, based on online social interactions, was introduced in 2014 in the Biometric Technologies laboratory at the University of Calgary. This type of behaviour-based identification based on monitoring online behaviour and writing styles have shown surprisingly accurate identification results [11, 13, 38]. The fourth type is soft biometrics such as gender, height, weight, eye colour, and hair color. These are not unique characteristics, but can help in authentication when combined with another physiological or behavioural trait.

7.2.3 *Unimodal and Multi-modal Biometric Systems*

The optimal biometric system is one having the properties of distinctiveness, universality, permanence, acceptability, collectability, and resistance to spoof attacks [39]. First biometric systems were identifying a user based on a single trait. A common example of a unimodal system is a mobile phone that uses a single fingerprint to unlock it. Unimodal systems are considered less reliable than multi-modal systems due to greater effect of noise, low quality data, similarity between classes, and intra-class differences [40]. Figure 7.3 demonstrates a generic architecture of a unimodal and multi-modal system.

It is impossible for a single biometric security system to satisfy all of these requirements despite tremendous progress in the biometric domain. However, as with many other fields, where a single-traits might fail, data aggregation and intelligent decision making can prevail. This has led to rise of multi-modal biometrics. In practically all scenarios, multi-modal architectures can provide a higher identification accuracy. A multimodal system can use one or more sensors (multi-sensorial system) for capturing instances of the same trait (right and left iris) or different traits (ear and face) [41]. A multimodal system may even be a multi-algorithmic biometric system that takes biometric input from a sensor and processes it using different algorithms. Multi-biometric systems combination with geometrical models, especially for face recognition has proven to be highly effective [42]. In some cases, multimodal architectures can approach zero Failure To Enroll (FTE) rate [43] and can be highly resistant against spoofing attacks [44]. However, a careful consideration of costs, training and types of data must be always given when deciding what multi-modal biometric system should be employed.

The advantages of multimodal biometric systems stem from the fact that there are multiple sources of information. The gain is in reliable recognition performance, increase in confidence in the decision made by multiple experts, and the greater level of assurance of a proper match during verification and identification. Increase in recognition performance of multimodal biometric systems is due to its ability to handle the noisy or poor quality data [4].

Development of a multibiometric system for security purposes is a difficult task. As with any unimodal system, the data acquisition procedure, sources of information, level of expected accuracy, system robustness, user training, data privacy and dependency on proper functioning of hardware and proper operational procedures impact directly the performance of a biometric system. The choice of biometric information that will be integrated must be made, information fusion methodology should be

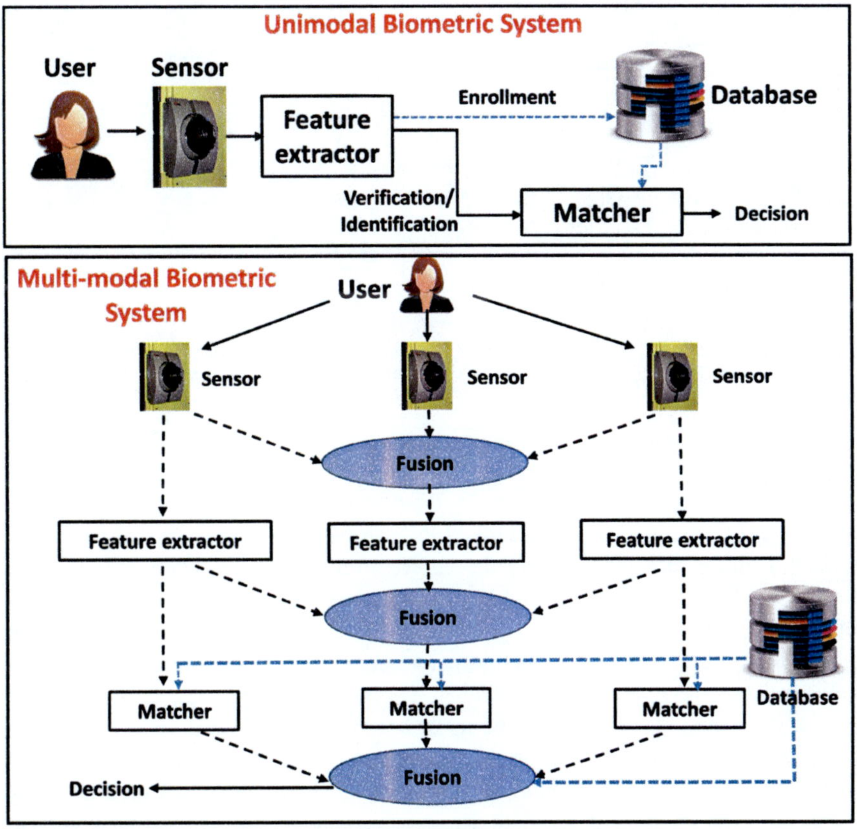

Fig. 7.3 Generalized unimodal and multi-modal biometric system architectures

selected, cost and benefit analysis should be performed, processing sequences must be developed, and system operators trained [4].

Since 2008, there has been an explosion in research on biometric information fusion types. Sanderson and Paliwal [45] proposed two main categories of multi-modal biometric information fusion: fusion before matching and fusion after matching. Fusion before matching contains sensor level fusion and feature level fusion, while fusion after matching contains match score level fusion, rank level fusion and decision level fusion. Figure 7.4 is a classification of biometric fusion techniques at various stages in a multimodal system. A new fusion mechanism called fuzzy logic fusion [46] was recently proposed. Fuzzy biometric fusion can be employed either in the initial stage, i.e. before matching occurs or in the latter stage, i.e. after matching occurs. Combining biometric traits can be facilitated at different time, starting from raw data acquisition to the final stage of match/non-match decision. For each type of information fusion, several algorithms rooted in statistics and confident decision-making, has been developed. For example, to get the consensus ranked list of most

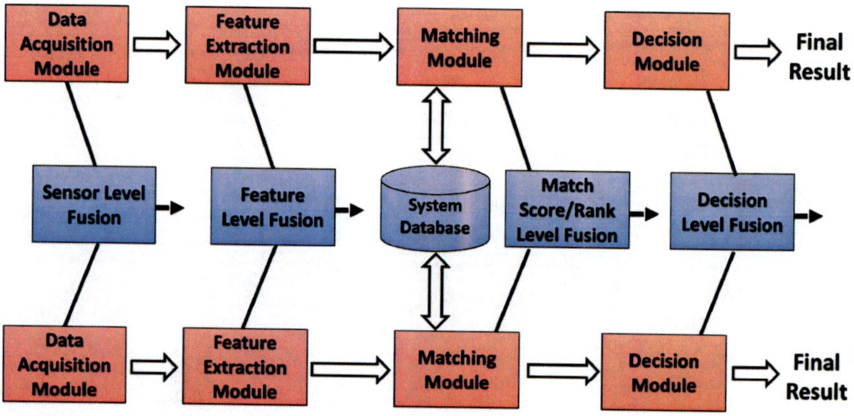

Fig. 7.4 Fusion mechanisms in multi-biometric systems

probable user identity, the initial ranked list can be integrated by the highest rank method, Borda count method, logistic regression method, Bayesian method, fuzzy method, or Markov chain method [47]. It is up to the end user of the biometric system (i.e. airport security, government agency, financial institution etc.) to determine the correct configuration and to choose biometric modalities. A good information fusion method allows the impact of less reliable sources be lowered compared to reliable ones, even in the presence of uncertainty or low performing modalities.

Before it made its way into biometric domain, information fusion was a key methodology in engineering and signal processing fields, multi-sensor processing and expert systems. A classical data driven information fusion approaches are also common in robotics, image processing and pattern recognition [4].

The origins of information fusion can be found in the neural network literature, since the work of combining neural network outputs appeared in 1965 [48]. Since then, the information fusion was successfully used in multimedia retrieval [49], multi-modal object recognition [50], multibiometrics [51] and video retrieval [52]. In summary, the benefit of fusion is that the influence of unreliable biometric traits can be lowered compared to reliable ones, while their inclusion still benefits the overall recognition accuracy.

7.2.4 Social Behavioral Biometrics and Privacy

A new type of biometric authentication that is based on online social data has been introduced in 2014 at the Biometric technologies laboratory and is called Social Behavioral Biometrics (SBB) [11]. It is based on a person's social interaction in on-line or off-line networks. The social feature set primarily includes (a) user's online temporal patterns and their frequency, (b) mode of communication through tweets, blogs, chats, (c) topic of interaction, interest areas, and d) online game playing strategies [11, 13]. Social biometrics has garnered interest in the biometric realm due to user communication and network patterns rooted in human interaction and psychology. A person's psychology drives gestures and emotions, whereas human interaction is dependent on friends, acquaintances, virtual networks, and communications.

Since 2014, many systems based on social behavior of users were developed. Thus, works [9, 11, 13, 53] utilized behavioral footprints and interaction patterns expressed through online communications. User profiles created through extracted behavioural features based on activity logs over a time frame were utilized to reliably authenticate users. Unobtrusive person recognition using multimodal fusion of behavioural patterns and gait was found to be successful in user identification [54]. Information sharing and collaboration websites such as Wikipedia were utilized in [38] to decipher the author's identity. Social biometrics also find applications in access control, risk assessment, and security [11, 13]. Most recent 2020 works [6, 22] studied the impact of writing profiles of users on Online Social Networks (OSNs) including Twitter to identify users based on their social behaviour. Predictions were made based on reply, retweet, shared weblink, trendy topics and temporal profiles to achieve high recognition accuracy, surpassing on closed data sets the performance of traditional biometric forms. Social behavioral biometric system can be further expanded with psychological traits and emotional state identification, opening new avenues for their exploration in human-computer interaction and social services domains [9].

Aesthetics refers to a person's preference for specific audio or visual content. Audio content includes songs and music whereas visual content includes pictures and videos [21]. Nowadays, search engines and recommendation algorithms target a person's preferences to display relevant advertisements. Surprisingly, aesthetics demonstrated potential for person identification in the biometric domain, by utilizing audio or visual preferences as a biometric feature set during authentication.

The concept of aesthetic biometrics emerged in 2012, with [55] exploring an identification system based on set of images a person liked. Aesthetic biometrics can be considered a subset of Social Behavioral Biometrics. They demonstrated that aesthetic biometrics have potential in identification and published a public dataset for future research. This research led to another work performed in 2014 which utilized greater contextual and visual categories of features [7]. In this work, authors optimized feature sets through a technique of multi-resolution counting grid. In 2017, Azam and Gavrilova obtained a higher accuracy by proposing a new feature extraction technique combined with Principal Component Analysis (PCA) for feature reduction [56]. Sieu and Gavrilova took it further in 2019, obtaining 95.1% identification accuracy [57] through composite features created by Gene Expression Programming (GEP). The latest research in 2020, by Bari, Sieu, and Gavrilova leveraged a deep learning Convolutional Neural Network (CNN) 'AestheticNet' for automatic aesthetic feature-set creation [21]. The three-stage network surpassed previous works by obtaining rank 1 accuracy of 97.73%.

As aesthetic identification as well as in general social behavioral biometrics involve analysis of personal choices and can be collected from publicly available data in online social networks, it is highly important to consider privacy and security issues in practical system development. There have been a number of approaches to ensure biometric data privacy, including authentication, encryption, random projection, template protection, network protection, and attack against biometric system detection [41, 58, 59]. An in-depth look at those mechanisms to mitigate privacy concerns is presented in this chapter in the section on deep learning in cancelable biometrics.

7.2.5 Deep Learning in Biometrics

Deep learning has shown to be a new and powerful tool in the domain of image processing and biometric recognition. It is a subset of Artificial Intelligence (AI) and Machine Learning (ML). It is inspired by the human brain's processing mechanism of interconnected neurons (neural networks). The word 'deep' refers to the many number of layers through which data is processed. Deep neural networks transform data into abstract representations which are very effective in classification and regression tasks [60]. Unlike traditional machine learning algorithms, deep neural architectures do not require manual setting of network parameters and can also work on labelled training data (supervised learning) or unknown/unlabelled data (unsupervised learning). Such networks utilize the technique of back-propagation to accordingly adjust network weights after each iteration so that the network learns better and improved accuracy is achieved.

Deep learning has shifted traditional biometric systems from classical biometric processing to cognitive intelligent authentication as indicated by advancements made in [5]. The earliest attempt in leveraging deep learning in biometrics was by [14] in 2011 and [15] in 2012, where a chaotic neural architecture was created for a multimodal biometric system. Advancements in computer processing power and networks made since 2012, offers significant opportunity to integrate deep learning in the biometrics field. Some recent research that applies deep learning to different areas of biometrics can be already found in video surveillance [16], risk analysis and prediction [17], privacy-sensitive applications [18], real-time emergency response systems [19], physio-rehabilitation medical practice [20], online recommender systems [21], online social networks [22], and other domains [8].

Popular deep learning models such as Convolutional Neural Networks (CNNs), Recurrent Neural Networks (RNNs), Long Short Term Memory (LSTMs) models, and Multi-Layer Perceptrons (MLPs) have shown to be efficient in biometric processing [61]. CNNs have shown to be powerful in learning unique type of feature representations from input data, especially visual data [62, 63]. Convolutional neural networks are preferred for biometric image recognition due to minimal preprocessing, convolution ability, shared weights, translation invariance, and low network complexity [64].

Convolutional segment consists of a set of layers or filters and a kernel window which parses over the input images. The parsing of the images with respect to each filter creates an activated or convolved image which is processed further in forthcoming layers. Pooling layers essentially reduce dimensions of image activations. This is done in order to reduce parameters and hence computational complexity. The types of pooling that can be employed are maximum pooling, minimum pooling, average pooling, and adaptive pooling [61]. Pooling can help prevent overfitting (when the network overlearns training data and degrades learning performance) and, in turn, increase accuracy on the test set. The combination of convolutional and pooling layers help a deep neural network to discern high level details of the images. The dropout layer is generally utilized subsequently after every convolutional and pooling combination. The dropout layer sets the values of random neuron weights to zero to help the CNN to learn the data more uniformly. The final layers of a CNN may consist of fully connected layers such as Flatten, Dense, and SoftMax. Flatten is used to flatten the tensor values to a one-dimensional array to pass it to the Dense layer. The Dense layer is a highly interconnected with each neuron's output depending on inputs from all the neurons of the previous layer. Softmax is the final layer or decision making layer which generates probability values of each class/user in the dataset.

Neural networks such as a CNN has been recently utilized for biometric feature extraction due to their data learning capabilities, automatic feature extraction, and dimension reduction abilities [18]. A recent survey of biometric recognition using deep learning conducted in [75] found that deep learning was an effective tool for speech, face, gait, iris, and signature recognition. Some examples of recent research that apply deep learning to biometric authentication are [16, 18, 67, 68, 76]. Papers [18] and [68] explored application of deep learning through convolutional neural networks for cancelable biometrics. CNNs were leveraged on standalone and cloud-based biometric systems for feature extraction from irises to create cancelable or revocable biometric templates. In [16], authors proposed a deep learning neural network for Kinect-based gait recognition in which the neural network was trained using dynamic joint relative cosine dissimilarity and joint relative triangle area and achieved high accuracy. Another recent publication [67] explored supervised application of deep learning on fingerprint recognition to create an Automatic Fingerprint Identification System (AFIS). Deep learning has also shown to be effective in facial recognition, with work [76] presenting a comprehensive review of the recent developments in deep facial recognition. It presented different network architectures and loss functions, along with facial processing methods of one-to-many augmentation and many-to-one normalization. Deep learning has also found application in class attendance systems for student facial detection [71]. Paper [77] presented a systematic review of speech recognition using Deep Neural Networks. In this review, it was found that traditional Mel Frequency Cepstral Coefficients (MFCC) for classical classifiers such as Hidden Markov Model (HMM) and Gaussian Mixture Model (GMM) were widely used as feature extraction methods of speech signals. The authors of [77] recommended using deep Convolutional Neural Networks (CNNs), Recurrent Neural Networks (RNNs), and Long Short Term Memory (LSTM) for speech signal processing. A comprehensive review of prior to 2018 research integrating deep learning methods with biometric domain can be found in [78]. Table 7.1 lists works that appeared after 2019 and utilize deep learning techniques for biometric authentication.

Table 7.1 Literature review of recent deep learning applications in biometric domain

Authors and paper	Year	Deep learning technique	Trait(s)
Abeer and Talal [65]	2021	CNN and image augmentation for person recognition	Gait
Tumpa, Luchak, and Gavrilova [22]	2021	Convolutional Neural Network for online social behavior	Twitter communication
Boucherit, Zmirli, Hentabli, and Rosdi [66]	2020	Deeply fused Convolutional Neural Network	Fingervein
Sudhakar and Gavrilova [18]	2020	Convolutional Neural Networks for Cancelable Biometrics	Irises and fingerveins
Wani, Bhat, Afzal, and Khan [67]	2020	CNN for Automated Fingerprint Identification System (AFIS)	Fingerprint
Bari, Sieu, and Gavrilova [21]	2020	AestheticNet CNN	Aesthetic Image Biometrics
Sudhakar and Gavrilova [68]	2020	Parallelized Convolutional Neural Networks for Cancelable Biometrics on the Cloud	Irises and fingerveins
Bari and Gavrilova [16]	2020	Deep Neural Network with Adam optimizer	Gait
Nada and Heyam [69]	2020	Fusion of three CNNs for multimodal biometrics	Iris, face, and fingervein
Bari and Gavrilova [70]	2019	Multi-Layer Perceptron (MLP)	Gait
Khan, Harous, Hassan, Ghani Khan, Iqbal, and Mumtaz [71]	2019	Convolution Neural Network and edge computing	Face
Minaee and Abdolrashidi [72]	2019	Residual Convolutional Neural Network (RCNN)	Iris
Wu, Tao, and Xu [73]	2019	Occluded face recognition using Convolutional Neural Network (CNN)	Face
Minaee, Azimi, and Abdolrashidi [74]	2019	Convolutional Neural Network (CNN)	Fingerprint

7.3 Deep Learning in Social Behavioral Biometrics

The previous section provided a comprehensive overview of biometric research domain, unimodal and multi-modal biometric system architectures, information fusion methods and emerging biometric modalities. It identified social behavioral biometric, aesthetics and biometric privacy as important topics for biometric security and cybersecurity. It also introduced deep learning concepts in the context of biometric security research. The following sections present an in-depth look into the design of multi-modal biometric systems, their performance and applications in other domains of research. We start this discussion with social behavioral biometric system design.

7.3.1 *Research Domain Overview of Social Behavioral Biometrics*

The concept of identifying users based on their social interactions and communication was introduced by Sultana et al. along with a formal definition of Social Behavioral Biometrics (SBB) [11]. The authors proposed five unweighted networks generated from retweets and reply acquaintances, shared hashtags, and URLs to identify users. In [53], another system was introduced using a set of features to identify users. The method generated weighted networks from the online social interactions and temporal behavior of the users. The integration of online social networking behavior of the users with the Internet of Things (IoT) infrastructure was proposed in [79]. The authors used data from built-in sensors and collected statistics of different social networks for continuous verification on smart devices. Saleema and Thampi proposed SBB features based on the cognitive psychological personality theories of dispositions and temperament in the context of online social network users [80]. This work proved the stability and uniqueness of the proposed SBB features to ensure the quality of the biometric templates. In [81], a machine learning-based SBB system, online social interaction-based weighted networks, temporal behavior and writing style of the users are incorporated for user identification. In [82], the same authors analyzed the impact of existing social biometric features, namely linguistic profile, temporal profile, reply network, retweet network, hashtag network and URL network on SBB. In [83], Twitter data is mined to recognize members of organizations based on the users' social networking relationships. In [84], fraudulent activity detection is detected based on user browsing behavior. Those works demonstrate potency of social data analysis in a variety of modern application domains.

7.3.2 *Social Behavioral Biometric Features*

In this day and age, users leave a huge amount of behavioral trails that can be used to identify them in online social networks. From these behavioral trails, three types of features can be obtained [11]: (a) Knowledge-based features, (b) Style-based features, and (c) Frequency-based features.

Knowledge-based features help to discover the person's behavioral information on online social networks regarding profile attributes and networks. Users provide attribute-based information during their profile creation on the online social network that includes username, affiliation, job, gender, date of birth, personal interest, webpage link, geographic location, address, etc. Network-based information is generated from the communicating behavior and shared contents of the users.

Style-based features are generated from the writing samples of the users. As everybody has their own writing pattern and choice of words, the style-based features hold the information of the user's writing style on online social networks. For example, how frequently a user uses emoticons, the most preferred emoticon for online social networking posts, user's preferred vocabulary set, use of punctuation, abbreviations, frequent spelling mistakes contribute to those features. Any kind of linguistic features obtained from the written contents of the users is considered as style-based features.

Frequency-based information is the statistical temporal features that are extracted from the users' online social network profiles. Statistical temporal information helps to reveal users' tweeting and networking behavior. A summary of the SBB features has been provided in Table 7.2.

7.3.3 General Architecture of Social Behavioral Biometrics System

Social Behavioral Biometric (SBB) features can be used in a unimodal system as well as in multimodal systems. In a unimodal system, the identification and verification of the users are performed using the SBB features only, while information fusion is still used to combine results from different SBB networks. In a multimodal system, SBB features are integrated with other biometric modalities, such as fingerprint,

Table 7.2 Summary of Social Behavioral Biometric (SBB) features

SBB features	Extracted information from OSN
Knowledge-based features	**Attributed-Based:** Username, date of birth, job, affiliation, education, gender, personal interest, geographic location, etc. **Network-Based:** Reply network, retweet network, hashtag network, URL network, follower network, followee network, etc.
Style-based features	Users' preferred emoticons, vocabulary set, punctuation, abbreviations, frequent spelling mistakes, character styling, etc.
Frequency-based features	Average probability of tweeting per day, average probability of tweeting per hour, seven days interval period, seven days tweeting period, average probabilities of original tweet per day, average probabilities of retweet per day, average probabilities of reply/mention per day, average probability of tweeting per week, etc.

gait, face, etc. to perform the identification and verification of the users. The general architecture of a Social Behavioral Biometric system is demonstrated in Fig. 7.5.

At first, the data obtained from the online social network is processed and divided into training and testing modules. Six networks, namely writing profile, reply network, retweet network, hashtag network, URL network and temporal profile are generated from both training and testing modules.

The writing profile is generated from the self-written tweets and replies of the user. The user's preferred vocabulary set is constructed using a feature extraction algorithm, Term Frequency-Inverse Document Frequency (TF-IDF) [85] and the Multinomial Naïve Bayes algorithm [86] is used to perform the classification.

The reply network is generated from the list of acquaintances whom a user replies or mentions frequently. The nodes are created from the extracted list of users and the edges are formed between the users based on the reply and mention relationship. The weights are calculated according to the frequency of the relationship and the edges between frequently replied and mentioned nodes gain higher weights. The list of retweeted acquaintances is parsed from the dataset whose tweets a user retweets often to generate the retweet network. A lot of users find it convenient to combine

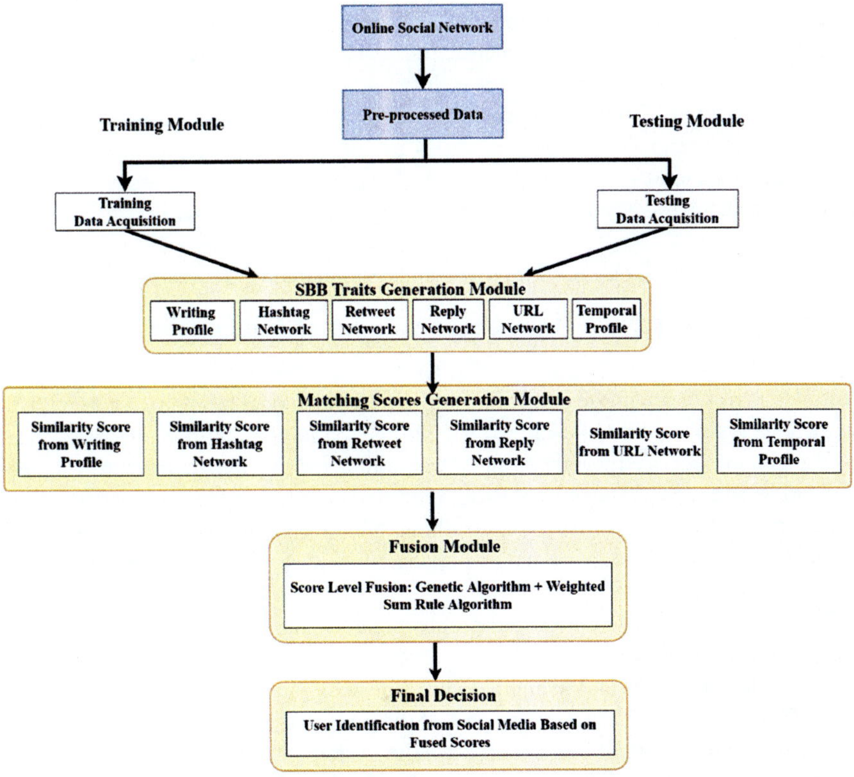

Fig. 7.5 General architecture of a Social Behavioral Biometric system

the weblinks with the tweets instead of writing the whole idea to share with the followers. The list of shared weblinks represents the user's preferred domains that contain the sharing pattern of that user. The shared URLs are parsed from the written contents of the dataset.

Twitter users use hashtags to categorize and connect their posts with the trends. These hashtag words can be considered as the main topic of the tweets. A trendy topic network or hashtag network can be generated from this hashtag relationship. Temporal profiles of users reveal their posting patterns in the social network. The temporal profile of each user can be created by extracting the features from the timestamps of user's profile, such as the average probability of tweeting per day, the average probability of tweeting per hour, average probability of tweeting per week, seven days interval period, seven days tweeting period, average probabilities of original tweet, retweet, and reply/mention per day, etc.

The SBB traits are fused using the score level or rank level fusion algorithm. The weights for all networks are chosen using the Genetic algorithm. Finally, the users of online social media are identified based on the fused scores.

The proposed SBB system can be combined with other physical or behavioral biometric systems to improve the system performance in identifying and verifying the users of the virtual world to enhance recognition performance. Figure 7.6 illustrates the general architecture of a multimodal SBB system. This system combines social behavioral and physiological biometric traits. The social behavioral profiles and soft biometrics are extracted from the online social networks and the physiological features are obtained from any physiological trait of the users. The scores obtained by social behavioral and physiological modules are combined using decision, score level or rank level fusion to arrive to the final authentication decision. Special attention should be given to privacy of user data, especially in the case of a multi-modal system.

7.3.4 Comparison of Rank and Score Level Fusion

Fusion is an important phase of any multi-trait or multi-modal biometric system [26]. Score level fusion and rank level fusion are highly popular in multi-modal biometric systems. A number of experiments have been conducted to compare the performance of the unimodal Social Behavioral Biometric (SBB) system using score level fusion algorithms and rank level fusion algorithms. This research was conducted at the Biometric technologies laboratory and appeared in [81]. Main findings are briefly summarized below.

The experiments were conducted on proprietary dataset of 250 users, with 200 tweets collected for each user [81]. The score level fusion algorithms: Median rule, Product rule, Sum rule and Weighted sum rule were applied to fuse all six SBB traits. Table 7.3 presents the performance of the proposed model in terms of average accuracy, precision, recall and f-measure. The system achieved the highest rank-1 recognition rate of 99.45% when utilized the GA-based weighted sum rule score

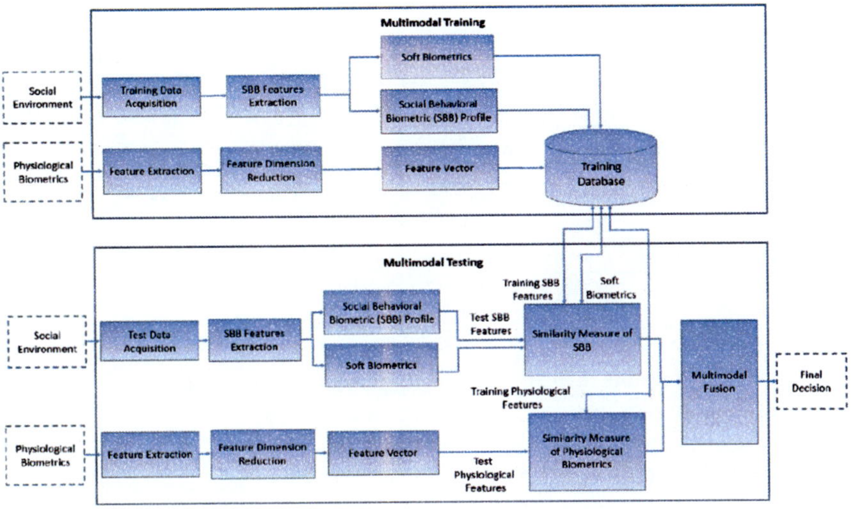

Fig. 7.6 General architecture of a multi-modal biometric system using social behavioral and physiological biometric traits

Table 7.3 Performance of the score level fusion algorithms on the SBB system [81]

Algorithm (%)	Accuracy (%)	Precision (%)	Recall (%)	F-measure (%)
Weighted sum rule + GA	99.45	99.17	99.45	99.26
Weighted sum rule	97.72	96.58	97.72	96.96
Sum rule	94.88	92.48	94.88	93.26
Product rule	89.90	85.52	89.90	86.92
Median rule	89.70	85.70	89.70	86.94

level fusion algorithm. The weights were assigned by the genetic algorithm after 200 generations.

The second experiment investigated the performance of rank level fusion algorithms on the SBB system by comparing the rank level fusion algorithms, namely Modified Highest Rank (MHR) and Weighted Borda Count (WBC). Table 7.4 illustrates the performance of the SBB system in terms of average accuracy, precision, recall and f-measure for different rank level fusion algorithms. The weight combination 0.54, 0.40, 0.03, 0.02, and 0.01 for writing, retweet, reply, shared weblinks, and trendy topic network, respectively was used in WBC_1, which achieved the highest rank-1 accuracy of 84.92% and rank-10 accuracy of 95.23%. All other weight combinations for rank level fusion algorithms, and Modified Highest Rank (MHR), did not perform as well. Thus, it was concluded that weighted sum rule and genetic algo-

Table 7.4 Performance of the rank level fusion algorithms on the SBB system [81]

Algorithm (%)	Accuracy (%)	Precision (%)	Recall (%)	F-measure (%)
WBC_1	84.92	79.02	84.92	80.79
WBC_2	84.79	78.75	84.79	80.57
WBC_3	84.72	78.68	84.72	80.50
WBC_4	82.37	75.59	82.37	77.60
WBC_5	80.36	73.03	80.36	75.16
MHR	73.51	65.28	73.51	67.49

rithm combination results in the highest performance for social behavioral biometric system based on tweets.

7.3.5 Deep Learning in Social Behavioral Biometrics

Previous section investigated information fusion methods combined with classical machine learning approaches for social behavioral biometric authentication system. In recent years, deep learning has emerged as a powerful classification methodology that can achieve very high accuracy when trained on large and complex datasets [87]. As millions of users are spending significant amount of time in online social networks and produce new data through their communication and social interactions, a vast amount of data is being generated everyday. Incorporating deep learning with Social Behavioral Biometrics (SBB) opens new frontiers to sensibly mine this data, with the purpose of improving processes, increasing user convenience and preventing unwanted risks. In this section, we present a general architecture of a SBB system using a deep learning architecture. The data obtained from the online social networks can be divided into two types: textual data and image data. Textual data includes tweets, replies, retweets, shared weblinks, hashtags, temporal information associated with the texts, etc. Image data includes the images attached with the shared posts, caption or title of the images, temporal information about the images, etc. The proposed architecture consists of two modules. One module contains a deep learning architecture to analyze the textual data and another module contains deep neural networks to analyze the image data. The system is trained with both modules separately and stores the extracted features in the database. In the testing phase, the system generates the features from the test dataset and matches them with the stored templates to measure the matching scores. The similarity scores obtained from both modules are fused to get the final identification report of the users.

Figure 7.7 presents the architecture of the proposed deep learning-based Social Behavioral Biometric system. In the text analysis module, the input is the textual data obtained from the social networks. The input data is passed through a deep neural network for feature extraction. The network is trained and tested to generate the

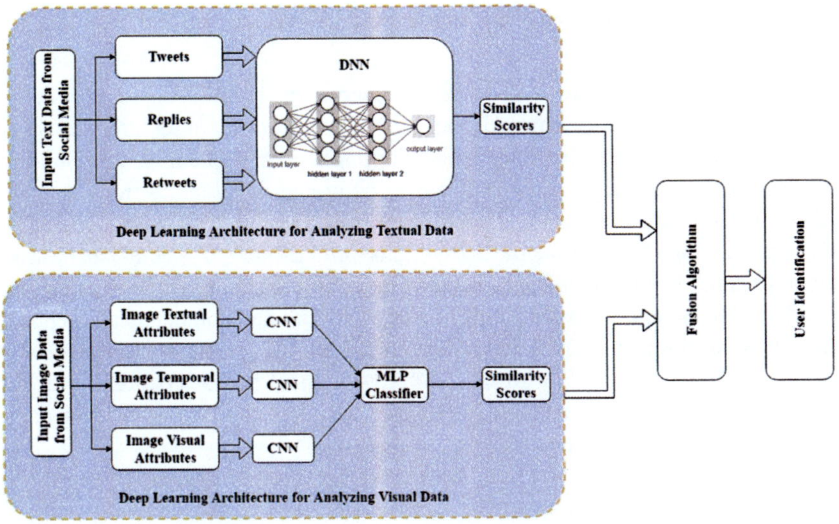

Fig. 7.7 Deep learning-based architecture of the proposed Social Behavioral Biometric (SBB) system

similarity scores. The Convolutional Neural Network (CNN) and Recurrent Neural Network (RNN) are commonly used in text and image analysis [88–90]. The output of the deep neural network is the similarity scores which will be the input for the fusion module.

In the image network, three types of attributes are used to describe the associated image. The first attribute is textual, such as the image title and image caption; the second attribute is temporal, such as the posting time, date and frequency; the third attribute is visual, which are the pixels representing the image. Each of the three attributes is fed into a separate Convolutional Neural Network (CNN) as each attribute represents a different type of data input. The CNNs were chosen over Recurrent Neural Networks (RNNs) for textual classification because of three reasons:

1. CNN's ability to extract local and position-invariant features, such as specific sentiments, names of places etc.
2. RNN's ability to understand sequential information is unnecessary for our length-limited textual and temporal attributes, causing potential overfitting when compared to CNN's [91].
3. Classifying temporal data has also been proven to be successful by CNNs, eliminating the need for more complex and computationally expensive RNNs [92].

The features learned from the three separately trained CNNs are aggregated and passed through Multilayer Perceptron (MLP) classifier for multi-modal deep learning. The similarity scores are then integrated with other SBB features as part of a multi-modal system.

7.3.6 Summary and Applications

Social Behavioral Biometrics is a powerful biometric trait for ensuring security in the cyberworld. This section has provided an overview of the current research trends as well as the future research direction of this emerging research domain. Traditional and deep learning architectures for social behavioral biometric system were presented, and comparison between rank and score level fusion methods was conducted. The proposed unimodal and multimodal Social Behavioral Biometric systems can be used in various applications, including continuous user authentication, anomaly detection, recommender systems, assisted living, robotics and human-computer interactions. The proposed unimodal person identification system using online social behavior can efficiently be integrated with existing physiological and behavioral biometric systems. Investigating most suitable information fusion methods for such integration can be a topic of future studies.

7.4 Deep Learning in Cancelable Biometrics

The previous section discussed information fusion and deep learning concepts in multi-modal social behavioral biometric systems. We established that information privacy is of enormous importance. The current section introduces the concept of cancelable or revocable biometrics to keep the users data safe. The multi-modal cancelable biometric system for irises and fingerveins is described. This standalone cancelable biometric system is further extended as a cloud-based architecture employing parallelized deep neural networks. The section concludes by discussing applications for deep learning based cancelable biometrics.

7.4.1 Biometric Privacy and Template Protection

Nowadays, hacking, information leakage, and data malpractices have become a common threat, thus security of biometric data is of utmost importance. Biometric data is a highly sensitive data which is unique to a person. It is essential that this sensitive data is not compromised.

It is essential to keep biometric templates well protected from intruders or an unauthorized access in a biometric system. Template protection schemes are commonly classified into two main categories—biometric cryptosystems and cancelable biometrics [93, 94].

Biometric cryptosystems is a combination of cryptography and biometrics [95]. The basis for biometric cryptosystems are key generation schemes and key binding schemes. In key generation schemes, the cryptographic key is generated from the biometric itself. This technique is utilized where it is not safe to separately store cryptographic keys. The main disadvantage of biometric cryptosystems is that the biometric templates are not protected and an adversary may be able to generate the crypto key using the unprotected templates. A better way would be to combine cryptosystems with cancelable biometrics, thereby protecting the biometric template as in recent work [96]. Key binding schemes work by binding the cypto key with the user's biometric. Common examples of key binding schemes are Fuzzy Commitment Schemes (FCS) and Fuzzy Vault Schemes (FVS).

FCS was initially proposed by Jules and Wattenberg [97], which are a combination of Error Correcting Codes (ECC) with cryptography. FCS by [97] had the possibility of a false reject if the biometric was captured with noise [98] and made it susceptible to information leakage [99, 100]. Fuzzy Vault Schemes (FVS) is the second type of key binding scheme. The vault has the hidden key of the cryptosystem which is only unlocked when genuine biometrics are presented [101]. According to [102], however, such scheme suffer from varying biometric samples that may reduce matching accuracies. Moreover, if biometrics are stored locally, there may be injection attack that will force the release of cryptographic keys.

The other popular template protection scheme is Cancelable Biometrics (CB) [103], which helps to overcome some of the challenges mentioned above. To protect biometric features stored in the biometric repository, cancelable biometrics emerged in 1998 [104]. Researchers realized that storage of biometric features without protection could allow hackers to reverse engineer the original samples. In those days, storage of data in a centralized database also added to the vulnerability.

The idea of cancelable biometrics is to transform raw biometric traits or computed biometric features into a different representation using certain transforms [103]. The transforms should not be invertible. The term cancelable or revocable refers to the property of the template being removed or reconstructed if necessary. Also, the cancelable templates stored in the biometric database are difficult to reverse engineer due to the non-invertible transformation applied. Another important advantage is that the cancelable template generation scheme can vary across multiple applications—which means one user biometric can be used to generate multiple CB templates for multiple applications [103].

According to [105], a cancelable biometric template should conform to certain essential properties. First and foremost it should be cancelable or revocable with an ability to generate a new template in its place if needed. Second, it should be non-invertible or computationally difficult to reverse engineer. Third, it should be distinct for a user. Uniqueness of cancelable templates within a system and between systems is very essential. Lastly, it is important that the cancelable biometric transformation doesn't degrade system accuracy.

7.4.2 Unimodal and Multi-modal Cancelable Biometrics

A cancelable biometric systems can be unimodal or multimodal depending on the number of biometric modalities used to generate the CB template. In the forthcoming paragraphs, a review of commonly used cancelable biometric techniques found in literature is presented. We classify cancelable template generation into two categories—based on salting techniques (adding passwords, patterns, or noise to biometrics) and general non-invertible transforms [106–108]. Cancelable methods of Random Permutations, Biometric Salting, Non-invertible Geometric Transforms, Bio-hashing, Bio-filters, and Random Projection are discussed. Table 7.5 reports cancelable biometric schemes utilized between 1998 to 2020 along with biometric modalities.

Non-invertible geometric transforms: Transformation of raw biometrics using geometric transforms such as polar transforms, sector transforms, and grid based transforms constitute non-invertible geometric transforms. In work [108], multiple minutia were mapped to a common region in the transformed domain. This made it challenging for an attacker to reverse engineer original fingerprint minutia positions. Non-invertible geometric transforms need to still address issues of first order transforms, user-specificity (discriminability), and performance.

Biometric salting: It is one of the oldest and classical technique to generate cancelable templates by mixing data, passwords, patterns, or noise termed as 'salt'. The level of security depends on the amount of salt added to the biometric, but too much 'salt' may reduce the reliability of the match. There also exists a risk of separately storing the additive data of passwords, patterns, or noise. A way to improve template protection is by a combination of transform-based approaches with salting techniques, which improves cancelability and discriminability [107]. In work [106], biometric salting using userkeys (user-specific token IDs) and Log Gabor filters contribute in creating a cancelable and user-specific template. Another work [111] employs error correction codes like Reed-Solomon codes to fuse neural networks and obtain user specific keys for biometric cancelability. It is essential that reverse engineering the original biometric should not be possible even if an adversary gains possession of the cancelable template and userkey. Also, the salt should be safely stored. If this additive data is compromised or modified, erroneous biometric authentication can jeopardize the system [114].

Table 7.5 Biometric template protection schemes from 1998 to 2020

Authors and paper	Year	Template protection scheme	Trait(s)
Kumar and Rawat [109]	2020	Random permutations	Face, iris, and ear
Sudhakar and Gavrilova [68]	2020	Random projection	Iris and fingerveins
Kho, J. Kim, I. J. Kim, and Teoh [110]	2019	Partial Local Structure (PLS) descriptor and Permutated Randomized Non-Negative Least Square (PR-NNLS)	Fingerprint
Soliman, Amin, and Samie [59]	2019	Random projection	Iris
Chauhan and Sharma [99]	2019	Fuzzy Commitment Scheme (FCS)	Iris
Sarkar, Singh, and Bhaumik [96]	2018	Biocryptosystem key generation scheme	Fingerprint
Kaur and Khanna [106]	2017	Biometric salting using Log Gabor filters and token IDs	Face, palmprint, palmvein, and Fingervein
Talreja, Valenti, and Nasrabadi [111]	2017	Deep networks and error correcting codes	Face and iris
Deshmukh and Balwant [58]	2017	Local Binary Pattern (LBP) and random projection	Palmprint
Paul and Gavrilova [41]	2014	Random cross folding and random projection	Face and ear
Syarif, Ong, Teoh, and Tee [112]	2014	Bio-Hashing—Most Intensive Histogram Block Location (MIHBL)	Fingerprint
Feng, Yuen, and Jain [107]	2010	Random projection and Fuzzy Commitment Scheme (FCS)	Face
Ratha, Chikkerur, and Connell [108]	2007	Non-invertible geometric transforms	Fingerprint
Juels and Sudan [101]	2006	Fuzzy vault scheme	Fingerprint
Savvides, Kumar, and Khosla [113]	2004	Biometric filters	Face
Clancy, Kiyavash, and Lin [98]	2003	Smart card based biocryptosystem	Fingerprint
Juels and Wattenberg [97]	1999	Fuzzy Commitment Scheme	Any modality
Soutar, Roberge, Stoianov, Gilroy, and Kumar [104]	1998	Biometric encryption—Bioscrypt	Fingerprint

Random permutations: In random permutations, a biometric is divided into sectors of rows and columns which are randomly shuffled. Work [109] devised a 'random permutation locality preserving projection' technique on face, iris, and ear datasets. Most random permutations techniques do not degrade authentication accuracy because they only involve biometric rearrangement operation. The downside is that noise present during biometric acquisition may get magnified due to the shuffling. It is required that the complexity of permutations prevents decryption using brute-force attacks.

Bio-filters: Biometric encryption method using kernels or filters constitute Bio-filters. A user-specific token or PIN is utilized to generate a kernel filter. For example, work [113], proposes to encrypt biometrics to produce a Minimum Average Correlation Energy (MACE) filter. The MACE filter is then used to authenticate a user based on the correlation between it and the encrypted test biometric sample obtained from the user. It was found that the raw biometrics were safeguarded from attacks because during authentication the images are never decrypted. Bio-filters should be applied thoughtfully because it is essential that the multiple transformations using random kernels, encrypted biometric templates, and encrypted MACE filters do not reduce the authentication capacity/accuracy of the system.

Bio-hashing: The bio-hash concept is based on user specific tokens to generate orthogonal pseudo-random vectors forming a bio-hash [115]. In bio-hashing, similar biometrics should have similar hash values and different biometrics should not have similar hashes. There should not be a major impact on hash values when the biometric template is tilted or changed positionally (location variance) [112].

Random projection: One of the most popular techniques in biometric template protection schemes has been random projection. Random projection is based off the Johnson-Lindenstrauss lemma [116], in which the biometric points are projected from a higher dimensional space to a lower dimensional subspace in such a way that the distance between them is maintained. Many recent papers such as [41, 58, 59] have utilized random projection. Work [59] employed sectorization and random projection on the CASIA iris dataset to achieve high accuracy, while work [58] utilized a local binary pattern along with random projection for palmprint protection. A mix of random crossfolding and random projection approaches has also shown to achieve high accuracy and discriminability while preserving template privacy as demonstrated in [41]. The effectiveness of random projection and its ability to combine transformation and salting approaches, has made it a popular technique [59, 110].

7.4.3 Deep Learning Architectures for Cancelable Multi-modal Biometrics

Standalone Cancelable Biometric Architecture Cancelable biometric research conducted at the Biometric Technologies laboratory on multi-modal systems employing deep learning has yielded good results. In this subsection, a standalone cancelable biometric architecture utilizing Convolutional Neural Networks (CNNs) is discussed along with experimental results. The proposed system is a multi-instance biometric verification system—it utilizes multiple different instances of the same modality to verify a user. The research has been published in [18, 68].

The proposed cancelable biometric system operates in three phases of biometric feature extraction, cancelable template generation, and user verification. In the biometric feature extraction phase, multi-instance biometrics are pre-processsed and its features extracted through the use of a CNN. The cancelable template generation phase provides cancelability and non-invertibility of the template through the technique of random projection. The third phase or user verification is carried out by a new Multi-Layer Perceptron architecture (MLP) to yield accurate verification. Figure 7.8 presents an architecture of the deep learning-based cancelable biometric system.

Biometric Feature Extraction The feature extraction phase is comprised of two stages—biometric pre-processing and feature extraction using convolutional neural network. Multi-instance irises (left and right iris) and fingerveins (middle and index fingervein) have been utilized for the proposed cancelable biometric system from three datasets [117–119]. Irises and fingerveins remain permanent over time and have unique and intricate patterns making them well-established biometric traits that offer high accuracy [120, 121]. Pre-processing was conducted through rescaling, pixelating, and normalization of raw biometric images. Infra-red fingervein images and RGB iris images were obtained with an extracted Region Of Interest (ROI) in the data acquisition phase before the pre-processing stage. RGB images were grey scale normalized from pixel range (0–255) to range (0–1) to help with illumination or lighting differences and aid in CNN convergence for the next phase. Figure 7.9 presents an architecture of the designed compact 6-layered Convolutional Neural Network (CNN).

The developed CNN utilizes smaller number of layers and parameters when compared to standard classical models of VGG16 and VGG19, and also uniquely extracts biometric features from the final *Dense* layer. As can be seen from Fig. 7.9, it is constructed using four convolutive layers and two dense layers. It also uses one-fourth of the trainable parameters of VGG16 and VGG19 (Proposed CNN parameters—

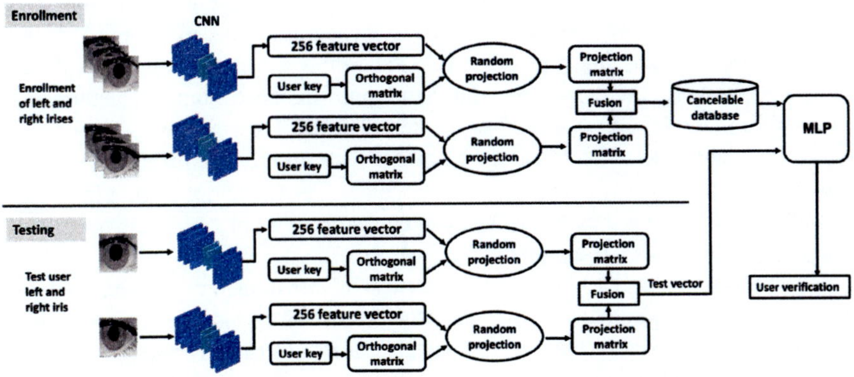

Fig. 7.8 Architecture of the deep learning-based cancelable biometric system

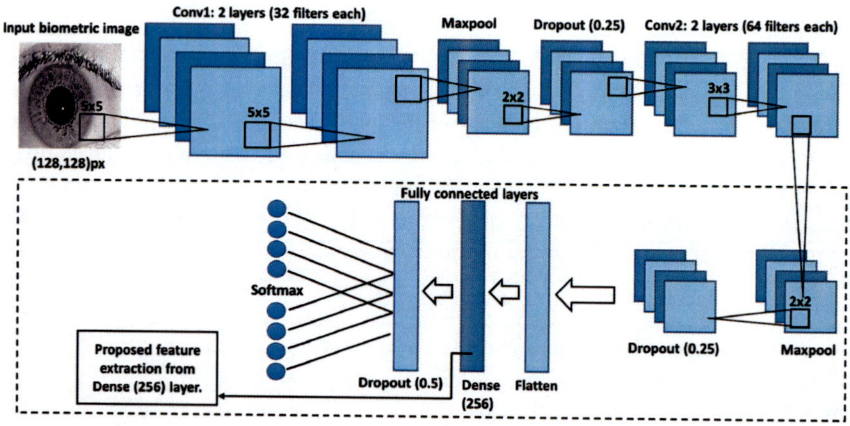

Fig. 7.9 Architecture of the compact 6-layered Convolutional Neural Network (CNN)

16,913,587, VGG16 parameters—65,972,256, VGG19 parameters 71,281,952). It outperforms VGG16 and VGG19 due to reduced overfitting problem due to less number of trainable parameters. Developed CNN performs automatic dimension reduction of features thereby avoiding use of techniques like Principle Component Analysis (PCA) or Linear Discriminant Analysis (LDA) for dimension reduction of features. Another advantage is that the CNN be used for different types of modalities and is positionally invariant (tilts, location of image do not degrade recognition performance). Optimizer RMSprop (Root Mean Square propagation). RMSprop solves the problem of Adagrad's aggressive and radically diminishing learning rate [122]. RMSprop is also quicker than SGD [123] in reaching global minimum of the loss function. Loss function used is categorical cross entropy that is suitable for the multi-class scenario [124]. The ReLU activation function has been chosen because of advantages over Hyperbolic Tangent (TanH) and Sigmoid activation functions. ReLU minimizes the vanishing gradient problem [125]. ReLU converges to the global minimum faster than TanH and Sigmoid. Finally, a ReduceLROnPlateau learning rate is opted due to better performance over constant learning rate, exponential decay, and other adaptive learning rates. Five fold validation is used to train and validate data (as per standard biometric procedure) and multi-instance biometric features are extracted from the final Dense layer of the neural network. In the next phase of cancelable template generation, extracted biometric features are transformed to create a non-invertible Cancelable Biometric (CB) template. The CB template should also be revocable which means a new template can be issued in its place in case necessary, rendering the previous CB template void.

Random projection (RP) [116] is used for cancelable transformation and it helps to maintain the Euclidean distance between the points before and after the projection, preserving its statistical properties and has also been successfully used in previous works [41, 58, 59] for biometric cancelability. A userkey K is used to generate orthogonal matrix O multiplied to both the feature matrices that generates projection

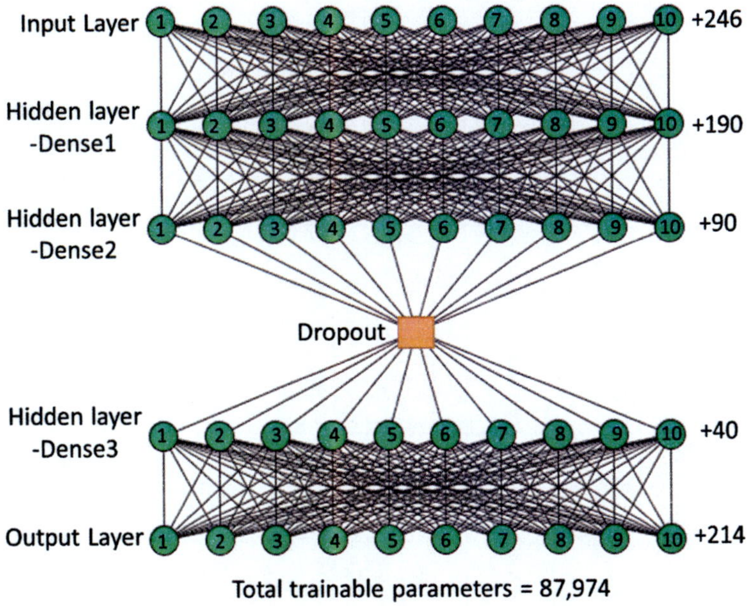

Fig. 7.10 Multi-Layer Perceptron (MLP) architecture for cancelable biometric verification

which are subsequently fused. The procedure renders final cancelable templates ($CB = CB_1, CB_2, CB_3...CB_n$) non-invertible, unlinkable, and cancelable.

The third phase or user verification phase is based on a Multi-Layer Perceptron architecture. MLP is known for its "efficiency in classification, regression, and mapping functions owing to its ability to distinguish data that is not linearly separable" [126]. MLP also outperformed machine learning models such as Support Vector Machine (SVM), k-Nearest Neighbor (k-NN), Decision Tree (D-Tree), and Naive Bayes (NB) algorithms on all three fingervein and iris datasets. Figure 7.10 displays MLP architecture that is compact and is made up of only 3 hidden layers (200, 100 and 50 fully connected neurons, respectively). The Adam optimizer is chosen due to its better performance in models with sparse gradients. The ability of mapping non-linear N-dimensional input signal to M-dimensional output signal makes the MLP suitable as a matching module.

Cloud Cancelable Biometric Architecture The standalone cancelable biometric system can be extended to a cloud-based cancelable system to accommodate surge in user enrollment, data storage, and computation power [18]. Cloud technology is "a way of providing infrastructure/platform/software as a service over the internet on the basis of pay as you use" [127]. It allows to extend the standalone cancelable system to a Biometrics-as-a-Service (BaaS) through cloud technology. The cloud also offers multiple processors for parallel computation of the proposed deep learning models reducing computational time.

Table 7.6 CNN performance: accuracy and loss over three trails on IITD iris dataset, MMU iris dataset, and FV-USM fingervein dataset. Acronyms: t.loss = training loss, t.acc = training accuracy, v.loss = validation loss, and v.acc = validation accuracy

	Run 1				Run 2				Run 3				Average			
	Acc and loss: 20 epochs				Acc and loss: 20 epochs				Acc and loss: 20 epochs				Acc and loss: 20 epochs			
Dataset	t.loss	t.acc	v.loss	v.acc	t.loss	t.acc	v.loss	v.acc	t.loss	t.acc	v.loss	v.acc	t.loss	t.acc	v.loss	v.acc
IITD	0.20	0.937	0.60	0.913	0.21	0.929	0.61	0.90	0.19	0.939	0.59	0.918	0.20	0.935	0.60	0.910
MMU	0.20	0.935	0.18	0.810	0.18	0.936	0.18	0.802	0.23	0.928	0.21	0.805	0.20	0.933	0.19	0.805
FVUSM	0.50	0.93	0.25	0.921	0.49	0.92	0.23	0.92	0.48	0.940	0.25	0.93	0.49	0.931	0.24	0.923

The cloud system is as a client-server model which is an extension of the standalone model. The few differences are: (a) Biometric crossfolding (which is fundamentally superimposing multiple instance of the biometric to form a cross-fold), (b) Embedding the crossfold in a QR code through steganography, and (c) Parallelization of the deep neural networks. Cross folding was carried out using binary matrices and its complementary matrix generated using a user's key. The binary and complement matrices were then multiplied to the multi-instance biometric feature matrices and fused to form a cross fold. Steganography was utilized to hide the cross fold in QR code generated by Python module—qrcode 6.1 [128] to transmit it over a transmission medium. Steganography for data hiding has been popularly used for biometric transmission through a communication medium, for example in work such as [129]. The proposed cloud system was built on Amazon Elastic Cloud compute EC2 G3 instance. Data was parallelized using the 4 T M60 GPUs by training using bath sizes of 128 (32 samples per GPU). Data parallelization meant that each GPU utilized a blueprint of the CNN to extract features from data divided between them. Data parallelism was found to increase processing speed of CNN training.

7.4.4 Performance of Cancelable Biometric System

The results presented in this section were previously published in journal papers [18] and [68]. All experiments conducted were carried out on three publicly available datasets of [117] (IITD iris dataset), [119] (MMU iris dataset), and [118] (FV-USM fingervein dataset). Accuracies of stand-alone and cloud-based cancelable multimodal system are identical.

Proposed CNN performance is reported in Table 7.6 which evaluates training and validation accuracy for 20 epochs over three trials. The final model weights and biases are chosen from the trial exhibiting optimal accuracy during subsequent feature extraction of biometric data. From Table 7.6, it can be noted that the CNN achieves high training accuracy of 93.5%, training loss of 0.20, validation accuracy of 91%, and validation loss of 0.60 on the IITD dataset. Proposed CNN also achieves training accuracy of 93.3%, training loss of 0.20, validation accuracy of 80.5%, and validation loss of 0.19 on the MMU dataset. For the FV-USM fingervein dataset, it achieves training accuracy of 93.1%, training loss of 0.49, validation accuracy of 92.3%, and validation loss of 0.24.

The introduced feature extraction using deep learning achieves better verification accuracy than classical feature extraction techniques (Log Gabor, Gabor with PCA (Principle Component Analysis), HOG (Histogram of Oriented Gradients), average and raw pixel intensities) given in Table 7.7. It also achieves lowest Equal Error Rate (EER is a threshold value to predetermine false accepts and false rejects [26]).

User verification is carried out using a new Multi-Layer Perceptron (MLP) architecture as in the standalone model. It has been employed for user verification because of its superior performance over traditional machine learning algorithms, lightweight architecture, and non-linear mapping property. Table 7.8 is a comparison of proposed

Table 7.7 Comparing Equal Error Rate (EER) and recognition accuracy of deep learning feature extraction against classical methods on three datasets [68]

Dataset	Method	EER	Accuracy (%)
IITD	5 × 5 Blocks of Avg Pixel Intensities	0.62	36
	Raw Pixel Intensities	0.50	50
	HOG	0.64	33
	Log Gabor [117]	0.38	63
	CNN Feature Extraction	**0.12**	**90**
MMU	5 × 5 Blocks of Avg Pixel Intensities [130]	0.86	15
	Raw Pixel Intensities [130]	0.76	22
	Gabor + PCA	0.51	52
	Log Gabor	0.42	62
	CNN Feature Extraction	**0.15**	**80**
FV-USM	5 × 5 Blocks of Avg Pixel Intensities	0.61	38
	Raw Pixel Intensities	0.59	40
	Gabor + PCA	0.39	58
	Log Gabor	0.06	91
	CNN Feature Extraction	**0.05**	**92.9**

Table 7.8 Comparison of five classifiers on cancelable biometric data for the standalone cancelable biometric system [18]

	EER IITD	EER MMU	EER FV-USM
Decision tree	0.35	0.45	0.27
Gaussian naive bayes	0.33	0.42	0.27
k-nearest neighbor	0.15	0.20	0.04
CNN + SVM	0.12	0.15	0.04
CNN + MLP	**0.04**	**0.14**	**0.01**

deep learning system against classical Machine Learning (ML) models such as SVM, kNN, DTrees, and Naive Bayes. Proposed system obtained lowest EER of 0.04, followed by SVM (EER 0.12) and DTrees (EER 0.35) for the IITD dataset. It also obtained lowest EER of 0.14 and 0.01 for MMU and FV-USM datasets, respectively. The speed-up of cloud system due to parallelization is given in Table 7.9.

The cloud-based system improved time efficiency of the neural network processing due to parallelization over multiple processors. The multiple GPUs provided by

Table 7.9 Speed and accuracy comparison of stand-alone cancelable biometric system versus cloud-based cancelable biometric system [18]

Dataset	Platform	CNN time (s)	MLP (s)	Accuracy(%)
IITD	Cloud [131]	–	–	95.27
	Standalone	31.8	41.38	98
	Cloud	**23.2**	**10**	**98.5**
MMU	Standalone	8.5	6.5	92.5
	Cloud	**7**	**5**	**92.5**
FV-USM	Standalone	39.38	28.55	99.55
	Cloud	**30**	**13**	**99.55**

AWS cloud help to increase speed by data parallelism on the datasets of MMU, IITD, and FV-USM. A final comparison of the proposed system to recent works in literature utilizing the same datasets has been made. Results are compiled in Table 7.10 along with year of research, techniques used, and recorded accuracy.

First, we compare recent research on the IITD iris dataset [131–133] is made against the developed system. Work [132] by Janani and Revathi in 2017, utilized iris localization, normalization, and feature extraction using Gabor filters on the IITD data to obtain 94.1% verification accuracy. Another 2017 work which also utilized cancelable biometric technique of Random Projection on irises [131] by Punithavathi, Geetha, and Shanmugam reported 95.27% verification accuracy. Work [133] by Omran and AlShemmary published in 2020, explored the use of deep learning in iris verification. AlexNet convolutional neural network (pretrained CNN developed by Alex Krizhevsky) and IRISNet (CNN proposed by them) were utilized for IITD segmented and unsegmented iris data. AlexNet recorded 95.09% verification accuracy, while IRISNet model obtained 95.98% verification accuracy with CNN training time of 14.26 min on segmented (normalized) irises and 97.32% verification accuracy with CNN training time of 16.05 min on unsegmented irises. IRISNet similarly split data into 80% training and 20% testing for 20 epochs as in our proposed work. Methodology utilizing CNN, MLP, and cancelable biometrics was able to further improve verification accuracy to 98% (standalone platform) with much lower CNN training time of 31.8 s and 98.5% (cloud platform) with CNN training time of 23.2 s. Work [130] by Andy Zeng on MMU iris dataset utilized Histogram of Oriented Gradients (HOG) and Support Vector Machine (SVM) on textons (micro units in images to perceive texture) to obtain 88.1% and 92% verification accuracy, respectively. Developed architecture utilizing CNN, MLP, and cancelable biometrics obtained verification accuracy of 92.5% on MMU data for both standalone and cloud platforms. Current state of the art research on the FV-USM fingervein dataset include [134–136]. Hu, Ma, and Zhan in 2018 utilized 2-Dimensional Principal Component Analysis (2DPCA) along with rotate invariant and uniform Local Binary Pattern (LBP) to obtain 97.15% and 98.10% verification accuracy, respectively on FV-USM fingervein data [134]. Another work by Das, Piciucco, Maiorana, and Campisi in

Table 7.10 Comparison of other methods with the developed cancelable system in terms of verification accuracy

Dataset	Author(s) and paper	Year	Technique(s)	Accuracy (%)
IITD	Janani and Revathi [132]	2017	GIRIST (Grus Iris Tool) based on iris localization, normalization, and feature extraction-Gabor filter	94.1
	Punithavathi, Geetha, and Shanmugam [131]	2017	CB system of random projection on iris features	95.27
	Omran and AlShemmary [133]	2020	AlexNet	95.09
	Omran and AlShemmary [133]	2020	CNN IRISNet with normalized iris images	95.98
	Omran and AlShemmary [133]	2020	CNN IRISNet for iris images without segmentation	97.32
	Standalone CB system	**2020**	**Deep leaning (CNN, MLP) with cancelable biometrics**	**98**
	Cloud CB system	**2020**	**Deep leaning (CNN, MLP) with cancelable biometrics**	**98.5**
MMU	Andy Zeng [130]	2018	Histogram of Oriented Gradients (HOG)	88.1
	Andy Zeng [130]	2018	Support Vector Machine (SVM) and textons	92
	Standalone CB system	**2020**	**Deep leaning (CNN, MLP) with cancelable biometrics**	**92.5**
	Cloud CB system	**2020**	**Deep leaning (CNN, MLP) with cancelable biometrics**	**92.5**
FV-USM	Hu, Ma, and Zhan [134]	2018	Rotate invariant Local Binary Pattern (LBP) and Two-Dimensional Principal Component Analysis (2DPCA)	97.15
	Hu, Ma, and Zhan [134]	2018	Uniform Local Binary Pattern (LBP) and Two-Dimensional Principal Component Analysis (2DPCA)	98.10
	Das, Piciucco, Maiorana, and Campisi [135]	2019	CNN	97.53
	Das, Piciucco, Maiorana, and Campisi [135]	2019	CNN and CLAHE (Contrast Limited Adaptive Histogram Equalization)	97.05
	Avci, Kocakulak, and Acir [136]	2019	CNN	98.44
	Standalone CB system	**2020**	**Deep leaning (CNN, MLP) with cancelable biometrics**	**99.55**
	Cloud CB system	**2020**	**Deep leaning (CNN, MLP) with cancelable biometrics**	**99.55**

2019 [135] utilized convolutional neural networks on the FV-USM dataset. They obtained 97.05% and 97.53% accuracy with and without CLAHE (Contrast Limited Adaptive Histogram Equalization). Another 2019 work [136] by Avci, Kocakulak, and Acir also utilized CNN on FV-USM data to obtain 98.44% verification accuracy. The cloud-based and standalone systems utilizing CNN, MLP, and cancelable biometrics obtain the highest verification accuracy of 99.5% as can be seen from Table 7.10. Thus, application of deep learning to cancelable biometrics can considerably improve accuracy and reliability of biometric authentication.

7.4.5 Summary and Applications

To summarize, this section explored the application of deep learning to cancelable biometrics. A discussion on biometric system vulnerabilities and various template protection schemes was presented. Cancelable biometrics has been a widely used security mechanism to safeguard raw biometric templates.

Overall the proposed system achieved a very high verification accuracy. This was possible due to the power of deep learning models for precise feature extraction and dimension reduction, coupled with a lighter model architecture containing less parameters which helped in accurately predicting values (avoiding overfitting and underfitting of the neural networks). Presented cancelable biometric system also outperformed existing research works that employ the same datasets utilized here, in terms of verification accuracy. The deep learning architecture performed better than classical machine learning models such as Support Vector Machine, K-Nearest Neighbour, D-Tree, and Naïve Bayes algorithms.

The standalone architecture was modified to a cloud-based architecture designed to be offloaded to the cloud as a client-server model. The biometric engine, cancelable database, and deep learning modules of the were hosted on the cloud server to be utilized as a cost-effective service (BaaS—Biometrics as a Service). Training time of the deep learning models reduced considerably due to data parallelization through multiple GPUs provided on the Amazon Web Service (AWS) cloud platform, while maintaining high verification accuracy.

There are a number of promising new directions of research that arise from this work. System performance in differing environments with low lighting, occlusion, and blurry input images can be studied. Another area of research scope lies in the deep learning architecture employed. For signal-based biometric modalities such as speech and gait, Recurrent Neural Networks (RNNs) and Long-Short-Term-Memory models (LSTMs) could be investigated due to their ability to utilize present and previous set of input values. Finally, other methods instead of random projection can be explored for biometric template protection.

7.5 Applications and Open Problems

7.5.1 User Authentication and Anomaly Detection

Social Behavioral Biometric features hold the information of users' behavior on the online social networks [11]. Nowadays, account credentials are used for granting a user access to their account. If the credentials get compromised, the social behavioral information can be used as a secondary authentication method to retrieve the account. In case of continuous user authentication, social behavioral traits can be repeatedly sampled over some period of time, and if their pattern undergoes a significant and a sudden change, user may need to re-identify to their account.

Anomaly detection refers to a situation when the data deviates from its regular behavior. In the online world, it is highly important to continuously monitor the vast amount of data to discover network intrusions and fraudulent activities. Social Behavioral Biometric systems can help to detect anomaly in user behaviors and create a quick response if the account is compromised. Aside from direct user authentication from social interactions, other auxiliary information such as emotional user state, aesthetic preferences or their psychological traits, can be analyzed. This research can find applications in recommended systems, human-computer interactions, and robotic domains.

7.5.2 Access Control

Biometric verification system for access control can include cancelable biometric. Cancelability of biometric templates adds to protection of biometric data through non-invertible transformations that can be revoked or cancelled if user information is compromised [18]. Moreover, the cloud-based cancelable biometric system can be provided as an authentication service on a pay-per-use basis with potentially unlimited storage capacity and computational power on the cloud. Biometric authentication services can be accessed over the internet using web APIs for user verification.

Cancelable biometric multi-modal system can include different behavioral biometric modalities such as speech, gait, aesthetic preferences, or communication expressed through images, texts, or tweets in a social media context. Investigation of performance of different behavioral biometrics and effect of data quality on overall system accuracy can be another interesting research direction.

7.5.3 Robotics

Machine learning and biometric emotion recognition emerged as some of the most promising research methodologies in Robotics. Recent research includes applying machine learning methods in the context for Robot-assisted surgeries (RAS) to model cognitive surgical robots employing surgical skill and competence [137]. Other focus areas are improving human-robot interactions, such as developing a rich categorization of emotional expressions in humans [138]. Usage of machine learning has been applied to one of the most important problems in robotics—motor skills learning—such as in a task of learning table tennis from a physical interaction with a human [139]. These works focus on how discrete and rhythmic tasks can be learned by combining both imitation and reinforcement learning.

Robots that can autonomously navigate new circumstances has been the long-standing vision of robotics and machine learning. Multiple fields need further exploration and refinement pertaining to human-robot interaction in the context of a biometric research. In surgical settings, there needs to be further research on how to adapt robot systems reliably and safely to new situations, based on changing circumstances and human behavior. Another area of research that can benefit from further exploration is having robots perform hospitality industry tasks, such as provide help with hotel check in, or serve as assistants for individuals with disabilities.

7.5.4 Assisted Living

The human life expectancy is increasing with advancements in medicine and will require innovative solutions to keep up with establishing effective and affordable long-term health care systems. It has been shown that in-home monitoring and smart technologies reduce hospital re-admissions and mortality rates [140]. Machine learning has been implemented in wearable devices to gather data and analyze it to detect behavioral pattern changes in aging individuals [141]. Subtle, but meaningful behavior changes (such as an increase in sedentary activities), can help predict onset of more serious health problems and result in earlier interventions. Significant adverse effects, such as falls, can result in a notification to caregivers for immediate help. Machine learning can also help humanize senior living experience by detecting loneliness through speech pattern analysis and engaging caregivers and family as needed [142].

Future research can include combining biometric-based remote monitoring with fostering meaningful social connections in senior communities through smart social networks. With limited hospital and nursing homes infrastructure, development of smart home systems that are safe, accessible and can provide additional benefits of remote monitoring is one of the fast growing domains of research.

7.5.5 Mental Health

High incidence rate of mental health illnesses and the need for effective methods has increased the use of information fusion and machine learning to detect, diagnose and develop treatment plans for patients [143]. This research creates a critical opportunity for early intervention in children who are identified as at-risk to develop mental health symptoms as adolescents [144]. Most studies focus on predicting the onset of mental health illnesses, including detection and diagnosis. The most common mental health illnesses studied that utilize machine learning methods include depression, schizophrenia and Alzheimer's disease [145].

Sensors in smartphone devices, wearables, and social networks are continuously collecting data which may be used towards identifying personal behavior, thoughts, emotions and personality traits. Mental health applications with machine learning, social behavioral biometrics and multimodal sensing can utilize sensor data for monitoring mental health of individuals, such as depression, anxiety, and stress [146].

Although in early stages, there is a big opportunity for using personal device sensors as clinical tools for conducting mental health research and monitoring at-risk individuals. Future works may include personalized treatment plans and measuring their effectiveness with the help of machine learning to assist professionals in providing mental health care.

7.5.6 *Education*

The incorporation of computers and technology into the education system has caused the greatest change in access, availability and delivery of education. Machine learning has allowed for optimization, customization and personalization of learning experiences for students. As more education activities are moving towards digital formats, this enables data collection and analysis for conducting studies. Such studies have been done to assess correlation between learning patterns and behaviors with academic performance in both formal and non-formal degrees [147].

Machine learning systems can also allow teachers to grade with greater speed and accuracy, in some cases with minimal human intervention. Studies have shown effective employment of automated grading in open-ended questions, resulting in good inter-rater agreement between machine learning grading system and the human instructor [148]. Other research evolved around plagiarism, such as detecting paraphrasing in written works [149].

Machine learning algorithms that can provide feedback to students for their presentation content and oral skills have also been developed. For example, a software presentation system may have a feature that autonomously provides real-time feedback and a report to help user to improve presentations [150].

Future research can include classroom monitoring systems based on biometric traits for curated personalized learning experiences taking into account student personalities or suggesting custom sitting arrangements. Other avenues to explore include creating human-computer interaction based on social behavioral biometrics, that can provide meaningful debates with students to help them practice their negotiation, communication and influential skills.

7.6 Summary

This book chapter provided a comprehensive overview of major developments in machine learning methods for biometric research in stand-alone, multi-modal and cloud-based security systems. It introduced the biometric security research, proceeded with defining unimodal and multi-modal architectures, surveyed both traditional and emerging biometric domains, and introduced the concepts of biometrics on the cloud. Deep learning methodology was discussed in the context of biometric security systems. The chapter focused on two major directions of emerging biometric research: social behavioral systems and cancelable multi-modal systems using deep learning. It listed key application domains of importance to industries and general public, including cybersecurity, border control, continuous authentication, defense, risk mitigation, smart homes, e-health, assisted living and education.

Future potent directions of research will be focused on ensuring data privacy in the context of social behavioral biometrics [151], investigating hybrid architectures based on a combination of traditional features with deep learning models, and ensuring seamless translation of research to real-time safety crucial applications. These open problems provide rich avenue for further advancement of the vibrant research on the intersection of machine learning and biometrics.

Acknowledgements Authors would like to thank NSERC DG program grant RT731064 and NSERC ENGAGE grant EGP522291 for partial support of this project, as well as all members of the Biometric Technologies laboratory at the University of Calgary and all collaborators for their valuable discussions during this chapter preparation.

References

1. A. Jain, A. Ross, S. Prabhakar, An introduction to biometric recognition. IEEE Trans. Circuits Syst. Video Technol. **14**(1), 420 (2004)
2. S. Yanushkevich, M. Gavrilova, P. Wang, S. Srihari, Image pattern recognition: synthesis and analysis in biometrics. Ser. Mach. Percept. Artif. Intell. **67**, 423 (2007)
3. M. Gavrilova (ed.), *Computational Intelligence: A Geometry-based Approach* (Springer Engineering book series Studies in Computational Intelligence (Springer, Berlin, 2009)
4. M. Gavrilova, M. Monwar, Multimodal biometrics and intelligent image processing for security systems, in *IGI Global* (2012)
5. Y. Wang, B. Widrow, L. Zadeh, Howard, S. Wood, V. Bhavsar, G. Budin, C. Chan, R. Fiorini, M. Gavrilova, D. Shell, Cognitive intelligence: Deep learning, thinking, and reasoning by brain-inspired system. Int. J. Cogn. Inform. Nat. Intell. (IJCINI) **10**(4), 1–20 (2016)
6. S. Tumpa, M. Sultana, P. Kumar, S. Yanushkevich, Y. Orly, H. Jison, M. Gavrilova, Social behavioral biometrics in smart societies, in *Advancements in Computer Vision Applications in Intelligent Systems and Multimedia Technologies, IGI Global* (2020), pp. 1–24
7. C. Segalin, A. Perina, M. Cristani, Personal aesthetics for soft biometrics: a generative multi-resolution approach, in *International Conference on Multimodal Interaction* (2014), pp. 180–187
8. M. Gavrilova, F. Ahmed, H. Bari, R. Liu, T. Liu, Y. Maret, B. Sieu, T. Sudhakar, Multi-modal motion capture based biometric systems for emergency response and patient rehabilitation. Res. Anthol. Rehabil. Pract. Ther. **32**, 653–678 (2021)

9. M. Gavrilova, Decoding intricacies of human nature from social network communications, in *Script-Based Semantics: Foundations and Applications. Essays in Honor of Victor Raskin* (2020), pp. 269–277
10. R. Yampolskiy, M. Gavrilova, Artimetrics: biometrics for artificial entities. IEEE Robot. Autom. Mag. **19**, 48–58 (2012)
11. M. Sultana, P. Paul, M. Gavrilova, A concept of social behavioral biometrics: motivation, current developments, and future trends, in *International Conference on Cyberworlds* (2014), pp. 271–278
12. S. Tumpa, A. Gavrilov, O. Duran, F. Zohra, M. Gavrilova, Quality estimation for facial biometrics, in *Innovations, Algorithms, and Applications in Cognitive Informatics and Natural Intelligence, IGI Global* (2020), pp. 298–320
13. M. Sultana, P. Paul, M. Gavrilova, Mining social behavioral biometrics in Twitter, in *International Conference on Cyberworlds* (2014), pp. 293–299
14. K. Ahmadian, M. Gavrilova, A novel multi-modal biometric architecture for high-dimensional features, in *International Conference on Cyberworlds* (IEEE, Banff, Canada, 2011), pp. 9–16
15. M. Gavrilova, K. Ahmadian, Dealing with biometric multi-dimensionality through novel chaotic neural network methodology. Int. J. Inf. Technol. Manag. Indersci. **11**(1–2), 18–34 (2012)
16. H. Bari, M. Gavrilova, Artificial neural network based gait recognition using Kinect sensor. IEEE Access **7**(1), 162708–162722 (2019)
17. M. Gavrilova, F. Ahmed, H. Bari, R. Liu, T. Liu, Y. Maret, B. Sieu, T. Sudhakar, *Multi-modal motion capture based biometric systems for emergency response and patient rehabilitation, in Design and Implementation of Healthcare Biometric Systems* (USA, IGI Global, Hershey, PA, 2018), pp. 160–184
18. T. Sudhakar, M. Gavrilova, Cancelable biometrics using deep learning as a cloud service. IEEE Access **8**, 112932–112943 (2020)
19. Y. Maret, D. Oberson, M. Gavrilova, Real-time embedded system for gesture recognition, in *International Conference on Systems, Man, and Cybernetics (SMC)* (IEEE, Japan, 2018), pp. 30–34
20. F. Ahmed, H. Bari, B. Sieu, J. Sadeghi, J. Scholten, M. Gavrilova, Kalman filter-based noise reduction framework for posture estimation using depth sensor, in *International Conference on Cognitive Informatics and Cognitive Computing (ICCI-CC)*. (IEEE, Italy, 2019), pp. 150–158
21. H. Bari, B. Sieu, M. Gavrilova, AestheticNet: deep convolutional neural network for person identification from visual aesthetic. Vis. Comput. **36**(10–12), 2395–2405 (2020)
22. S. Tumpa, I. Luchak, M. Gavrilova, Behavioral biometric identification from online social media using deep learning, *Women in Data Science Conference (WiDS) Poster, Calgary, Canada*
23. C. Louis, Why your biometrics are your best password (2020), https://www.forbes.com/sites/louiscolumbus/2020/03/08/why-your-biometrics-are-your-best-password/#524cd91b6c01. Accessed Dec 2020
24. Verizon, Data breach investigations report (2019), https://www.nist.gov/system/files/documents/2019/10/16/1-2-dbir-widup.pdf. Accessed Dec 2019
25. Thales Group, Biometrics: authentication and identification (definition, trends, use cases, laws and latest news) - 2020 review (2020), https://www.thalesgroup.com/en/markets/digital-identity-and-security/government/inspired/biometrics. Accessed Dec 2020
26. A. Jain, A. Ross, K. Nandakumar, *Introduction to Biometrics* (Springer, Boston, MA, 2011). 978-0-387-77325-4
27. S. Bharadwaj, M. Vatsa, R. Singh, Biometric quality: a review of fingerprint, iris, and face. EURASIP J. Image Video Process. **2014**(1), 1–28 (2014)
28. P. Rot, Z. Emeršič, V. Struc, P. Peer, Deep multi-class eye segmentation for ocular biometrics, in *IEEE International Work Conference on Bio-inspired Intelligence (IWOBI), San Carlos* (2018), pp. 1–8

29. P. Schuch, S. Schulz, C. Busch, Survey on the impact of fingerprint image enhancement. IET Biom. **7**(2), 102–115 (2018)
30. R.S. Choras, A review of image processing methods and biometric trends for personal authentication and identification. Int. J. Circuits Syst. Signal Process. **10**, 367–376 (2016)
31. C. Wang, M. Gavrilova, Delaunay triangulation algorithm for fingerprint matching, in *International Symposium on Voronoi Diagrams in Science and Engineering (ISVD'06)* (2006), pp. 208–216
32. H. Fathee, O. Ucan, M. Jassim, O. Bayat, Efficient unconstrained iris recognition system based on CCT-like mask filter bank. Math. Probl. Eng. J. Hindawi **2019**(6575019), 10 (2019)
33. A. Kumar, A. Potnis, A. Singh, Iris recognition and feature extraction in iris recognition system by employing 2D DCT. Int. Res. J. Eng. Technol. (IRJET) **3**(12), 503–510 (2016)
34. S. Monisha, G. Sheeba, Gait based authentication with Hog feature extraction, in *International Conference on Inventive Communication and Computational Technologies (ICICCT), Coimbatore* (2018), pp. 1478–1483
35. B. Kumar, Biometric matching, in *Encyclopedia of Cryptography and Security*, ed. by H.C.A. Van Tilborg, S. Jajodia (Springer, Boston, 2011)
36. A. Jain, R. Bolle, S. Pankanti, *Biometrics: Personal Identification in Networked Society* (Kluwer Academic Publications, 1999). ISBN 978-0-7923-8345-1
37. F. Ahmed, H. Bari, M. Gavrilova, Emotion recognition from body movement. IEEE Access **8**, 11761–11781 (2020)
38. M. Sultana, P. Paul, M. Gavrilova, Social behavioral biometrics: an emerging trend. Int. J. Pattern Recognit. Artif. Intell. **29**(08), 1556013 (2015)
39. A.K. Jain, K. Nandakumar, A. Ross, 50 years of biometric research: accomplishments, challenges, and opportunities. Pattern Recognit. Lett. **79**, 80–105 (2016)
40. M. Ghayoumi, A review of multimodal biometric systems: fusion methods and their applications, in *IEEE/ACIS International Conference on Computer and Information Science (ICIS)* (Las Vegas, NV, 2015), pp. 131–136
41. P. Paul, M. Gavrilova, Rank level fusion of multimodal cancellable biometrics, in *International Conference on Cognitive Informatics and Cognitive Computing* (IEEE, London, 2014), pp. 80–87
42. Y. Luo, M. Gavrilova, P. Wang, Facial metamorphosis using geometrical methods for biometric applications. Int. J. Pattern Recognit. Artif. Intell. **22**(3), 555–584 (2008)
43. T. Danny, Unimodal biometrics vs. multimodal biometrics (2018), https://www.bayometric.com/unimodal-vs-multimodal/. Accessed Dec 2018
44. P. Wild, P. Radu, L. Chen, J. Ferryman, Robust multimodal face and fingerprint fusion in the presence of spoofing attacks. Pattern Recognit. **50**, 17–25 (2016)
45. C. Sanderson, K. Paliwal, Information fusion for robust speaker verification, in *European Conference on Speech Communication and Technology* (Alborg, Denmark, 2001), pp. 755–758
46. M. Monwar, M. Gavrilova, Y. Wang, A novel fuzzy multimodal information fusion technology for human biometric traits identification, in *International Conference on Cognitive Informatics and Cognitive Computing (ICCI-CC)* (IEEE, Banff, Canada, 2011), pp. 112–119
47. M. Gavrilova, M. Monwar, Markov chain model for multimodal biometric rank fusion. Signal Image Video Process **7**(1), 137–149 (2013)
48. K. Tumer, J. Gosh, Linear order statistics combiners for pattern classification, *Combining Artificial Neural Networks* (1999), pp. 127–162
49. Y. Wu, K. Chang, E. Chang, J. Smith, Optimal multimodal fusion for multimedia data analysis, in *ACM International Conference on Multimedia* (2004), pp. 572–579
50. L. Wu, P. Cohen, S. Oviatt, From members to team to committee - a robust approach to gestural and multimodal recognition. Trans. Neural Netw. **13**(4), 972–982 (2002)
51. N. Poh, S. Bengio, How do correlation and variance of base-experts affect fusion in biometric authentication tasks? IEEE Trans. Acoust. Speech Signal Process. **53**, 4384–4396 (2005)
52. R. Yan, A. Hauptmann, The combination limit in multimedia retrieval, in *ACM International Conference on Multimedia* (2003), pp. 339–342

53. M. Sultana, P. Paul, M. Gavrilova, User recognition from social behavior in computer-mediated social context. IEEE Trans. Hum.-Mach. Syst. **47**(3), 356–367 (2017)
54. S. Bazazian, M. Gavrilova, A hybrid method for context-based gait recognition based on behavioral and social traits. Trans. Comput. Sci. Springer, Berlin, Heidelberg **25**, 115–134 (2015)
55. P. Lovato, A. Perina, N. Sebe, O. Zandona, A. Montagnini, M. Bicego, M. Cristani, Tell me what you like and I'll tell you what you are: discriminating visual preferences on Flickr data, in *Asian Conference on Computer Vision* (2012), pp. 45–56
56. S. Azam, M. Gavrilova, Person identification using discriminative visual aesthetic, in *Canadian Conference on Artificial Intelligence* (2017), pp. 15–26
57. B. Sieu, M. Gavrilova, Biometric identification from human aesthetic preferences. Sensors **20**(4), 1133 (2020)
58. M. Deshmukh, M.K. Balwant, Generating cancelable palmprint templates using local binary pattern and random projection, in *International Conference on Signal-Image Technology & Internet-Based Systems (SITIS), Jaipur, India* (2017), pp. 203–209
59. R. Soliman, M. Amin, F. Abd El-Samie, A modified cancelable biometrics scheme using random projection. Ann. Data Sci. **6**, 223–236 (2019)
60. Y. LeCun, L. Bottou, Y. Bengio, P. Haffner, Gradient-based learning applied to document recognition. Proc. IEEE **86**(11), 2278–2324 (1998)
61. A. Shrestha, A. Mahmood, Review of deep learning algorithms and architectures. IEEE Access **7**, 53040–53065 (2019)
62. F. Sultana, A. Sufian, P. Dutta, Advancements in image classification using convolutional neural network, in *Fourth International Conference on Research in Computational Intelligence and Communication Networks (ICRCICN), IEEE* (2019), pp. 122–129
63. S. Yadav, S. Jadhav, Deep convolutional neural network based medical image classification for disease diagnosis. J. Big Data **6**(1), 1–18 (2019)
64. K. Simonyan, A. Zisserman, Very deep convolutional networks for large-scale image recognition. CVPR **1409**, 1556 (2014)
65. M. Abeer, H. Talal, Analysis and best parameters selection for person recognition based on gait model using CNN algorithm and image augmentation. J. Big Data **8**(1), 1–20 (2021)
66. I. Boucherit, M. Zmirli, H. Hentabli, B. Rosdi, Finger vein identification using deeply-fused convolutional neural network. J. King Saud Univ.-Comput. Inf. Sci. (2020). ISSN 1319-1578
67. M. Wani, F. Bhat, S. Afzal, A. Khan, Supervised deep learning in fingerprint recognition. *Advances in Deep Learning* (Springer, 2020), pp. 111–132
68. T. Sudhakar, M. Gavrilova, Deep learning for multi-instance biometric privacy. ACM Trans. Manag. Inf. Syst. (TMIS) **12**(1), 1–23 (2020)
69. A. Nada, H. Heyam, Deep learning approach for multimodal biometric recognition system based on fusion of iris, face, and finger vein traits. Sensors **20**, 1–17 (2020)
70. H. Bari, M. Gavrilova, *Multi-layer perceptron architecture for Kinect-based gait recognition, in Computer Graphics International Conference (CGI)* (Springer, Cham, Switzerland, 2019), pp. 356–363
71. M. Khan, S. Harous, S. Hassan, M. Ghani, R. Iqbal, S. Mumtaz, Deep unified model for face recognition based on convolution neural network and edge computing. IEEE Access **7**, 72622–72633 (2019)
72. S. Minaee, A. Abdolrashidi, DeepIris: iris recognition using a deep learning approach (2019), arXiv:1907.09380v1
73. G. Wu, J. Tao, X. Xu, *Occluded face recognition based on the deep learning, in Chinese Control And Decision Conference (CCDC)* (China, Nanchang, 2019), pp. 793–797
74. S. Minaee, E. Azimi, A. Abdolrashidi, Pushing the limits of fingerprint recognition using convolutional neural network. FingerNet **1907**, 12956 (2019)
75. S. Minaee, A. Abdolrashidi, H. Su, M. Bennamoun, D. Zhang, Biometric recognition using deep learning: a survey (2019), arXiv:1912.00271
76. M. Wang, W. Deng, Deep face recognition: a survey (2019), arXiv:804.06655v8

77. A. Nassif, I. Shahin, I. Attili, M. Azzeh, K. Shaalan, Speech recognition using deep neural networks: a systematic review. IEEE Access **7**, 19143–19165 (2019)
78. K. Sundararajan, D. Woodard, Deep learning for biometrics: a survey. ACM Comput. Surv. **51**(3), 1–34 (2018)
79. F. Anjomshoa, M. Aloqaily, B. Kantarci, M. Erol-Kantarci, S. Schuckers, Social behaviometrics for personalized devices in the internet of things era. IEEE Access **5**, 12199–12213 (2017)
80. A. Saleema, S. Thampi, User recognition using cognitive psychology based behavior modeling in online social networks, in *International Symposium on Signal Processing and Intelligent Recognition Systems* (2019), pp. 130–149
81. S. Tumpa, M. Gavrilova, Score and rank level fusion algorithms for social behavioral biometrics. IEEE Access **8**, 157663–157675 (2020)
82. S. Tumpa, M. Gavrilova, Linguistic profiles in biometric security system for online user authentication, in *IEEE International Conference on Systems, Man, and Cybernetics (SMC)* (2020), pp. 1033–1038
83. H. Si, Z. Chen, W. Zhang, J. Wan, J. Zhang, N.N. Xiong, A member recognition approach for specific organizations based on relationships among users in social networking Twitter. Futur. Gener. Comput. Syst. **92**, 1009–1020 (2019)
84. X. Ruan, Z. Wu, H. Wang, S. Jajodia, Profiling online social behaviors for compromised account detection. IEEE Trans. Inf. Forensics Secur. **11**(1), 176–187 (2015)
85. H. Aulia, E. Alva, I. Kho, G. Maulahikmah, M. Wahyu, Automated document classification for news article in Bahasa Indonesia based on term frequency inverse document frequency (TF-IDF) approach, in *International Conference on Information Technology and Electrical Engineering (ICITEE)* (2014), pp. 1–4
86. F. Eibe, B.R. Remco, Naïve bayes for text classification with unbalanced classes, in *European Conference on Principles of Data Mining and Knowledge Discovery* (2006), pp. 503–510
87. J. Kelleher, *Deep Learning* (MIT Press, 2019)
88. P. Liu, X. Qiu, X. Huang, Recurrent neural network for text classification with multi-task learning (2016), arXiv:1605.05101
89. S. Georgakopoulos, S. Tasoulis, A. Vrahatis, V. Plagianakos, Convolutional neural networks for toxic comment classification, in *Hellenic Conference on Artificial Intelligence* (2018), pp. 1–6
90. K. Ryczko, K. Mills, I. Luchak, C. Homenick, I. Tamblyn, Convolutional neural networks for atomistic systems. Elsevier Comput. Mater. Sci. **149**, 134–142 (2018)
91. Y. Wenpeng, K. Kann, Y. Mo, H. Schütze, Comparative study of CNN and RNN for natural language processing (2017), arXiv:1702.01923
92. P. Charlotte, I. Geoffrey, P. François, Temporal convolutional neural network for the classification of satellite image time series. Remote Sens. **11**(5), 523 (2019)
93. A. Jain, K. Nandakumar, A. Nagar, Biometric template security. EURASIP J. Adv. Signal Process. **2008**(113), 1–17 (2008)
94. V. Patel, N. Ratha, R. Chellappa, Cancelable biometrics: a review. IEEE Signal Process. Mag. **32**(5), 54–65 (2015)
95. A. Kholmatov, B. Yanikoglu, *Biometric cryptosystem using online signatures, in International Symposium on Computer and Information Sciences* (Springer, Berlin, Heidelberg, 2006), pp. 981–990
96. A. Sarkar, B. Singh, U. Bhaumik, Cryptographic key generation scheme from cancelable biometrics. Prog. Comput., Anal. Netw., Springer, Singapore **710**, 265–272 (2018)
97. A. Juels, M. Wattenberg, A fuzzy commitment scheme, in *ACM Conference on Computer and Communications Security* (1999), pp. 28–36
98. T. Clancy, N. Kiyavash, D. Lin, Secure smartcard-based fingerprint authentication, *ACM SIGMM Workshop on Biometrics Methods and Applications* (2003), pp. 45–52
99. S. Chauhan, A. Sharma, Improved fuzzy commitment scheme. Int. J. Inf. Technol. 1–11 (2019)
100. T. Ignatenko, F. Willems, Information leakage in fuzzy commitment schemes. IEEE Trans. Inf. Forensics Secur. **5**(2), 337–348 (2010)

101. A. Juels, M. Sudan, A fuzzy vault scheme. Des. Codes Cryptogr. **38**(2), 237–257 (2006)
102. U. Uludag, S. Pankanti, S. Prabhakar, A. Jain, Biometric cryptosystems: issues and challenges. Proc. IEEE **92**(6), 948–960 (2004)
103. D. Rachapalli, H. Kalluri, A survey on biometric template protection using cancelable biometric scheme, in *International Conference on Electrical. Computer and Communication Technologies (ICECCT), Coimbatore* (2017), pp. 1–4
104. C. Soutar, D. Roberge, A. Stoianov, R. Gilroy, B. Kumar, Biometric encryption using image processing. Opt. Secur. Count. Feit Deterrence Tech. **2**(3314), 178–188 (1998)
105. S. Kanade, D. Delacrétaz, B. Dorizzi, Cancelable biometrics for better security and privacy in biometric systems, in *International Conference on Advances in Computing and Communications* (Springer, Berlin, Heidelberg, 2011), pp. 20–34
106. H. Kaur, P. Khanna, Cancelable features using log-Gabor filters for biometric authentication. Multimed. Tools Appl. **76**(4), 4673–4694 (2017)
107. Y. Feng, P. Yuen, A. Jain, A hybrid approach for generating secure and discriminating face template. IEEE Trans. Inf. Forensics Secur. **5**, 103–117 (2010)
108. N. Ratha, S. Chikkerur, J. Connell, R. Bolle, Generating cancelable fingerprint templates. IEEE Trans. Pattern Anal. Mach. Intell. **29**, 561–572 (2007)
109. N. Kumar, M. Rawat, RP-LPP: a random permutation based locality preserving projection for cancelable biometric recognition. Multimed. Tools Appl. **79**, 2363–2381 (2020)
110. J. Kho, J. Kim, I. Kim, A. Teoh, Cancelable fingerprint template design with randomized non-negative least squares. Pattern Recognit. **91**, 245–260 (2019)
111. V. Talreja, M. Valenti, N. Nasrabadi, Multibiometric secure system based on deep learning, in *IEEE Global Conference on Signal and Information Processing (globalSIP)* (2017), pp. 298–302
112. M. Syarif, T. Ong, A. Teoh, C. Tee, Improved biohashing method based on most intensive histogram block location. Int. Conf. Neural Inf. Process. Springer, Cham **8836**, 644–652 (2014)
113. M. Savvides, B. Kumar, P. Khosla, Cancelable biometric filters for face recognition. Int. Conf. Pattern Recognit. **3**, 922–925 (2004)
114. A. Jin, L. Hui, Cancelable biometrics. Scholarpedia **5**(1), 9201 (2010)
115. A. Jin, D. Ling, A. Gohb, Biohashing: two factor authentication featuring fingerprint data and tokenised random number. Pattern Recognit. **37**(11), 2245–2255 (2004)
116. W. Johnson, J. Lindenstrauss, Extensions of lipschitz mappings into a hilbert space, *Contemporary Mathematics* (1984), pp. 186–206
117. A. Kumar, A. Passi, Comparison and combination of iris matchers for reliable personal identification, in *IEEE Computer Society Conference on Computer Vision and Pattern Recognition Workshops* (2008), pp. 1–7
118. M. Asaari, S. Suandi, B. Rosdi, Fusion of band limited phase only correlation and width centroid contour distance for finger based biometrics. Expert. Syst. Appl. **41**(7), 367–3382 (2014)
119. Multimedia-University. Iris database MMU database. pesona.mmu.edu.my/ ccteo/. Accessed December 2018
120. K. Shaheed, H. Liu, G. Yang, I. Qureshi, J. Gou, Y. Yin, A systematic review of finger vein recognition techniques. Inf. J. **9**(9), 213–242 (2018)
121. V. Nazmdeh, S. Mortazavi, D. Tajeddin, H. Nazmdeh, M. Asem, *Iris recognition: from classic to modern approaches, in Annual Computing and Communication Workshop and Conference (CCWC)* (NV, USA, IEEE, Las Vegas, 2019), pp. 981–988
122. S. Ruder, An overview of gradient descent optimization algorithms (2017), arXiv:1609.04747v2
123. E. Dogo, O. Afolabi, N. Nwulu, B. Twala, C. Aigbavboa, A comparative analysis of gradient descent-based optimization algorithms on convolutional neural networks, in *International Conference on Computational Techniques, Electronics and Mechanical Systems (CTEMS), Belgaum, India* (2018), pp. 92–99

124. J. Brownlee, How to choose loss functions when training deep learning neural networks. Mach. Learn. Mastery-Deep. Learn. Perform (2020), https://machinelearningmastery.com/how-to-choose-loss-functions-when-training-deep-learning-neural-networks/. Accessed May 2020
125. A. Krizhevsky, I. Sutskever, G. Hinton, ImageNet classification with deep convolutional neural networks. Commun. Assoc. Comput. Mach. **60**, 84–90 (2017)
126. H. Ramchoun, M. Idrissi, Y. Ghanou, M. Ettaouil, MLP: architecture optimization and training. Int. J. Interact. Multimed. Artif. Intell. **4**, 26–30 (2016)
127. A. Kaur, V. Singh, S. Gill, The future of cloud computing: opportunities, challenges and research trends, in *International Conference on I-SMAC (IoT in Social, Mobile, Analytics and Cloud), Palladam, India* (2018), pp. 213–219
128. Pypi. Pure Python QR Code generator QRCode 6.1. Pypi Version Jan 2019 https://pypi.org/project/qrcode/. Accessed June 2020
129. I. McAteer, A. Ibrahim, G. Zheng, W. Yang, C. Valli, Integration of biometrics and steganography: a comprehensive review. Technologies **7**(2), 34–56 (2019)
130. A. Zeng, Iris recognition (2018) cs.princeton.edu. Accessed Dec 2018
131. P. Punithavathi, S. Geetha, S. Shanmugam, Cloud-based framework for cancelable biometric system, in *IEEE International Conference on Cloud Computing in Emerging Markets (CCEM)* (2017), pp. 35–38
132. B. Janani, M. Revathi, Comparison of iris database performance using GIRIST. Int. J. Adv. Res. Trends Eng. Technol. (IJARTET) **4**(11), 479–485 (2017)
133. M. Omran, E.N. AlShemmary, An iris recognition system using deep convolutional neural network. J. Phys.: Conf. Ser. **1530**, 012159 (2020)
134. N. Hu, H. Ma, T. Zhan, A new finger vein recognition method based on LBP and 2DPCA, in *Chinese Control Conference (CCC), Wuhan* (2018), pp. 9267–9272
135. R. Das, E. Piciucco, E. Maiorana, P. Campisi, Convolutional neural network for finger-vein-based biometric identification. IEEE Trans. Inf. Forensics Secur. **14**(2), 360–373 (2019)
136. A. Avci, M. Kocakulak, N. Acir, Convolutional neural network designs for fingervein-based biometric identification, in *International Conference on Electrical and Electronics Engineering (ELECO), Bursa, Turkey* (2019), pp. 580–584
137. Z. Beiqun, R. Waterman, R. Urman, R. Gabriel, A machine learning approach to predicting case duration for robot-assisted surgery. J. Med. Syst. **43**(2), 1–32 (2019)
138. G. Jam, J. Rhim, A. Lim, Developing a data-driven categorical taxonomy of emotional expressions in real world human robot interactions, in *Companion of the 2021 ACM/IEEE International Conference on Human-Robot Interaction, NY, USA, New York* (2021), pp. 479–483
139. K. Mulling, J. Kober, O. Kroemer, J. Peters, Learning to select and generalize striking movements in robot table tennis. Int. J. Robot. Res. **32**(3), 263–279 (2013)
140. K. Blum, S. Gottlieb, The effect of a randomized trial of home telemonitoring on medical costs, 30-day readmissions, mortality, and health-related quality of life in a cohort of community-dwelling heart failure patients. J. Card. Fail. **20**(7), 513–521 (2014)
141. D. Naranjo-Hernández, L. Roa, J. Reina-Tosina, M. Estudillo-Valderrama, SoM: a smart sensor for human activity monitoring and assisted healthy ageing. IEEE Trans. Biomed. Eng. **59**(11), 3177–3184 (2012)
142. V. Badal, S. Graham, C. Depp, K. Shinkawa, Y. Yamada, L. Palinkas, H. Kim, D. Jeste, E. Lee, Prediction of loneliness in older adults using natural language processing: exploring sex differences in speech. Am. J. Geriatr. Psychiatry: Off. J. Am. Assoc. Geriatr. Psychiatry **S1064–7481**(20), 30479–6 (2020)
143. A. Thieme, D. Belgrave, G. Doherty, Machine learning in mental health: a systematic review of the HCI literature to support the development of effective and implementable ML systems. ACM Trans. Comput.-Hum. Interact. (TOCHI) **27**(5), 1–53 (2020)
144. A. Tate, R. McCabe, H. Larsson, S. Lundström, P. Lichtenstein, R. Kuja-Halkola, Predicting mental health problems in adolescence using machine learning techniques. PLoS One **15**(4), e0230389 (2020)
145. A. Shatte, D. Hutchinson, S. Teague, Machine learning in mental health: a scoping review of methods and applications. Psychol. Med. **49**(9), 1426–1448 (2019)

146. E. Garcia-Ceja, M. Riegler, T. Nordgreen, P. Jakobsen, K. Oedegaard, J. Tørresen, Mental health monitoring with multimodal sensing and machine learning: a survey. Pervasive Mob. Comput. **51**, 1–26 (2018)
147. J. Xu, K.H. Moon, M. Van der Schaar, A machine learning approach for tracking and predicting student performance in degree programs. IEEE J. Sel. Top. Signal Process. **11**(5), 742–753 (2017)
148. I. Ndukwe, B. Daniel, C. Amadi, A machine learning grading system using chatbots, in *International Conference on Artificial Intelligence in Education* (Springer, Cham, 2019), pp. 365–368
149. E. Hunt et al., Machine learning models for paraphrase identification and its applications on plagiarism detection, in *IEEE International Conference on Big Knowledge (ICBK), Beijing, China* (2019), pp. 97–104
150. H. Trinh, R. Asadi, D. Edge, T. Bickmore, RoboCOP: a robotic coach for oral presentations. ACM Interact. Mob. Wearable Ubiquitous Technol. **1**(2), 1–24 (2017)
151. T. Habibu, A. Sam, Assessment of vulnerabilities of the biometric template protection mechanism. Int. J. Adv. Technol. Eng. Explor. **5**(45), 243–254 (2018)

Marina L. Gavrilova is a Full Professor in the Department of Computer Science, University of Calgary, and a head of the Biometric Technologies Laboratory. Her publications include over 200 journal and conference papers, edited special issues, books and book chapters in the areas of image processing, pattern recognition, machine learning, biometric and online security. She is Founding Editor-in-Chief of LNCS Transactions on Computational Science Journal. Dr. Gavrilova has given over 50 keynotes, invited lectures and tutorials at major scientific gatherings and industry research centers, including Stanford University, SERIAS Center at Purdue, Microsoft Research USA, Oxford University UK, Samsung Research South Korea and others. Dr. Gavrilova serves as an Associate Editor for IEEE Access, IEEE Transactions on Computational Social Systems, Vis. Comput. and the International Journal of Biometrics, and was appointed by the IEEE Biometric Council to serve on IEEE Transactions on Biometrics, Behavior, and Identity Science Committee. She is a passionate promoter of diversity, equity and inclusion at her workplace and in the society as a whole.

Iryna Luchak is currently pursuing her M.Sc. in Computer Science at the University of Calgary, Canada, at the Human-Computer Interaction Laboratory. She recently obtained a B.Eng. degree in Computer Engineering from the University of British Columbia. Her research interests include machine learning, social computing and data analytics. Iryna has published two journal articles and presented at the Women in Data Science 2021 Conference.

Tanuja Sudhakar received the B.Tech. degree in information technology from Anna University, in 2016. She recently obtained the M.Sc. degree in computer science at the University of Calgary, Canada, under the Supervision of Prof. Marina Gavrilova. She worked as a Software Developer at Mphasis, a Blackstone Company, from August 2016 to August 2018. Her research interests include cancelable biometrics, biometric security, computer vision, and deep learning.

Sanjida Nasreen Tumpa is currently pursuing her M.Sc. in Computer Science at the University of Calgary, Canada, under the supervision of Prof. Marina L. Gavrilova. She received a B.Sc. degree in Computer Science and Engineering (CSE) from the Military Institute of Science and Technology (MIST), Bangladesh in 2014 and an M.Sc. degree in CSE from Bangladesh University of Engineering and Technology (BUET), Bangladesh in 2019. She served the Department of CSE, MIST, Bangladesh as a faculty member from 2015 to 2019. She has published over 15 journal and conference papers, in addition to three book chapters. Her research interests include social network analysis, natural language processing, and machine learning.

Chapter 8
Early Smoke Detection in Outdoor Space: State-of-the-Art, Challenges and Methods

Margarita N. Favorskaya

Abstract In recent decades, early smoke detection in outdoor environment is a hot topic due to its practical importance for a fire safety. Many researchers have contributed to this area since the 1990s. The chapter aims to follow the evolution of conventional image processing and machine learning methods based on the motion, semi-transparent, color, shape, texture and fractal features to deep learning solutions using various deep network architectures. Our experimental researches in this area have been conducted since 2010. This chapter reflects the original techniques of early smoke detection in complex outdoor scenes.

Keywords Smoke recognition · Smoke detection · Smoke segmentation · Machine learning · Deep learning · Features · Outdoor space · Complex scene

8.1 Introduction

Video smoke detection in outdoor space is often combines with video flame detection [1], but in the most cases video smoke detection as a factor of early detection of danger is a separate problem [2, 3]. Generally speaking, the dynamic and spatial properties of smoke and flame are different, resulting in different methods of extracting their low-level and high-level features. Such methods should provide high values of precision and low rates of false alarm. For smoke detection in outdoor space, low rates of false alarm are crucial. Thus, all algorithmic improvements in this field are being made to distinguish the smoke regions and regions looking like smoke in the wild but not cigarette smoke or vapour. Unfortunately, fire or wildfire can appear in sometime and somewhere people live or in nature by a wide spectrum of reasons, caused or not caused by a human. Therefore, interest in this topic has remained unchanged over the past decades.

M. N. Favorskaya (✉)
Institute of Informatics and Telecommunications, Reshetnev Siberian State University of Science and Technology, 31, Krasnoyarsky Rabochy ave, Krasnoyarsk 660037, Russian Federation
e-mail: favorskaya@sibsau.ru

M. Virvou et al. (eds.), *Advances in Selected Artificial Intelligence Areas*, Learning and Analytics in Intelligent Systems 24,
https://doi.org/10.1007/978-3-030-93052-3_8

Smoke as an object of research has a large variability in shape, color, semi-transparency, turbulence variance, boundary roughness and time-varying flicker effect at boundaries, as well as unstable motion [4]. Typical shooting artifacts such as low resolution, blur, adverse illumination and weather conditions complicate the problem. Thus, smoke detection in dark-time and nights remains a hot topic [5, 6]. The conventional approach for video smoke detection is to analyze the motion using block-matching algorithm or optical flow and chrominance features in a scene and detect all regions looking like smoke. It should be noted that smoke analysis can be done not only in spatial domain using the conventional filters and/or texture estimates, but also in the frequency domain applying digital wavelet transform.

One of the challenges is the semi-transparency of smoke, especially in the early phases of burning. During the main burning phases, smoke can be white, light gray, dark gray or black [7]. Such smoke is not a problem for detection. Obviously, the colored smoke shows burning phase. Smoke semi-transparency is more significant property affecting early smoke detection, but it is difficult to determine, for example in relation to atmospheric haze in the outdoor space. Local features such as turbulence, semi-transparency, flicker effect on boundaries and unstable motion help to verify the smoke regions.

Conventional techniques supposing handcrafted feature extraction allowed the problem to be decomposed into several sequential steps using methods of motion estimation and texture/edge analysis in a scene. When motion is reliably detected, many machine learning methods such as *k*-nearest neighbors, Fisher linear discriminant, Naive Bayes classifier, Support Vector Machine (SVM), AdaBoost [8, 9], decision tree classifier, boosted random forests [10], among others, can be employed. Markov models were also applied for smoke recognition. Machine learning methods prevailed until the 2010s. In recent years, the effectiveness of deep learning based smoke recognition has been proven by many researchers [11–13].

The main advantage of deep learning is the ability to create high-level features from low-level features and automatically classify them. The deep learning approach avoids complex image preprocessing and objectively provides higher values of precision due to many network parameters that are tuned during the training stage. Thus, Convolutional Neural Network (CNN) architecture similar to the AlexNet architecture has been used for smoke detection with high false alarm prediction [14]. The network, which was proposed in [15], improved CNN by a batch-normalized convolutional layer that helped to avoid overfitting and obtain good precision values. The application of Faster R-CNN for forest wildfire detection was investigated in [16]. In addition to CNN, Recurrent Neural Networks (RNNs) [17], Generative Adversarial Networks (GANs) [18] and some hybrid architectures have been successfully used for video smoke detection with low values of false alarm. However, the durable training stage and high hardware requirements increase the computational cost of the deep learning approach.

It should be noted that some intermediate solutions also exist. They are based on the handcrafted features and incredibly smaller neural network (for example, SqueezeNet uses only one-fiftieth of the AlexNet parameters but provides the same accuracy [19]).

In this study, we provide a brief overview of conventional machine learning and deep learning methods to address smoke detection in outdoor space. Smoke detection techniques can be divided into three categories: smoke recognition (smoke is present in the frame or not), smoke detection (the smoke location is indicated by a bounding box) and smoke segmentation (each pixel is classified as smoke or non-smoke pixel). All smoke detection techniques are being developed and used depending on the problem being solved. We show that conventional machine learning methods remain on the frontier of research, despite the fact that deep learning methods are very attractive way to get good results in complex outdoor scenes. We also describe our approaches for smoke detection over the past decade.

The rest of this chapter is structured as follows. Section 8.2 describes a problem statement and challenges. Conventional machine learning methods are introduced in Sect. 8.3, while Sect. 8.4 contains a description of deep learning methods. The proposed deep architecture for smoke detection in outdoor space is discussed in Sect. 8.5. Section 8.6 presents an overview of the available datasets. Section 8.7 gives comparative experimental results. Finally, conclusions are drawn in Sect. 8.8.

8.2 Problem Statement and Challenges

Fire and smoke are components of the burning process, and some researchers consider their behavior simultaneously. However, many publications discuss these topics separately due to different spatio-temporal features of smoke and fire. Moreover, smoke often appears earlier than the flame. Thus, smoke detection plays a crucial role in early fire alarms.

Smoke is a mixture of gaseous products of chemical combustion reactions. The combustion products of many organic and non-organic substances contain suspended solids (carbon black, oxides, salts, etc.). From this point of view, smoke is a disperse system of combustion products and air, consisting of gases, vapors and hot particles. Depending on the stage of burning process, combustion products and environment conditions, different chemical combustion reactions take place. The density of volatile combustion products is 3–5 times lower than the density of the ambient air. The convective flow of hot vapor–gas mixture of combustion products above the combustion zone arises continuously, and, at the same time, absorption of fresh air from below to the combustion zone occurs. It is well-known that burning of natural and polymer materials causes various chemical reactions. Therefore, the temperature, color and speed of smoke propagation for wildfire and industrial fire are different.

We roughly categorize the problem of smoke detection into far and close smoke detection. A wildfire is usually considered as far smoke detection problem with many interfering factors such as haze, fog, clouds and plants. Fires in urban environment are usually closer than wildfires due to distributed smoke fire surveillance. Note that the main interfering factors come down to the overlap of visual objects in the urban environment.

In the task of video smoke detection, still image I or video sequence $S = \{F_i\}$, $i = \overline{1, n}$, where n is the number of frames, are the input data. According to the three categories of smoke detection techniques, the output data *Out* are the following:

- Smoke recognition. For smoke/non-smoke image: $Out = \{I\}$, where $I = \{Smoke\ image/Non\text{-}smoke\ image\}$. For smoke/non-smoke video sequence: $\{Out_i | F_i \in S(F_i)\}$, $Out_i = \{F_i\}$, where $F_i = \{Smoke\ frame/Non\text{-}smoke\ frame\}$.
- Smoke detection. For smoke/non-smoke images: $Out = \{R_j\}$, $j = \overline{1, n_j}$, where $\{R_j\}$ is the set of smoke regions bounded in rectangles, n_j is the number of smoke regions. For smoke/non-smoke video sequence: $\{Out_i | F_i \in S(F_i)\}$, $Out_i = \{R_{ij}\}$, $j = \overline{1, n_j}$, where$\{R_{ij}\}$ is the set of smoke regions in frame F_i, n_j is the number of smoke regions bounded in rectangles.
- Smoke segmentation. For smoke/non-smoke images: $Out = \{P_{jk}\}$, $k = \overline{1, n_{jk}}$, where $\{P_{jk}\}$ is the set of pixels k in smoke region R_j, n_{jk} is the number of pixels k in smoke region R_j. For smoke/non-smoke video sequence: $\{Out_i | F_i \in S(F_i)\}$, $Out_i = \{P_{ijk}\}$, $k = \overline{1, n_{ijk}}$, where $\{P_{ijk}\}$ is the set of pixels k in smoke region R_j in frame F_i, n_{ijk} is the number of pixels k in smoke region R_j of frame F_i.

A special case is the detection of semi-transparent smoke that looks like fog/haze [20]. Smoke detection is also affected by the thickness of a smoke as it spreads in an outdoor space.

In conventional machine learning methods for video smoke detection, motion analysis is implemented at the first step of algorithms. Smoke surveillance usually involves static scenes. Therefore, the background model is of great importance. The main requirement for the background model is its adaptation for luminance conditions (including night-time), moving objects (especially for scenes with relatively small depth), weather conditions and atmospheric phenomena for scenes with large depth.

It is reasonable to create different background models for scene with small and large depths. In [21], a background subtraction model based on color information was proposed for scenes with slowly varying brightness. Suppose that camera noise in three color channels of RGB color space is normally distributed with the noise variance $\sigma_{Rn}^2, \sigma_{Gn}^2, \sigma_{Bn}^2$. For each pixel with coordinates (x, y), the mean and variance values of the intensity function $I(x, y)$ were calculated using a set of source frames. The parameters are updated in each pixel with coordinates (x, y) using Eqs. 8.1–8.2, where μ_t and σ_t^2 are the mean value and dispersion at time t, respectively, μ_{t+1} and σ_{t+1}^2 are the mean value and variance at time $t + 1$, α is the empirical constant.

$$\mu_{t+1}(x, y) = \alpha \mu_t(x, y) + (1 - \alpha) I_{t+1}(x, y) \tag{8.1}$$

$$\begin{aligned} \sigma_{t+1}^2(x, y) = \alpha\left[\sigma_t^2(x, y) + (\mu_{t+1}(x, y) - \mu_t(x, y))^2\right] \\ +(1 - \alpha)(I_{t+1}(x, y) - \mu_{t+1}(x, y))^2 \end{aligned} \tag{8.2}$$

Background model for scenes with large depth faces the problem of image dehazing. Haze and fog degrade the image due to the atmospheric absorption and scattering, and smoke recognition becomes difficult. Such images lose contrast and color rendering. The radiative transport equation describing the light passing through a scattering medium along its original course and distributed in other directions can be simplified using the so called air-light (or also as the veiling light) [22]. In computer vision, a simpler image formation model is often used:

$$I(x) = J(x)\, t(x) + A(1 - t(x)), \tag{8.3}$$

where **I** stands for the observed image, J is the surface radiance vector at the intersection point of the scene and the real-world ray corresponding to the pixel $\mathbf{x} = (x, y)$, **A** is the air-light color vector, $t(\mathbf{x})$ is the transmission along that ray.

Equation 8.3 explains the loss of contrasts due to haze resulting from averaging an image with constant color **A**. The first term on the right side of Eq. 8.3 is called direct attenuation and describes the scene radiance and its decay in the medium. The second term is called air-light and causes the scene color to shift. When the atmosphere is homogenous, the transmission *tr* can be defined by Eq. 8.4:

$$tr(\mathrm{x}) = e^{-\beta d(\mathrm{x})}, \tag{8.4}$$

where β is the scattering coefficient of the atmosphere, *d* is the scene depth.

The scattering coefficient indicates that the scene radiance is exponentially attenuated with the scene depth. The atmospheric scattering model contains three unknown parameters that determine an infinite number of solutions. In this regard, the image itself is often used to construct a light transmission map during haze removal or to determine the scene depth. In [22], a refined image formation model was formulated under the assumption that the transmission and surface shading are locally uncorrelated. The use of reliable statistics has increased the visibility of complex scenes containing different surface albedos. A simple but effective image prior (dark channel prior) to haze removal from a single input image was proposed in [23]. After removing the haze, the background model for scenes with large depth is built using conventional methods, for example, Gaussian mixture model.

Also two features appear at the boundaries of the smoke. The first one is a random flickering on the boundaries from frame to frame [24]. It was shown in [25] that the flickering frequency of smoke is 1–3 Hz, and for a flame this parameter is 10 Hz. Frequency analysis, in particular wavelet analysis, allows one to estimate this parameter on the boundaries of candidate regions.

The second parameter is a turbulence estimate based on fractal theory. Catrakis and Dimotakis [26] proposed to quantify the geometric complexity of the shape of surfaces using a dimensionless area-volume measure. Thus, for closed surfaces embedded in *d*-dimensional space ($d \geq 2$), the size can be measured by $V_d^{1/d}$, where V_d is the volume enclosed by the surface. For a given size, the sphere $S_{d,\mathrm{sph}}$ has the least surface area in terms of the sphere volume $V_{d,\mathrm{sph}}$:

$$S_{d,\mathrm{sph}} = \frac{1}{k_d} V_{d,\mathrm{sph}}^{(d-1)/d} \quad \text{with} \quad k_d = \frac{\Gamma^{1/d}(1 + d/2)}{d\pi^{1/2}}, \tag{8.5}$$

where $\Gamma(x)$ is the gamma function.

For any closed surface, the surface area S_d is bounded from below by Eq. 8.6.

$$S_d = \frac{1}{k_d} V_d^{(d-1)/d} \tag{8.6}$$

The resulting area-volume measure Ω_d is a measure of the complexity of the surface shape: Ω_d quantifies the deviation of the shape from its minimum value, i.e. it will be equal to one for spheres and unbounded from above:

$$1 \leq \Omega_d \equiv \frac{k_d S_d}{V_d^{(d-1)/d}} \leq \infty. \tag{8.7}$$

Fractal surfaces are correspond to $\Omega_d = \infty$. For 2D closed contours, the size measure becomes the square root of the enclosed area, i.e. $V_2^{1/2} = A^{1/2}$. Thus, 2D shape complexity is described by Eq. 8.8, where $P = S_2$ is the perimeter of the contour.

$$1 \leq \Omega_2 \equiv \frac{k_2 P}{A^{1/2}} \leq \infty \quad \text{with} \quad k_2 = \frac{1}{2\pi^{1/2}} \tag{8.8}$$

Parameter Ω_2 measures the bounding arc length per unit square root of the enclosed area. Based on Eqs. 8.5–8.8, one can evaluate the fractal properties of candidate regions.

Color and textural [27] properties are estimated in a conventional way taking into account possible complex cases of luminance [28].

Video smoke detection is inferior to particle-sampling based detectors in terms of false alarm rates. Thus, fire smoke sensors prevail in indoor space. For outdoor space, video smoke detection plays a significant role because the concentration of particles in the air is rapidly reduced by wind. Also, the sensors have a limited detection range.

The main challenges of video smoke detection are that smoke varies greatly in color, texture and shape during a burning process. It is difficult to achieve satisfactory results in smoke detection using conventional feature extraction methods: only specialized feature extraction methods or ensemble of classification methods provide higher precision and low false alarm results.

Another challenge relates to the complexity of smoke modelling due to variability of smoke density, scene illumination and interfering objects. Usually smoke detection algorithms use the spatial and motion information in an attempt to extract the color, texture, energy and dynamic features. Many researchers prefer to use less time-consuming features. This makes it impossible to process complex scenes. The main purpose of video smoke detection is to reduce the false alarm rates and missed detection rates.

8.3 Conventional Machine Learning Methods

Many algorithms have been proposed in this field since the 2000s. In [25], the classical approach was developed that involved a combination of motion, flicker, edge blurring and color features, which includes five steps:

- Moving pixels or regions are detected.
- High frequency content corresponding to the edges in these regions is checked.
- If the edges are not sharpened, then U and V channels are checked.
- Flicker analysis is executed using a temporal wavelet transform.
- Shape of the moving region is checked for convexity.

Another popular approach uses statistical estimates such as mean, standard deviation, skewness, kurtosis and entropy calculated in sub-bands of wavelet transform with following SVM classifier [29]. In [30], a dynamic texture for smoke detection was implemented using a surfacelet transform and a hidden Markov tree model. Random forest classifiers have also been a popular approach for smoke recognition [10]. However, the most famous family for extracting smoke features from video was Local Binary Patterns (LBPs) since the 2010s [31, 32]. In [33], a higher order Linear Dynamical System (h-LDS) descriptor was introduced focusing on smoke detection.

Many modifications of the LBPs have been proposed for solving various problems of computer vision especially face recognition and texture analysis. Suppose that a texture patch in a gray image is described by the joint distribution *Dis* of $(M + 1)$ pixels, where $M > 0$:

$$Dis = d(g_c, g_0, g_1, \ldots, g_{M-1}). \tag{8.9}$$

After subtracting the central pixel value from the neighborhood, it can be assumed that the central pixel value is independent on the differences, and Eq. 8.9 can be factorized as follows:

$$Dis \approx d(g_c)\ d(g_0 - g_c, g_1 - g_c, \ldots, g_{M-1} - g_c). \tag{8.10}$$

The first factor $d(g_c)$ as the intensity distribution in the central pixel contains useless information, while the second factor as the joint distribution of differences $d(g_0 - g_c, g_1 - g_c, \ldots, g_{M-1} - g_c)$ can be used to simulate the local texture. However, it can be difficult to reliably estimate such a multi-dimensional distribution. A vector quantization-based solution to this problem was proposed by Ojala et al. [34]. The generic LBP operator is derived from this joint distribution.

The LBP was introduced by Ojala et al. [35] as a binary operator robust to lighting variations with low computational cost and the ability of simple coding of neighboring pixels around the central pixel as a binary string or decimal value. The operator $LBP_{P,R}$ is calculated in the surrounding relative to a central pixel with the intensity g_c provided by Eq. 8.3, where P is the number of pixels in the neighborhood on radius

R. If $(g_i - g_c) \geq 0$, then $s_2(g_i - g_c) = 1$, otherwise $s(g_i - g_c) = 0$. Subscript "2" means a dimension equaled 2, while subscript "3" means a dimension equaled 3.

$$LBP_{P,R} = \sum_{i=0}^{P-1} s_2(g_i - g_c) \cdot 2^i \tag{8.11}$$

Let us consider LBP modifications for video smoke detection.

A rotation invariant measure $VAR_{P,R}$ as local spatial pattern and local contrast texture descriptor was proposed in [36]. This measure can be defined as:

$$VAR_{P,R} = \frac{1}{P}\sum_{i=0}^{P-1}\left(g_i - \frac{1}{P}\sum_{i=1}^{P-1} g_i\right)^2. \tag{8.12}$$

Joint distribution of $LBP_{P,R}$ and $VAR_{P,R}$ can better describe a local texture that using $LBP_{P,R}$ alone. However, $VAR_{P,R}$ has continuous values and needs to be quantized. The LBP Variance (LPBV) descriptor overcomes the problems of $LBP_{P,R}/VAR_{P,R}$ descriptor. The LBPV histogram is computed using Eqs. 8.13–8.14.

$$LBPV_{P,R}(k) = \sum_{i=1}^{N}\sum_{j=1}^{M} w\big(LBP_{P,R}(i, j), k\big) \quad k \in [0, K] \tag{8.13}$$

$$w\big(LBP_{P,R}(i, j), k\big) = \begin{cases} VAR_{P,R}(i, j) & LBP_{P,R}(i, j) = k \\ 0 & \text{otherwise} \end{cases} \tag{8.14}$$

One fruitful idea is to find shape-invariant features of smoke in videos. This idea is implemented in [37] as a four-stage heuristic approach. First, a feature vector was built concatenating three histograms (edge orientation, edge magnitude and LBP) and four density features (edge magnitude, LBP, color intensity and saturation). Second, the multi-scale partitions were obtained by fragmentation of a detection window into a set of small blocks. Third, statistic features, including mean, variance, skewness, kurtosis and seven Hu moments, were calculated for each partition. Fourth, the AdaBoost algorithm was used to select discriminative shape-invariant features from a feature pool. In [38], this approach was enhanced by dynamic estimation of smoke probability being detected by the AdaBoost detector.

The two main versions of 3D LBP called Volume LBP (VLBP) and Spatio-Temporal LBP (STLBP) are the well-known extensions of 2D LBP. The VLBP called LBP from Three Orthogonal Planes (LBP-TOP) analyzes information from three orthogonal planes XY, XT and YT, where T is the time axis [39]. The VLBP is computationally simple and easy to extend. However, information in some pixels can be duplicated and is counted twice. The STLBP collects information from adjacent frames relative the central pixel, making them more suitable for video analysis. 2D

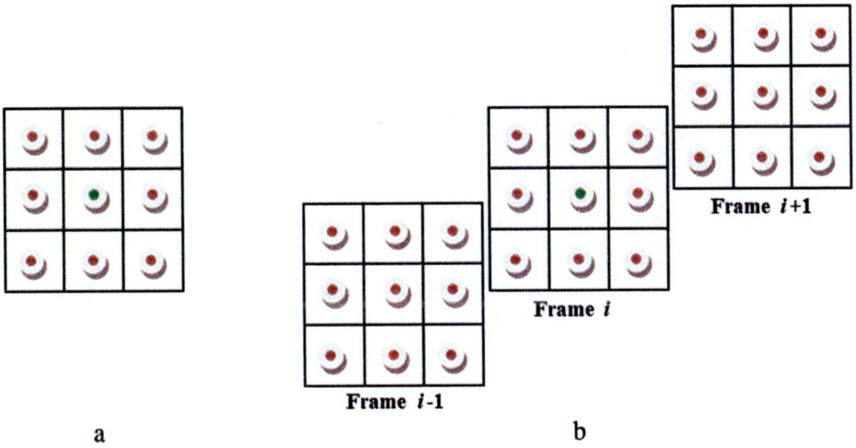

Fig. 8.1 LBP representation: **a** generic LBP in spatial domain ($R = 1$), **b** STLBP in spatio-temporal domain ($R = 1$)

generic and 3D STLBPs in the spatial and spatio-temporal domains, respectively, are depicted in Fig. 8.1 (green dot means central pixel).

In [40], the STLBP technique was applied not only to the original image as a set of intensity value map but also to the temporal brightness gradient map, normal flow map and Laplacian map, which were built based on four measures proposed in [41]:

- Pixel intensity $\mu_I(q_0, t_0)$ provided by Eq. 8.15, where $I(q, t)$ is the intensity value of pixel q in time instant t, $B(q_0, t_0, r_s, r_t)$ is a 3D cube centering at the point (q_0, t_0) with spatial radius r_s and temporal radius r_t.

$$\mu_I(q_0, t_0) = \iint\limits_{B(q_0, t_o, r_s, r_t)} I(q, t)\ dq\ dt \tag{8.15}$$

- Temporal brightness gradient $\mu_B(q_0, t_0)$ as a summation of temporal intensity changes around the point (q_0, t_0) provided by Eq. 8.16, where $B(q_0, t_0, r_s)$ is the spatial square centering at the point (q_0, t_0) with spatial radius r_s.

$$\mu_B(q_0, t_0) = \iint\limits_{B(q_0, t_o, r_s)} \frac{\partial I(q, t)}{\partial t}\ dq \tag{8.16}$$

- Normal flow $\mu_F(q_0, t_0)$ measures the motion of pixels in a direction perpendicular to the brightness gradient. Edge motion is an appropriate measure for the chaotic motion of dynamic textures. This measure can be calculated using Eq. 8.17.

$$\mu_F(q_0, t_0) = \iint\limits_{B(q_0, t_o, r_s)} \frac{\partial I(q, t)/\partial t}{\|\nabla I(q)\|}\ dq \tag{8.17}$$

- Laplacian $\mu_L(q_0, t_0)$ provides the local co-variance of the pixel intensities at the point (q_0, t_0) in the spatial–temporal domain (Eq. 8.18).

$$\mu_L(q_0, t_0) = \iint\limits_{B(q_0, t_o, r_s)} \Delta I(q, t)\ dq \tag{8.18}$$

The Kullback–Leibler (KL) divergence was chosen to compare histograms. It can be adapted to measure the distances between histograms in order to analyze the probability of occurrence of code numbers for compared textures [42]. First, the probability of occurrence of code numbers is accumulated in one histogram per image. Each bin in the histogram represents a code number. Second, the constructed histograms of test images are normalized. Third, the KL divergence is computed by Eq. 8.19, where $h \in 1, 2$ is the number of compared histograms $H(\cdot)$ and K is the total number of coded numbers.

$$D_{KL} = \sum_{h=1}^{2}\sum_{j=1}^{K} H(h, j) \log H(h, j) - \sum_{j=1}^{K} H_p(j) \log H_p(j)\ H_p(j) = \sum_{h=1}^{2} H(h, j) \tag{8.19}$$

In [43], High-order Local Ternary Patterns (HLTPs) were introduced. They were based on Local Ternary Patterns (LTPs) proposed by Tan and Triggs [44] with the indicating function $s_3(g_i, g_c, t)$ provided by Eq. 8.20, where t is the threshold for quantization.

$$s_3(g_i, g_c, t) = \begin{cases} +1 & g_i - g_c \geq t \\ 0 & |g_i - g_c| < t \\ -1 & g_i - g_c \leq -t \end{cases} \tag{8.20}$$

The high order directional derivatives are expressed as a 1D signal $f_i(u)$ along the ith direction, where u denotes for the local coordinates of resamplings according to the Taylor decomposition. Thus, the kth order LBP in 2D space was calculated as follows:

$$LBP^k_{P,R} = \sum_{i=1}^{P} s_2\left(f_i^{(k)}(0)\right) \cdot 2^i, \tag{8.21}$$

where $f_i^{(k)}(0)$ denotes the derivative of the kth order calculated at the point g_c for the direction i.

To reduce a complexity in 3D space, LTP 3-value pattern was decomposed into the upper LBP and lower LBP and 2D joint histograms of upper and lower LBPs were converted into a 1D vector for SVM classification. For noisy images, local magnitude patterns and local center patterns generated the HLTPs as concatenated histograms.

The original smoke feature extraction proposed in [45] was very close to deep learning. Thus, 3D local differences in scale space were the convolutions of image patches with Gaussian filters without downsampling. A sliding 3D sampling window in scale space extracted fine-to-coarse features. Many definitions such as "training", "layers", "backward propagation", "discriminative features" and "deep structure" were taken from deep learning theory. The authors argued that their efforts were focused on extracting smoke features, involving three steps: calculation of 3D differences, constructing projection matrices and feature maps and calculating within-map and between-map encodings. Then Taylor-like coefficients were applied to weight the histograms in the difference layers. In this sense, the proposed deep structure differed, for example, from PCA-Net [46]. The novelty of this approach was the combination of training and encoding methods for smoke recognition. However, the proposed approach was aimed only at feature extraction. This approach was time consuming and reasonable for small, unbalanced datasets.

In [47], conventional approach for smoke detection using a stationary camera based on background subtraction method and color, shape and edge features was boosted through parallel processing with a unified computational architecture of graphics processing unit. The algorithm was tested on multiple low and high resolution video sequences and demonstrated adequate processing time for a realistic range of frame sizes.

As the intermediate conclusion, we can mention that most of machine learning methods for smoke detection are better suited for image analysis than video analysis. The temporal dimension is poorly considered in classical recognition methods.

8.4 Deep Learning Methods

Deep learning methods use a different paradigm compared to conventional vision-based smoke detection methods. In deep neural networks, spatio-temporal features are extracted automatically during the training. Generally speaking, we know common recommendations for choosing an optimization method, deep network architecture and training set, but fine tuning of parameters is done during experiments. Also, the quantity and quality of input data are of great importance. Consider a development of deep learning methods for smoke detection, recognition and segmentation. In the first studies, only spatial features were taken into account. Features from both domains, spatial and temporal, were automatically extracted using more complex deep architectures.

Frizzi et al. [14] suggested CNN for detection of forest fire smoke, very close in architecture to the structure of the LeNet-5. The structure of this CNN is depicted in Fig. 8.2. Leaky Rectified Linear Units (ReLUs) with coefficient $\alpha = 1/3$ were used in convolutional and fully connected layers.

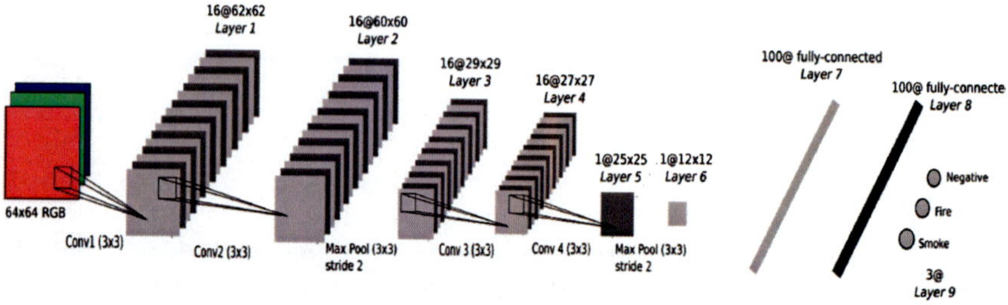

Fig. 8.2 CNN architecture for detection of forest fire smoke [14]

This CNN was trained with 27,919 RGB labeled images with sizes 64 × 64 pixels. The three subsets included 60% training images, 20% validation images and 20% test images. Stochastic Gradient Descent (SGD) with mini-batches of size 100 was applied for error minimization. The weights in the network were randomly initialized. The initial learning rate was 0.01 and momentum 0.9. The learning rate decreased by a factor 0.95 each 5 epochs. Dropout of 0.5 in the two fully connected layers allowed to avoid overfitting. The network was trained for approximately 100 cycles. Such network tuning provided 97.9% accuracy of smoke, flame and other images classification. However, this result was compared to other classifiers such as SVM and random forest.

The study discussed in [48] was very similar to [14]. The proposed CNN included 5 convolutionl layers followed by 3 fully connected layers according to the AlexNet architecture, but for two classes. The authors claimed that their approach achieved 99.4% detection rates with 0.44% false alarm rates on a large dataset.

Deep Normalization and Convolutional Neural Network (DNCNN) with 14 layers for smoke detection was suggested in [15]. This architecture is represented in Fig. 8.3.

In this end-to-end network, conventional convolutional layers were replaced on normalization and convolutional layers in order to accelerate the convergence speed of training and improve performance. Also, the input data were additionally generated in order to remove the imbalance of negative and positives samples. In DNCNN, the first eight layers were the normalization and convolutional layers, the next three pooling layers served for feature extraction, and the remaining three layers were fully connected layers for classification.

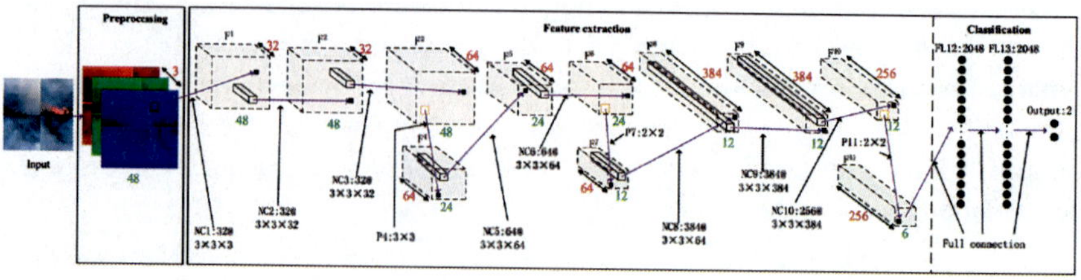

Fig. 8.3 DNCNN architecture for smoke detection [15]

Since smoke images are captured under varying illumination, the min–max normalization method [49] was adapted to eliminate the effects of illumination.

The main idea of the study in [50] was that CNN cannot provide features in the temporal domain. Smoke has a dynamic nature and permanently changes its shape. Thus, RNN is useful for extracting dynamic features from consecutive frames. The Inception-V4 architecture [51] was used because of its ability to run the sets of layers in parallel shown in Fig. 8.4, as well as the Xception, which is an "extreme" version of the Inception-V3 module [52]. This structure used the depthwise separable convolutions.

In [16], the Faster R-CNN was used for forest smoke detection. The Faster R-CNN was trained on a real forest smoke dataset, real smoke plus forest background dataset and simulative smoke plus forest background dataset.

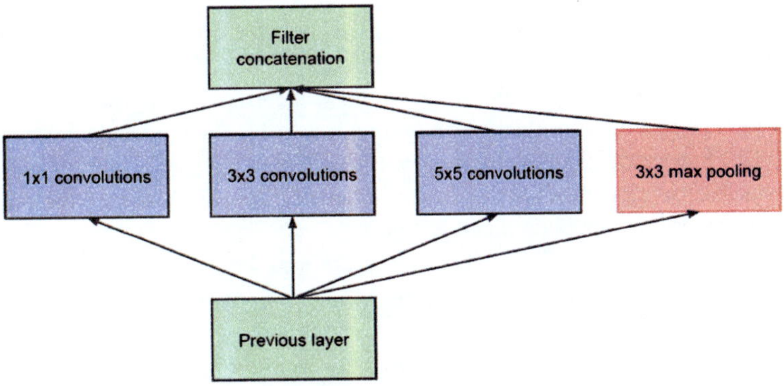

a

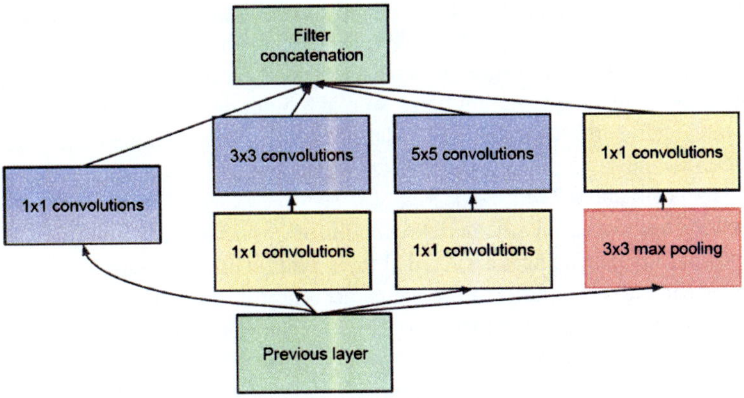

b

Fig. 8.4 Inception module: **a** naive version, **b** with dimension reductions

As shown in [18], the conditional GAN (cGAN) as an extension of the GAN in combination with the U-net can be used to find the foreground target (generative model) and calculate the higher dimensional differences between the generated smoke distribution and real data distribution (discriminative model). The cGAN architecture depicted in Fig. 8.5 is a huge U-net type network mapping a smoke frame onto a segmented image.

The network weights are optimized by adversary training between the generator and the discriminator. The network combines downsampling layers on the left side and upsampling layers on the right side and connects convolution with deconvolution. Each pixel is classified using a convolution network, and the final segmentation result is obtained using the deconvolution and pixel locating. In this study, the significant feature of smoke—motion was not used. At the same time, average of precision achieved 0.96 that can be explained by the overfitting on the dataset used: 5,194 training images and 283 test images were randomly selected from 5,477 images.

A deep smoke segmentation network divided into the coarse and fine paths was proposed in [53]. The coarse path as an encoder-decoder of a Fully Convolutional Network (FCN) [54] extracted the global context information of smoke and generated a coarse segmentation mask. The fine path as a FCN encoder-decoder, which extracted the fine spatial details of smoke, was shallower than the coarse path network. A very small network combined the results of both paths. The FCN architecture with two paths and skip layers is depicted in Fig. 8.6.

Path 1 gained global context information to generate a coarse smoke segmentation map. It utilized the first five blocks of the VGG 16 network, which contained 13 convolutional layers and 4 max-pooling layers as a decoder part, and the second five blocks involving 9 convolutional layers and 5 upsampling layers. Path 2 also had an encoder-decoder structure including 7 convolutional layers and 2 max-pooling layers in the encoding part, as well as 4 convolutional layers and 2 upsampling layers in the decoding part. Path 2 captured rich local information regarding blurred edges and semi-transparent property of smoke. The Fusion network produces the final prediction map. This end-to-end network was tested on a synthetic smoke dataset in order to avoid manual labelling fuzzy smoke boundaries. Quantitative evaluation experiments using three synthetic test datasets with different background images computed two measures: mean intersection over union and mean squared error with best values 71.04 and 0.2745, respectively.

An interesting aspect of smoke detection in foggy surveillance environment was developed in [55]. The authors argued that their method used a light-weight architecture considering all necessary requirements (accuracy, running time and deployment feasibility in smart cities) compared to the AlexNet, GoogleNet and VGG architectures. The MobileNet V2 was chosen as the basic architecture, in which 1,000 classes were changed to four classes: "smoke", "non-smoke", "smoke with fog" and "non-smoke with fog". The structure of this intelligence-assisted smoke detection system for foggy surveillance environment is depicted in Fig. 8.7, while details of a single block are shown in Fig. 8.8.

An energy-efficient system based on deep CNNs for early smoke detection in both normal and foggy Internet of Things (IoT) environments was developed in [56]. This

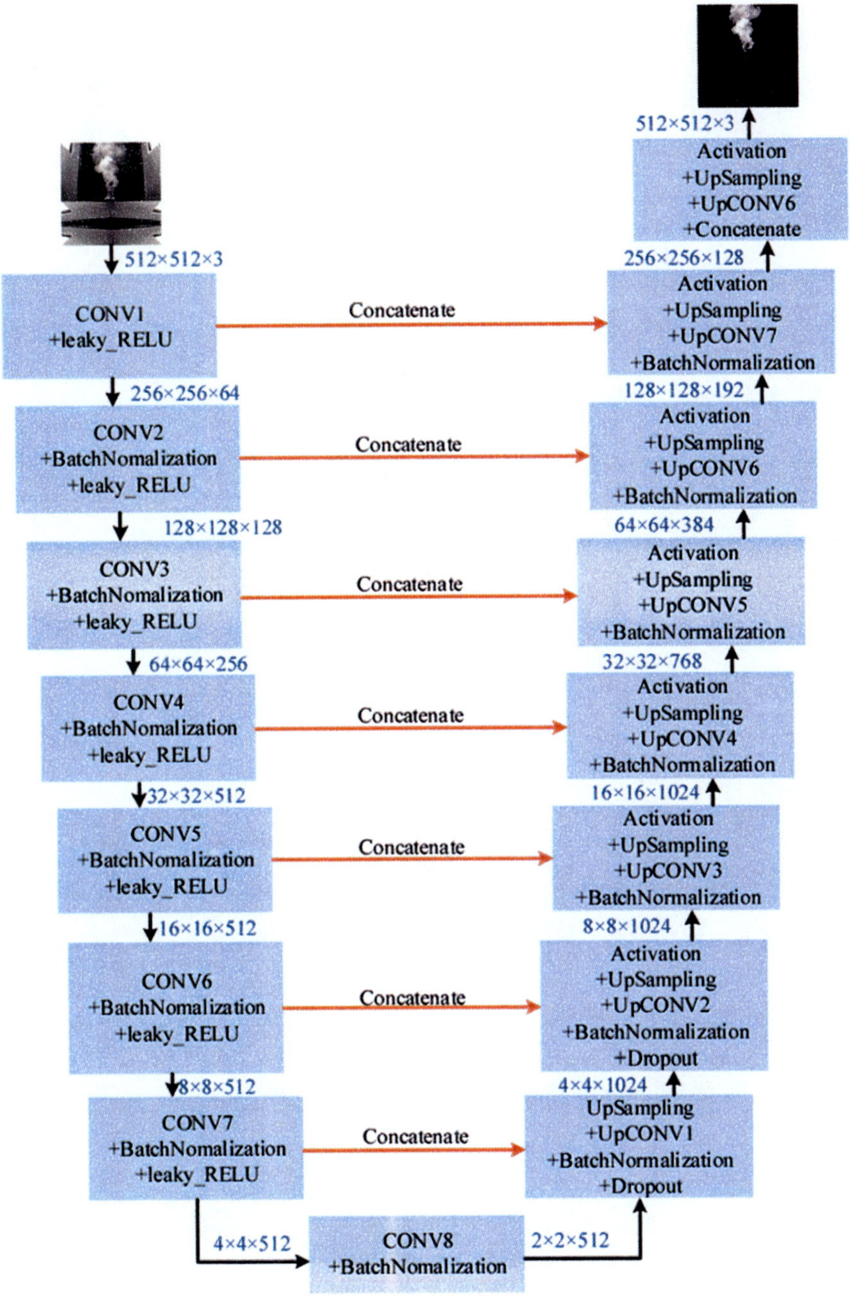

Fig. 8.5 cGAN architecture for smoke segmentation [18]

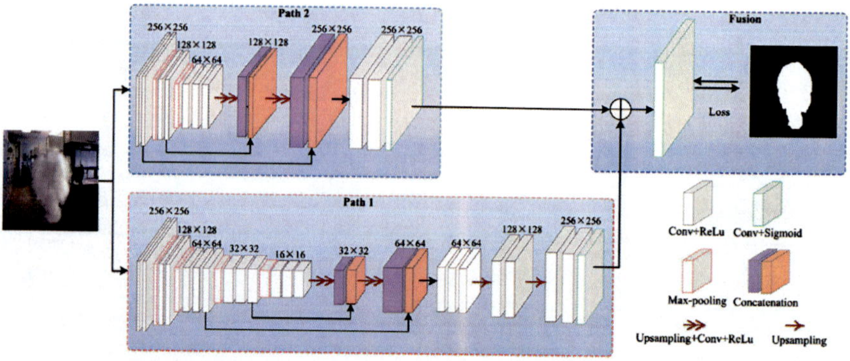

Fig. 8.6 FCN architecture for smoke segmentation [53]

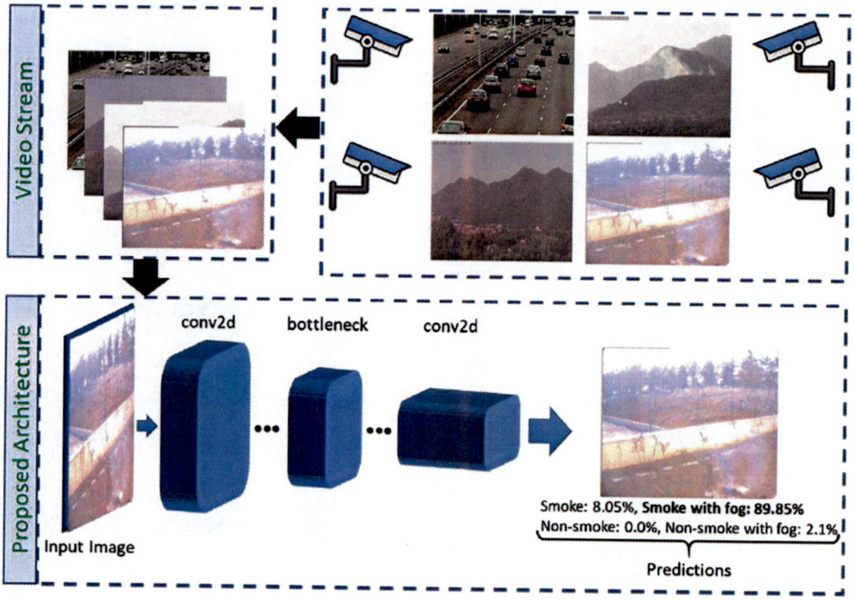

Fig. 8.7 Architecture of intelligence-assisted smoke detection for foggy surveillance environment [55]

method was implemented on the VGG-16 architecture, providing better performance in terms of accuracy, false alarms rate and efficiency in smart cities for early detection of smoke in normal and foggy IoT environments.

An attention-based deep fusion CNN for smoke detection in the fog environment was suggested in [57]. The VGG16-based CNN architecture includes an attention mechanism combining spatial and channel attention, as well as feature-level and decision-level fusion modules. Feature-level fusion is based on the feature pyramid network structure, and classification is implemented using decision-level fusion. For

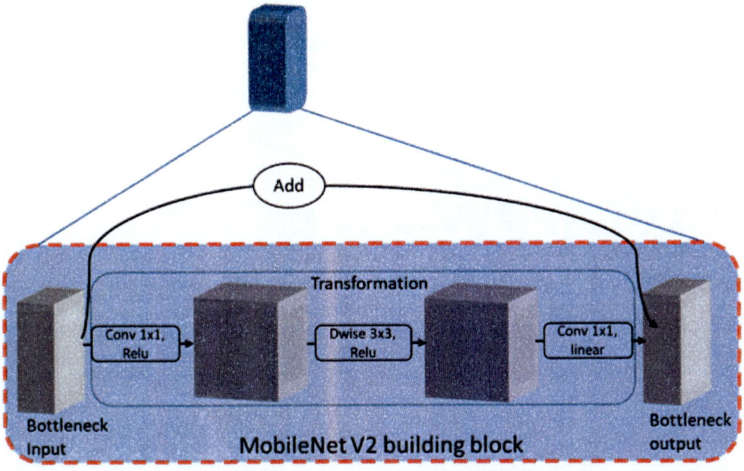

Fig. 8.8 Details of a single block of architecture depicted in Fig. 8.7

the experiments, a self-created fog smoke dataset with diverse positive and hard negative samples was collected. The accuracy of the proposed method with smoke and smoke with fog reached 90.7% and 92.0%, respectively.

Video smoke detection method based on deep saliency network was suggested in [58]. In this end-to-end deep saliency network, a deep feature map was combined with a saliency map to predict the presence of smoke in an image. The Region Proposed Network (RPN) used in this study was suggested in [59]. Its outputs were the candidate boxes with confident scores, and the objectness score of each pixel was normalized to [0, 255]. The objectness function was taken from [60].

For each bounding box B_i with a confidence score b^i generated by the RPN in an image, its confidence score b^i was added to all the pixels in the bounding box. The objectness score s_p has the form of Eq. 8.22, in which confidence scores from all bounding boxes weighted by the Gaussian function for smoothness:

$$s_p = \left[\sum_{i=0}^{N} b_i^2 I(p \in B_i) \exp(-\lambda d(pb_i)) \right]^{1/2}, \tag{8.22}$$

where $d(p, B_i)$ is the normalized distance between pixel p and the centre of bounding box B_i (the total number is N). $I(p\ B_i)$ indicates whether p is inside B_i.

The RPN contained a master branch (pixel-level CNN) and a partner branch for predicting existence. The final smoke saliency map is the fusion of pixel-level and object-level saliency maps through experimental comparison. A feature map and a saliency map were combined to predict smoke, as shown in Fig. 8.9.

The encoder architecture was based on 13 convolutional layers of the VGG16. The Recurrent Convolutional Layer (RCL) efficiently incorporated local contexts

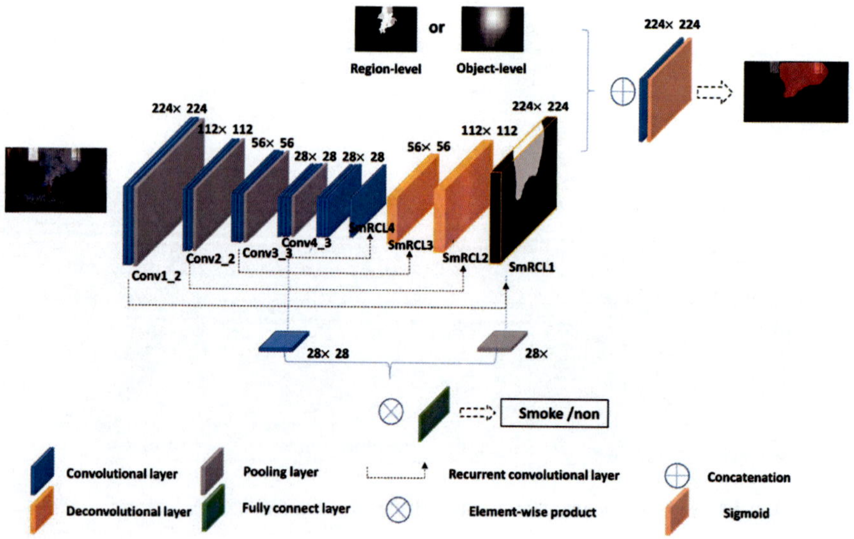

Fig. 8.9 Architecture of deep saliency network [58]

into feature maps in the encoder to refine the details of the saliency map in the decoder (SmRCL1, SmRCL2, SmRCL3 and SmRCL4). The detailed architecture of the RCL is depicted in Fig. 8.10.

The final pixel-level saliency map had the same resolution as the original input image. The object-level saliency map was generated based on the proposals produced by the RPN, and the region-level saliency map was generated using a superpixel-based approach. It was found from experiments that a fusion of object-level and pixel-level saliency maps worked better than a combination of region-level and pixel-level saliency maps.

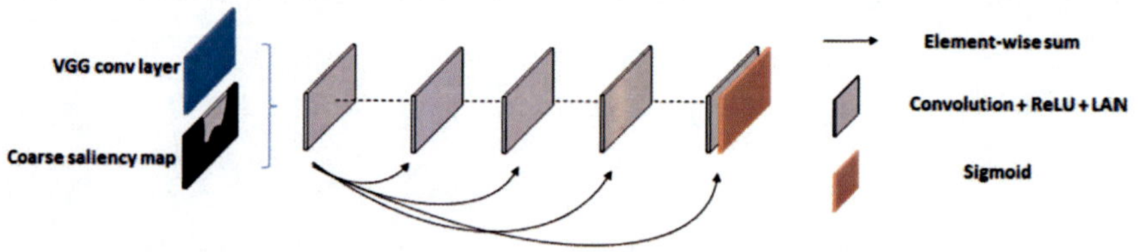

Fig. 8.10 Detail architecture of RCL

8.5 Proposed Deep Architecture for Smoke Detection

Consider the proposed architecture based on CNN and Long Short-Term Memory (LSTM) blocks. Our Weaved Recurrent Single Shot Detector (WRSSD) architecture considers both spatial and temporal domains using convolutional and recurrent layers, respectively.

The RNNs have been successfully applied to process various sequences because they can keep latent state from one input to the next one. A fragment of neural network A, which takes the input value x_t and returns the output value h_t, is depicted in Fig. 8.11. Feedbacks allow to transfer information from one step of RNN to another step. The RNN can be considered as several copies of the same network, each part of which transfers information to the next copy.

Most RNNs are currently based on the LSTM blocks. The main components of LSTM are three types of nodes called gates: input gate, forget gate and output gate. Two vectors, the input data vector x_t and the vector of the hidden state h_{t-1} obtained from the hidden state in the previous step, are entered into the LSTM. The typical LSTM architecture is shown in Fig. 8.12.

The first step in the LSTM block is to make a decision to forget information. This decision is implemented by a sigmoid layer called the forget gate. Its output is the number u_t in the range [0...1] in each cell state c_{t-1}. Unit means storing information, while zero means information forgetting. The second step is to decide about information reservation in a cellular state. This step has two procedures. First, a sigmoid sub-layer called the input gate determines the updated values i_t. Second, a sub-layer of hyperbolic tangent creates a vector of candidates for new values $\widetilde{c_t}$. This vector can be added to the current state. In the third step, an update is created for the state based on two previous procedures. The state update is calculated using Eq. 8.23.

$$c_t = u_t * c_{t-1} + i_t * \widetilde{c_t} \tag{8.23}$$

Sigmoid sub-layer forms a resulting value defined by Eq. 8.24.

$$o_t = \sigma(W[h_{t-1}, x_t] + b_0) \tag{8.24}$$

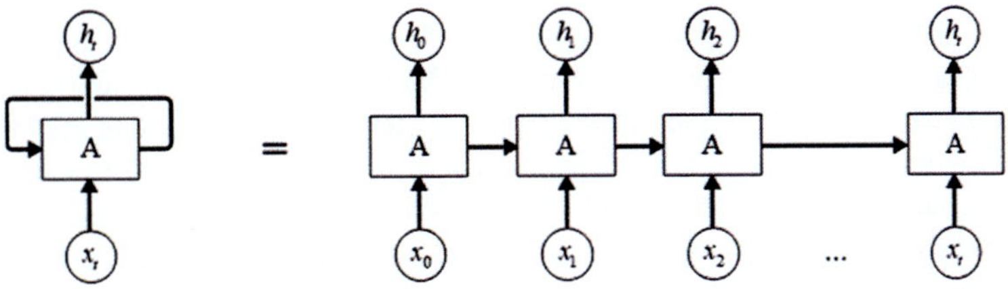

Fig. 8.11 Full presentation of RNN

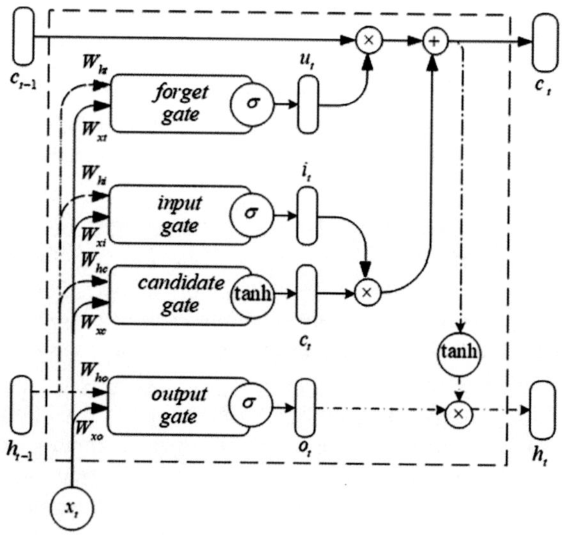

Fig. 8.12 Structure of LSTM block

Then the cellular state passes through a sub-layer of hyperbolic tangent and is multiplied by the output of sigmoid sub-layer:

$$h_t = o_t * \tanh(c_t). \tag{8.25}$$

The LSTM blocks can be incorporated into CNN architecture. The feature map of the previous frame is used as the input state of the LSTM block, and the feature map of the current frame is the output state of the LSTM block. In order to ensure high precision of object detection, it is necessity to determine the place of the LSTM blocks in CNN architecture. It should be noted that Liu and Zhu [61] proposed more efficient bottleneck LSTM blocks incorporating Single Shot Detector (SSD) architecture with depthwise separable convolutions, which immediately reduce the required computation by 8 to 9 times compared to previous definitions.

To detect smoke in videos, the SSD architecture depicted in Fig. 8.13 was chosen, and one of its types—MobileNet [62] has been used as the basic architecture due to its capability to process data in real-time.

Frames with different sizes reduced to 300 × 300 pixels are entered in the first convolutional layer, represented as three color channels. The resulting one-channel feature maps are fed to the input of a sequence of MobileNet convolutional layers. The MobileNet uses a combination of two different convolution operations: depthwise and pointwise convolutions. Depthwise convolution performs convolution on each channel separately. For an image with three channels, convolution creates an output image that also has three channels. Depthwise convolution is accompanied by a 1 × 1 convolution called pointwise convolution. The combination of these two convolutions reduces computational costs.

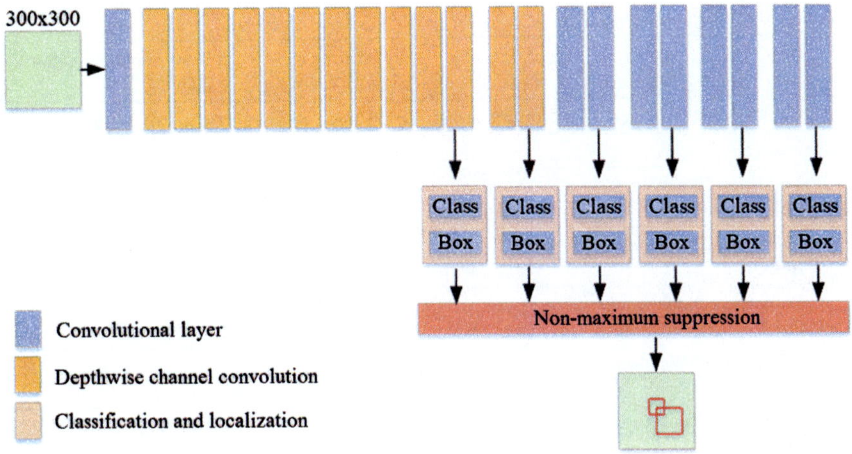

Fig. 8.13 Basic SSD architecture

After each convolutional layer, batch normalization is applied to increase productivity and stabilize operations of neural network. The main idea of the model is a data scaling. Data is normalized to values in the range from 0 to 1. The received data is entered into the input of the ReLU activation function represented in standard form:

$$f(x) = \max(0, x), \tag{8.26}$$

where x is the input value.

The problem of detecting rectangles bounding objects is solved using anchors. Bounding rectangles with different proportions and sizes are evenly spaced in the frame. The displacements of the rectangle centers, the values for changing the width and height of rectangles, and also the classes of bounded objects are determined using convolutional classification and localization layers that receive different feature maps. After predicting the placement of rectangles, a non-maximum suppression procedure is applied in order to determine the best rectangles. The network output is a set of coordinates of bounded rectangles, a set of object classes and the probabilities of belonging to certain classes. An object can be classified into one of three classes: low density smoke (semi-transparent smoke), medium density smoke and dense smoke. The complete architecture of the proposed WRSSD is presented in Table 8.1.

A convolutional neural network identifies the topological features of frames. The LSTM blocks have been added to the network structure in order to consider the temporal features. The use of LSTM convolutional blocks is aimed to modify the feature maps taking into account the feature maps resulting from the processing of previous frames. The SSD network operation scheme with the LSTM blocks is depicted in Fig. 8.14.

Table 8.1 Architecture of WRSSD

No	Layer	Filter size	Stride	Sizes of output feature map
1	Conv, BatchNorm, ReLU	3 × 3 × 3 × 32	1 × 2 × 2 × 1	1 × 150 × 150 × 32
Basic network				
2	Depthwise, BatchNorm, ReLU	3 × 3 × 32 × 1	1 × 1 × 1 × 1	1 × 150 × 150 × 32
	Pointwise, BatchNorm, ReLU	1 × 1 × 32 × 64	1 × 1 × 1 × 1	1 × 150 × 150 × 64
3	Depthwise, BatchNorm, ReLU	3 × 3 × 64 × 1	1 × 2 × 2 × 1	1 × 75 × 75 × 64
	Pointwise, BatchNorm, ReLU	1 × 1 × 64 × 128	1 × 1 × 1 × 1	1 × 75 × 75 × 128
4	Depthwise, BatchNorm, ReLU	3 × 3 × 128 × 1	1 × 1 × 1 × 1	1 × 75 × 75 × 128
	Pointwise, BatchNorm, ReLU	1 × 1 × 128 × 128	1 × 1 × 1 × 1	1 × 75 × 75 × 128
5	Depthwise, BatchNorm, ReLU	3 × 3 × 128 × 1	1 × 2 × 2 × 1	1 × 38 × 38 × 128
	Pointwise, BatchNorm, ReLU	1 × 1 × 128 × 256	1 × 1 × 1 × 1	1 × 38 × 38 × 256
6	Depthwise, BatchNorm, ReLU	3 × 3 × 256 × 1	1 × 1 × 1 × 1	1 × 38 × 38 × 256
	Pointwise, BatchNorm, ReLU	1 × 1 × 256 × 256	1 × 1 × 1 × 1	1 × 38 × 38 × 256
7	Depthwise, BatchNorm, ReLU	3 × 3 × 256 × 1	1 × 2 × 2 × 1	1 × 19 × 19 × 256
	Pointwise, BatchNorm, ReLU	1 × 1 × 256 × 512	1 × 1 × 1 × 1	1 × 19 × 19 × 512
8	Depthwise, BatchNorm, ReLU	3 × 3 × 512 × 1	1 × 1 × 1 × 1	1 × 19 × 19 × 512
	Pointwise, BatchNorm, ReLU	1 × 1 × 512 × 512	1 × 1 × 1 × 1	1 × 19 × 19 × 512
9	Depthwise, BatchNorm, ReLU	3 × 3 × 512 × 1	1 × 1 × 1 × 1	1 × 19 × 19 × 512
	Pointwise, BatchNorm, ReLU	1 × 1 × 512 × 512	1 × 1 × 1 × 1	1 × 19 × 19 × 512
10	Depthwise, BatchNorm, ReLU	3 × 3 × 512 × 1	1 × 1 × 1 × 1	1 × 19 × 19 × 512
	Pointwise, BatchNorm, ReLU	1 × 1 × 512 × 512	1 × 1 × 1 × 1	1 × 19 × 19 × 512
11	Depthwise, BatchNorm, ReLU	3 × 3 × 512 × 1	1 × 1 × 1 × 1	1 × 19 × 19 × 512

(continued)

Table 8.1 (continued)

No	Layer	Filter size	Stride	Sizes of output feature map
	Pointwise, BatchNorm, ReLU	1 × 1 × 512 × 512	1 × 1 × 1 × 1	1 × 19 × 19 × 512
12	Depthwise, BatchNorm, ReLU	3 × 3 × 512 × 1	1 × 1 × 1 × 1	1 × 19 × 19 × 512
	Pointwise, BatchNorm, ReLU	1 × 1 × 512 × 512	1 × 1 × 1 × 1	1 × 19 × 19 × 512
13	Depthwise, BatchNorm, ReLU	3 × 3 × 512 × 1	1 × 2 × 2 × 1	1 × 10 × 10 × 512
	Pointwise, BatchNorm, ReLU	1 × 1 × 512 × 1024	1 × 1 × 1 × 1	1 × 10 × 10 × 1024
14	Depthwise, BatchNorm, ReLU	3 × 3 × 1024 × 1	1 × 1 × 1 × 1	1 × 10 × 10 × 1024
	Pointwise, BatchNorm, ReLU	1 × 1 × 1024 × 1024	1 × 1 × 1 × 1	1 × 10 × 10 × 1024
Additional convolutional layers				
15	Conv, BatchNorm, ReLU	1 × 1 × 1024 × 256	1 × 1 × 1 × 1	1 × 10 × 10 × 256
16	Conv, BatchNorm, ReLU	3 × 3 × 256 × 512	1 × 2 × 2 × 1	1 × 5 × 5 × 215
17	Conv, BatchNorm, ReLU	1 × 1 × 512 × 128	1 × 1 × 1 × 1	1 × 5 × 5 × 128
18	Conv, BatchNorm, ReLU	3 × 3 × 128 × 256	1 × 2 × 2 × 1	1 × 3 × 3 × 256
19	Conv, BatchNorm, ReLU	1 × 1 × 256 × 128	1 × 1 × 1 × 1	1 × 3 × 3 × 128
20	Conv, BatchNorm, ReLU	3 × 3 × 128 × 256	1 × 2 × 2 × 1	1 × 2 × 2 × 256
21	Conv, BatchNorm, ReLU	1 × 1 × 256 × 64	1 × 1 × 1 × 1	1 × 2 × 2 × 64
22	Conv, BatchNorm, ReLU	3 × 3 × 64 × 128	1 × 2 × 2 × 1	1 × 1 × 1 × 128
Classification and localization No 1				
23	Conv	1 × 1 × 512 × 12	1 × 1 × 1 × 1	1 × 19 × 19 × 12
24	Conv	1 × 1 × 512 × 12	1 × 1 × 1 × 1	1 × 19 × 19 × 12
Classification and localization No 2				
25	Conv	1 × 1 × 1024 × 24	1 × 1 × 1 × 1	1 × 10 × 10 × 24
26	Conv	1 × 1 × 1024 × 24	1 × 1 × 1 × 1	1 × 10 × 10 × 24
Classification and localization No 3				
27	Conv	1 × 1 × 512 × 24	1 × 1 × 1 × 1	1 × 5 × 5 × 24
28	Conv	1 × 1 × 512 × 24	1 × 1 × 1 × 1	1 × 5 × 5 × 24
Classification and localization No 4				
29	Conv	1 × 1 × 256 × 24	1 × 1 × 1 × 1	1 × 3 × 3 × 24
30	Conv	1 × 1 × 256 × 24	1 × 1 × 1 × 1	1 × 3 × 3 × 24

(continued)

Table 8.1 (continued)

No	Layer	Filter size	Stride	Sizes of output feature map
Classification and localization No 5				
31	Conv	1 × 1 × 256 × 24	1 × 1 × 1 × 1	1 × 2 × 2 × 24
32	Conv	1 × 1 × 256 × 24	1 × 1 × 1 × 1	1 × 2 × 2 × 24
Classification and localization No 6				
33	Conv	1 × 1 × 128 × 24	1 × 1 × 1 × 1	1 × 1 × 1 × 24
34	Conv	1 × 1 × 128 × 24	1 × 1 × 1 × 1	1 × 1 × 1 × 24

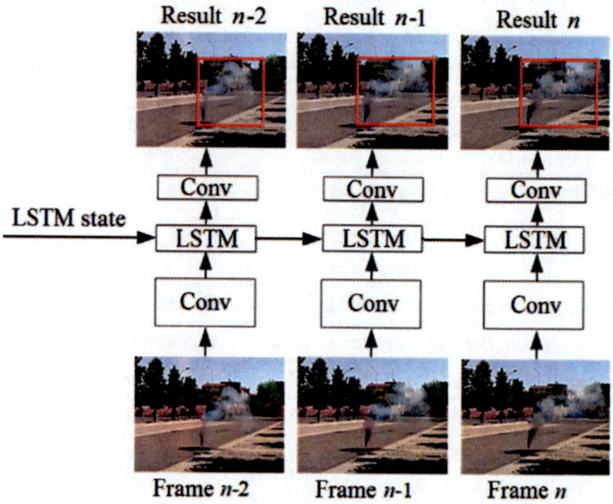

Fig. 8.14 SSD network operation scheme with the LSTM blocks

Two feature maps with size $W \times H \times M$ obtained after the set of convolutional layers and with size $W \times H \times N$ obtained from the output of the LSTM block at the previous step are entered in the LSTM block. These feature maps are combined through the channels forming the $W \times H \times (M + N)$ feature map, which is fed into the inputs of the forget gate, input gate and output gate. The output of the LSTM block is feature maps that are passed to the next convolutional layers and also to the LSTM block in the next step.

Creating a feature map using the LSTM block is time consuming. For this reason, the LSTM blocks are placed after low resolution feature maps. It is also necessary to select the optimal number of the LSTM blocks providing the highest values of precision. During the experiments, several architectures were tested, when:

- LSTM block was located after the thirteenth convolutional layer.
- LSTM blocks were located after the twelfth and thirteenth convolutional layers.
- LSTM blocks were located after the sixth, twelfth and thirteenth convolutional layers.

The neural network was trained using a gradient descent algorithm and began its work by finding the correspondences between the detected rectangles and sample rectangles. If the intersection area of these rectangles exceeds 50%, then the output $z_{ij}^{p} = 1$, otherwise $z_{ij}^{p} = 0$ [63]. The next step is to evaluate the error function. The network parameters are then changed to minimize the error function.

Error function of rectangles localization is calculated using Smooth L1-loss as a combination of L1-loss and L2-loss:

$$
\begin{gathered}
L_{loc}(z, l, g) = \sum_{i \in Pos}^{N} \sum_{m \in \{cx, cy, w, h\}} z_{ij}^{p} Smooth_{L1}\left(l_i^m - \hat{g}_j^m\right), \\
\hat{g}_j^{cx} = \frac{g_j^{cx} - d_i^{cx}}{d_i^w} \qquad \hat{g}_j^{cy} = \frac{g_j^{cy} - d_i^{cy}}{d_i^h} \\
\hat{g}_j^{=} \log\left(\frac{g_j^w}{d_i^w}\right) \qquad \hat{g}_j^h = \log\left(\frac{g_j^h}{d_i^h}\right)
\end{gathered}
\tag{8.27}
$$

where l is the detected rectangle, g is the sample rectangle, d is the anchor, w and h are the width and height of rectangle, respectively, cx and cy are the shifts of the anchor origin with coordinates (x, y), Pos is the set of rectangles, for which $z_{ij}^{p} = 1$, N is the correspondence number of rectangles.

The error function of classification is calculated using Eq. 8.28, where c is the probability of classification, Neg is the set of rectangles, for which $z_{ij}^{p} = 0$. Class "0" is associated with background objects.

$$
L_{conf}(z, c) = -\sum_{i \in Pos}^{N} z_{ij}^{p} \log\left(\hat{c}_i^p\right) - \sum_{i \in Neg} \log\left(\hat{c}_i^0\right) \tag{8.28}
$$

The resulting loss function is calculated using Eq. 8.29, where α is the weight coefficient.

$$
L(z, c, l, g) = \frac{1}{N}\left(L_{conf}(z, c) + \alpha L_{loc}(z, l, g)\right) \tag{8.29}
$$

To avoid overfitting, the WRSSD was initially trained without recurrent elements. Then, the LSTM blocks were added to the pre-trained WRSSD, and training continued. The dataset includes around 1,500 images and short videos with smoke and around 1,300 images without smoke selected from the datasets ViSOR [64], Bilkent [65], DynTex [66], as well as free Internet resources. This dataset was divided into the training, validation and testing sets with proportions 60%, 20% and 20%, respectively. Training samples for the WRSSD with the LSTM blocks involves the batches consisting of 10 consecutive frames.

Paths to images and videos, coordinates of bounded rectangles and their classes are written in XML files, which are then converted into a format available for training a neural network using open library TensorFlow 0.8.

8.6 Datasets

Public smoke and fire datasets have been collected since the 2010s. The most famous of them are the following:

- Database of Bilkent University [65]. This is a directory of sample videos containing smoke and smoke/fire flows in open or large spaces.
- DynTex [66]. The total contents of the DynTex database are over 650 sequences, but a small fraction of them is related to videos with smoke. This database includes 7 types of dynamic textures such as waving/oscillating motion, directed motion, turbulent/irregular motion, oscillating motions, directed motions, irregular motions and direct appearance change. Smoke and fire are in the medium category.
- Video Smoke Detection [67]. This dataset includes smoke/non-smoke images divided into 4 sets with different ratios of smoke/non-smoke images and smoke/non-smoke videos. "Traffic", "Basketball yard" and "Waving leaves" are concerned to the non-smoke videos.
- ViSOR [64]. ViSOR (Video Surveillance Online Repository) contains 13 categories, among which videos for smoke detection contain 14 videos with 25,570 total frames. Various levels of annotation can be adopted such as smoke event detection only, events plus bounding boxes of the smoke and events plus annotations. Both ground truth and automatic annotations are provided.
- Smoke Detection Dataset [68]. This dataset is one of the most recent, formed in July 2012 at the University of Salerno. It consisted of 149 videos, each of approximately 15 min, with a total duration of over 35 h. This dataset is available for testing both smoke detectors and fire detectors, as it contains red houses in a wide valley, mountains at sunset, sun reflections in a camera, several smokes and clouds. The videos are of good quality and useful for experimenting with deep learning models.
- V-MOTE Database [69]. The V-MOTE project is the incorporation of vision at the nodes of Wireless Sensor Network (WSN) using power efficient hardware. To demonstrate the successful application of WSN for environmental control and monitoring, videos with dynamic textures and very early detection of forest fires by means of visual smoke detection are collected and available for downloading.
- Wildfire Observers and Smoke Recognition [70]. The dataset includes two image databases. The first database contains a collection of non-segmented wildfire smoke images captured from both the ground and the air. The second database is a collection of wildfire smoke images that have been manually segmented by a human reference observer into three classes defined as smoke, maybe smoke

Fig. 8.15 Examples of frames: **a** with smoke, **b** without smoke

and no-smoke. The purpose of this resource is to create a comprehensive dataset of wildfire smoke images available to all researchers. The training set contains only 49 original images with wildfire smoke, and the testing set has also other 49 original images with wildfire smoke. Both sets were manually segmented. Videos are stored as consecutive images saved in a folder. The package contains 5 smoke sequences and 1 no-smoke sequence with a total 256 images.

Available datasets vary widely in volumes and quality of videos. The Smoke Detection Dataset [68] was formed for experiments with deep neural models, while other datasets are more suitable for conventional machine learning methods. Some frames with and without smoke from the datasets mentioned above are depicted in Fig. 8.15.

Existing smoke datasets include thousand samples. This means that the researchers are limited in images and videos for deep network training, especially for forest fires detection. In these cases, researchers try to combine deep learning and conventional feature extraction to recognize the fire smoke areas. In [71], spatial features were extracted by CNN within Caffe framework, but dynamic features were not considered for most situations.

We can recommend several ways to solve this problem:

- Generation of samples of synthetic smoke images based on background extraction from real images [72].
- Augmentation.
- GAN application to generate the samples based on real images.
- Extension of existing smoke datasets manually as a traditional approach to data collection.

Let us consider in detail the first three ways. Often, synthetic images have visible artifacts despite a wide variety of shapes, backgrounds and lighting conditions of images with a modelling object. And smoke is no exception. Xu et al. [72] proposed

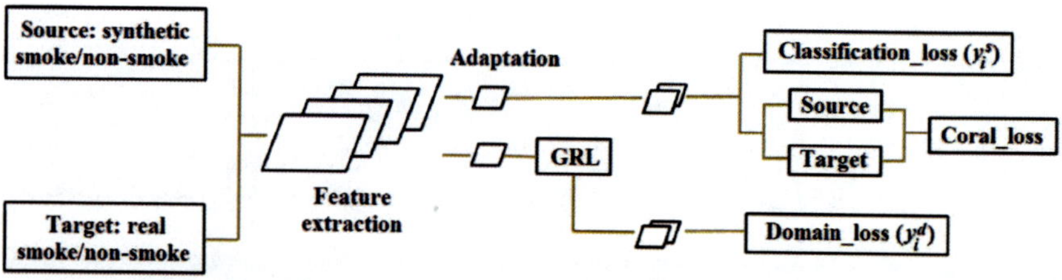

Fig. 8.16 CNN architecture for generating synthetic smoke images [72]

to create synthetic smoke images taking into account statistical distribution of the smoke properties greater than the smoke visibility. For this, CNN architecture similar to the AlexNet architecture depicted in Fig. 8.16 was proposed. This CNN contains the shared feature extraction layers, feature adaptation layers and three loss function layers. The source dataset included synthetic smoke and non-smoke images, and target dataset involved real smoke and non-smoke images. The feature distributions of synthetic and real smoke images is confusing with a combination of supervised and unsupervised adaptations simultaneously. Fuzzy semi-transparent smoke was modelled on background images collected from ImageNet. The output of this network is two datasets – smoke and non-smoke images.

An adaptation layer was added to prevent overfitting using the source dataset. The adaptation layer projected the source and target distributions into a lower dimension space for better classification.

The softmax loss as the Classification_loss L_s of the label y_i^s is defined by Eq. 8.30, where y_i^s indicates whether the input sample x_i is a smoke image or not, α_i is the predicted probability of the input sample x_i on the label y_i^s, N is the size of a batch, set as 64.

$$L_s = \frac{1}{N}\sum_{i=1}^{N}\log[\text{softmax}(\alpha_i)] \tag{8.30}$$

The hinge loss as the Domain_loss L_d of the label y_i^d is estimated by Eq. 8.31:

$$L_d = \frac{1}{N}\sum_{i=1}^{N}(\max(0, 1-\sigma(l_i = k)t_{ik}))^p, \tag{8.31}$$

where y_i^d indicates whether x_i is a real or synthetic image, l_i is the predict label, t_{ik} is the predicted probability of x_i on the label k,

$$\sigma(condition) = \begin{cases} 1 & \text{if } k = y_i^d \\ -1 & \text{otherwise} \end{cases}.$$

During training, the classification error L_s was minimized, while the domain loss L_d was maximized to confuse the distributions between synthetic and real smoke images through the gradient reversal layer [73].

Classification and domain objectives are roughly equalizing the distributions of feature extracted from synthetic and real smoke images in a statistical distribution: the correlation of their features cannot be guaranteed. The authors added a Coral_loss layer L_{coral} to flatten the second-order statics of the source and target feature distributions [74] defined by Eq. 8.32, where C_S and C_T are the covariance matrices of the source and target feature representations of d-dimension.

$$L_{coral} = \frac{1}{4d^2} \|C_S - C_T\|_F^2 \tag{8.32}$$

The joint loss function was defined by Eq. 8.33, where the loss weight α_{label}, β_{domain} and γ_{coral} determined a contribution of each type of three loss functions to optimization.

$$L = \alpha_{label} \cdot L_s + \beta_{domain} \cdot \varphi \cdot L_d + \gamma_{coral} \cdot L_{coral} \tag{8.33}$$

The rectangular region of smoke in the synthetic smoke image was mentioned before the synthetic smoke image was generated, while the smoke region in the real image was manually selected. Thus, the prepared dataset is available for supervised learning.

Augmentation includes color changes and affine transformations (translation, rotation, scaling, flipping and shearing), for example, with the following parameters [75]:

- Random rotation in the range [–45°, + 45°].
- Random flip of the image followed by horizontal alignment.
- Random resizing in the range [0.75, 1.25] or scaling up to 25% zooming in/out.
- Random cropping of images.
- Random gamma shift in the range [–25.5, + 25.5], when all color values are shifted together.
- Random shift in each color channel in the range [–25.5, + 25.5] (10% of maximum color value).

The GANs were proposed by Goodfellow et al. [76] in 2014. This is a generative model consisting of a generator, which captures the data distribution and generates simulated samples, and a discriminator, which estimates a probability that a sample was obtained from the training data rather than the generator. A capability of discriminator for data feature extraction is upgraded via an adversarial training process. This means that the GANs can be used for unsupervised representation learning with unlabeled data. GANs have many impressive applications, one of which is a generation of examples for image datasets. Xie et al. [77] proposed a temporally coherent generative model providing super-resolution fluid flows modelling. Four-dimensional physics fields (3D volumetric data and temporal dimension) have been synthesized by GAN

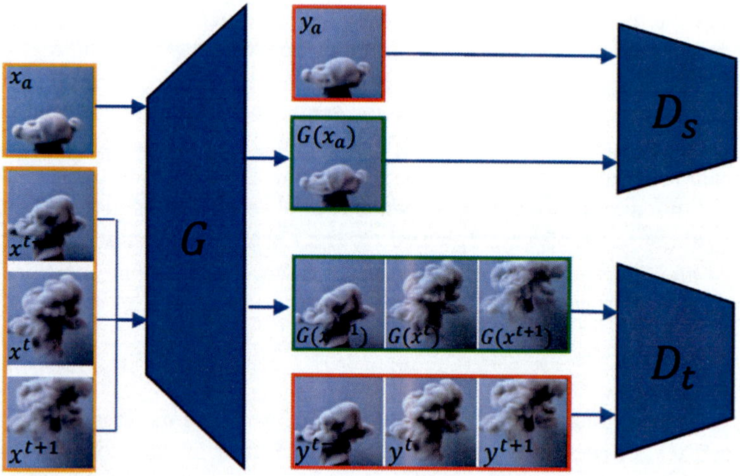

Fig. 8.17 High level overview of GAN architecture for physics fields' synthesis [77]

using a novel temporal discriminator that takes into account the velocity and vorticity of inputs. High level overview of this approach is depicted in Fig. 8.17. Generator G is guided during training by two discriminator networks, one of which focuses on the space D_s, while another focuses on the temporal aspect D_t. At runtime, both of these networks are discarded, and only the generator network G is evaluated.

The discriminator is in the form of a simple binary classifier that is trained in a supervised manner to reject generated data ($D(G(x)) = 0$) and accept real data ($D(y) = 1$). For training, the loss for the discriminator is given by sigmoid cross entropy for the two classes "generated" and "real":

$$\begin{aligned} L_D(D, G) &= \mathrm{E}_{y \sim p_y(y)}\left[-\log D(y)\right] + \mathrm{E}_{x \sim p_x(x)}\left[-\log(1 - D(G(x)))\right], \\ &= \mathrm{E}_m\left[-\log D(y_m)\right] + \mathrm{E}_n\left[-\log(1 - D(G(x_n)))\right] \end{aligned} \quad (8.34)$$

where n is the number of drawn inputs x, m is the number of real data samples y.

The notation $y \sim p_y(y)$ is used to denote the samples y being drawn from a corresponding probability data distribution p_y. The continuous distribution $L_D(D, G)$ gives the mean of discrete samples y_n and x_m. The generator is trained to "fool" the discriminator into accepting its samples and generates output that is close to real data. In practice, this means that the generator is trained to control the discriminator result. Instead of directly application of the negative discriminator loss (Eq. 8.35), GANs typically apply the following function:

$$L_G(D, G) = \mathrm{E}_{x \sim p_x(x)}\left[-\log(D(G(x)))\right] = \mathrm{E}_n\left[-\log(D(G(x_n)))\right]. \quad (8.35)$$

All known approaches to extend datasets can help in the learning process. However, there are no theoretical estimates of the required volume of datasets. They are highly dependent from the network architecture, in other words, on a number of training parameters.

8.7 Comparative Experimental Results

In the literature, we can find different types of scores' captions with approximately the same calculations. Thus, the original comparison of deep learning and encoding methods is given in [45]. The authors proposed to study the multi-scale and multi-order features from 3D local differences using their own dataset with smoke and non-smoke images [78]. The datasets are described in Table 8.2.

Ten comparison methods with twelve versions of features were tested, among which are the following:

- Multi-scale and multi-order features from 3D local differences [45].
- Original LBP without mapping [25].
- Sub-oriented histograms of LBP [78].
- Completed modeling of LBP operator [79].
- Multi-channel decoded LBP with decoder and adder [80].
- Pairwise rotation invariant co-occurrence LBP [81].
- HLTP and HLTP based on the magnitudes of noise removed derivatives and the values of the central pixels [43].
- Discriminant face descriptor [82].
- Dense micro-block difference [83].
- Deep normalization and convolutional neural network [15].

The dimensions of the compared methods are unbalanced, from 62 to 20,480, and it is unclear why the method with the least number of features provided very close results in Detection Rate (DR), False Alarm Rate (FAR) and ERror Rate (ERR), sometimes superior to other methods in FAR estimates. For the dataset used, DR, FAR and ERR estimates are in the ranges [79.3–98.7], [0.99–7.34] and [1.26–10.0], respectively. Deep normalization and CNN performed the worst results in this experiment.

Table 8.2 The image datasets for training and classifying [78]

Dataset	Number of smoke images	Number of non-smoke images	Total number of images	Stage
Set1	552	831	1,383	Training
Set2	688	817	1,505	Testing
Set3	2,201	8,511	10,712	Testing
Set4	2,254	8,363	10,617	Testing

To the best of our knowledge, it is better to use conventional machine learning estimates such as accuracy, precision, recall, F-measure, false positive ratio and false negative ratio [55]. In terms of smoke recognition, the false positive ratio and false negative ratio are often referred to as the False Acceptance Ratio (FAR) and False Rejection Ratio (FRR).

Our team has been studying the problem of early smoke detection since 2010 and has implemented several original algorithms based on a feature approach (motion, color, texture, edge and fractal analysis [8]), LPB approach [40] and deep learning approach. For the experiments, different datasets [65, 66] and [70] with varied total duration of videos were used. Average estimates for early smoke detection (semi-transparent smoke) such as True Recognition (TR), FAR and FRR values, are shown in Table 8.3.

In additional, six models, including one model with hand-crafted features and five deep learning models, were implemented in order to obtain the estimates and compare the results:

- Model based on conventional spatio-temporal features (Model 1).
- SSD model without motion verification (Model 2).
- SSD model with motion verification (Model 3).
- WRSSD model with one LSTM block located after 13th layer (Model 4).

Table 8.3 Average results for early smoke detection (semi-transparent smoke) in our experiments, 2010–2019

Basic methods for detection and classifying	Type of smoke	Averaged TR, %	Averaged FRR, %	Averaged FAR, %
Feature-based and clustering, 2010	Semi-transparent	70.3–89.1	10.9–29.3	5.3–8.6
LBP and Kullback–Leibler divergence, 2015	Semi-transparent	56.1–77.1	25.8–66.2	4.8–43.9
LBP and Kullback–Leibler divergence, 2015	Dense	99.0–99.8	0.25–1.91	0.13–1.5
LBP and SVM, 2015	Semi-transparent and dense	72.8–100	0–27.2	2.9–12.7
LBP and Random forests	Semi-transparent and dense	81.5–100	0–20.0	1.36–8.5
LBP and Boosted random forests, 2016	Semi-transparent and dense	85.7–100.0	0.0–16.8	0.8–6.3
CNN with LSTM blocks (WRSSD), 2019	Semi-transparent and dense	91.5–100	0.0–10.9	0.9–4.7

Table 8.4 Average results for early smoke detection using Model 1–Model 6

Model	Number of analyzed frames per sec	Averaged TR, %	Averaged FRR, %	Averaged FAR, %
Model 1	2	82.7	20.3	16.5
Model 2	8	88.0	16.2	8.1
Model 3	8	89.1	15.1	8.1
Model 4	7	92.0	12.1	6.3
Model 5	7	92.5	12.0	5.9
Model 6	5	94.5	10.2	3.7

- WRSSD model with two LSTM blocks located after the 12th and 13th layers (Model 5).
- WRSSD model with three LSTM blocks located after the 6th, 12th and 13th layers (Model 6).

The experiments were conducted using a PC with Intel Core i5-7300HQ, 2.5 GHz, NVIDIA GeForce GTX 1050, RAM 6 Gb. The obtained results are shown in Table 8.4.

The speed of algorithms based on all types of deep architectures exceeds significantly the speed of image processing with handcrafted smoke features. However, the temporal costs increase when LSTM blocks are added to the neural network architecture. The precision of smoke detection using a neural network is higher than the precision of smoke detection based on image processing methods. Checking the movement after the neural network is applied can reduce FRR values. Extraction of temporal features using the LSTM blocks decreases FAR values and improves a precision of smoke detection. The precision of smoke detection based on the WRSSD with two LSTM blocks is not much better than a precision of smoke detection by the WRSSD with one LSTM block. The use of WRSSD with three LSTM blocks improves smoke detection, but it is time consuming.

The output of neural networks is a list of object classes with probabilities. If the probability value exceeds the predetermined threshold, then the classification is performed correctly. A convenient way to visualize the precision values at different thresholds is to use ROC-curves. The percentage of FRR and FAR depends on the given threshold value. The optimal threshold value in the range [0…1] can be selected to minimize FRR or FAR. For each threshold value, the true positive ratio and false positive ratio are plotted on OX and OY axes, respectively. The ROC-curves depicted in Fig. 8.18 show the classification results for SSD and WRSSD.

The conducted experiments show that smoke with difference densities in video scenes with variable depth can be successfully detected. Smoke with high optical density can be easy detected using the algorithms based on the spatio-temporal features, extracted manually or automatically. Precision of semi-transparent smoke detection is worse. Thus, the precision results vary over time due to changes in

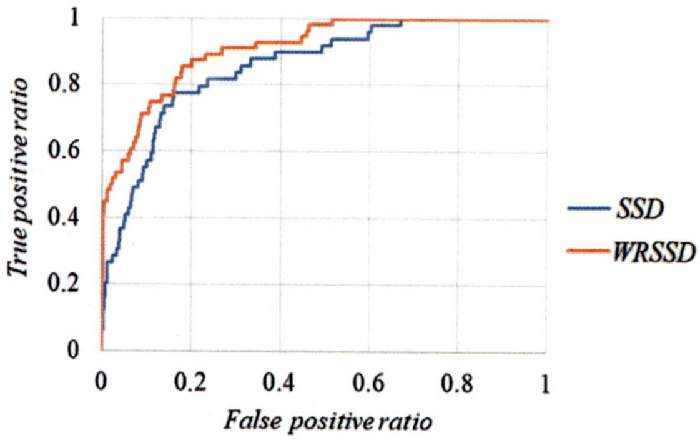

Fig. 8.18 ROC-curves for SSD and WRSSD models

smoke density. Objectively, deep learning methods require more careful parameters optimization compared to conventional machine learning methods.

8.8 Conclusions

Smoke in outdoor space as a complex object for detection and recognition has attracted much attention of researchers since the 2000s. This is a good example showing a substantial development of classification techniques in recent decades. Fisher linear discriminant, *k*-nearest neighbors, Naive Bayes classifier, SVM, AdaBoost, decision tree classifier, boosted random forests and Markov models were the main set of machine learning methods applied to extract handcrafted features in the spatio-temporal domain. The latest generation of deep learning models provides a built-in mechanism for automatically extracting of low-level features and, at the same time, high-level features classification. Deep networks address the efforts of researchers to model optimization, big data collection and durable training process. However, objectively the modern deep networks show better classification results even in difficult cases of smoke detection caused by its inner properties and outdoor impacts.

References

1. D. Han, B. Lee, Flame and smoke detection method for early real-time detection of a tunnel fire. Fire Safety J. **44**(7), 951–961 (2009)
2. V. Vipin, Image processing based forest fire detection. Int. J. Emerging Technol. Adv. Eng. **2**(2), 87–95 (2012)
3. C.Y. Lee, C.T. Lin, C.T. Hong, M.T. Su, Smoke detection using spatial and temporal analyses. Int. J. Innov. Comput. Inf. Control **8**(6), 1–11 (2012)
4. P. Barmpoutis, K. Dimitropoulos, N. Grammalidis, Smoke detection using spatio-temporal analysis, motion modeling and dynamic texture recognition, in *22nd European of the Signal Processing Conference* (2014), pp. 1078–1082

5. O. Gunay, K. Tasdemir, U. Toreyin, A.E. Cetin, Video based wildfire detection at night. Fire Safety J. **44**, 860–868 (2009)
6. C.-C. Ho, M.-C. Chen, Nighttime fire smoke detection system based on machine vision. Int. J. Precis. Eng. Manuf. **13**, 1369–1376 (2012)
7. G. Miranda, A. Lisboa, D. Vieira, F. Queiros, C. Nascimento, Color feature selection for smoke detection in videos, in *12th IEEE International Conference on Industrial Informatics* (2014), pp. 31–36
8. M. Favorskaya, K. Levtin, Early video-based smoke detection in outdoor spaces by spatio-temporal clustering. Int. J. Reason.-Based Intell. Syst. **5**(2), 133–144 (2013)
9. H. Kim, D. Ryu, J. Park, Smoke detection using GMM and Adaboost. Int. J. Comput. Commun. Eng. **3**(2), 123–126 (2014)
10. M. Favorskaya, A. Pyataeva, A. Popov, Spatio-temporal smoke clustering in outdoor scenes based on boosted random forests. Procedia Comput. Sci. **96**, 762–771 (2016)
11. S. Khan, K. Muhammad, T. Hussain, J.D. Ser, F. Cuzzolin, S. Bhattacharyya, Z. Akhtar, V.H.C. de Albuquerque, DeepSmoke: deep learning model for smoke detection and segmentation in outdoor environments. Expert Syst. Appl. **182**, 115125.1–115125.10 (2021)
12. H. Liu, F. Lei, C. Tong, C. Cui, L. Wu, Visual smoke detection based on ensemble deep CNNs. Displays **69**, 102020.1–102020.10 (2021)
13. Y. Jia, W. Chen, M. Yang, L. Wang, D. Liu, Q. Zhang, Video smoke detection with domain knowledge and transfer learning from deep convolutional neural networks. Opt. – Int. J. .Light Electron Opt. **240,** 166947.1–166947.13 (2021)
14. S. Frizzi, R. Kaabi, M. Bouchouicha, J.M. Ginoux, E. Moreau, Convolutional neural network for video fire and smoke detection, in *42nd Annual Conference of the IEEE Industrial Electronics Society* (2016), pp. 18429–18438
15. Z. Yin, B. Wan, F. Yuan, X. Xia, J. Shi, A deep normalization and convolutional neural network for image smoke detection. IEEE Access **5**, 18429–18438 (2017)
16. Q. Zhang, G. Lin, Y. Zhang, G. Xu, J. Wang, Wildland forest fire smoke detection based on Faster R-CNN using synthetic smoke images. Procedia Eng. **211**, 441–446 (2018)
17. M. Yin, C. Lang, Z. Li, S. Feng, T. Wang, Recurrent convolutional network for video-based smoke detection. Multi. Tools Appl. **8**, 1–20 (2018)
18. Y. Jia, H. Du, H. Wang, R. Yu, L. Fan, G. Xu, O. Zhang, Automatic early smoke segmentation based on conditional generative adversarial networks. Optik – Int. J. Light Electron Opt. **193**, 162879.1–62879.13 (2019)
19. Y. Peng, Y. Wang, Real-time forest smoke detection using hand-designed features and deep learning. Comput. Electron. Agric. **167**, 105029.1–105029.18 (2019)
20. H. Tian, W. Li, P.O. Ogunbona, L. Wang, Detection and separation of smoke from single image frames. IEEE Trans. Image Process. **27**(3), 1164–1177 (2018)
21. M. Favorskaya, D. Pyankov, A. Popov, Motion estimations based on invariant moments for frames interpolation in stereovision. Procedia Comput. Sci. **22**, 1102–1111 (2013)
22. R. Fattal, Single image dehazing. ACM Trans. Graphics **27**(3),72.1–72.9 (2008)
23. K. He, J. Sun, X. Tang, Single image haze removal using dark channel prior. IEEE Trans. Pattern Anal. Mach. Intell. **33**(12), 2341–2353 (2011)
24. B.U. Toreyin, Y. Dedeoglu, U. Gueduekbay, Computer vision based method for real-time fire and flame detection. Pattern Recogn. Lett. **27**(1), 49–58 (2006)
25. B.U. Toreyin, Y. Dedeoglu, A.E. Cetin, Wavelet based real-time smoke detection in video, in *13th European Signal Processing Conference* (2005), pp. 1–4
26. H.J. Catrakis, P.E. Dimotakis, Shape complexity in turbulence. Phys. Rev. Lett. **80**(5), 968–971 (1998)
27. M. Favorskaya, M. Damov, A. Zotin, Intelligent method of texture reconstruction in video sequences based on neural networks. Int. J. Reason.-Based Intell. Syst. **5**(4), 223–236 (2013)
28. M. Favorskaya, A. Pakhirka, A way for color image enhancement under complex luminance conditions, in *Intelligent Interactive Multimedia: Systems and Services, SIST*, vol. 14, ed. by T. Watanabe, J. Watada, N. Takahashi, R.J. Howlett, L.C. Jain (Springer, Berlin, 2012), pp. 63–72

29. J. Gubbi, S. Marusic, M. Palaniswami, Smoke detection in video using wavelets and support vector machines. Fire Safety J. **44**, 1110–1115 (2009)
30. W. Ye, J. Zhao, S. Wang, Y. Wang, D. Zhang, Z. Yuan, Dynamic texture based smoke detection using surfacelet transform and HMT model. Fire Safety J. **73**, 91–101 (2015)
31. F. Yuan, Video-based smoke detection with histogram sequence of LBP and LBPV pyramids. Fire Safety J. **46**, 132–139 (2011)
32. H. Maruta, Y. Iida, F. Kurokawa, Smoke detection method using local binary patterns and AdaBoost, in *IEEE International Symposium on Industrial Electronics* (2013), pp. 1–6
33. K. Dimitropoulos, P. Barmpoutis, N. Grammalidis, Higher order linear dynamical systems for smoke detection in video surveillance applications. IEEE Trans. Circuits Syst. Video Technol. **27**(5), 1143–1154 (2017)
34. T. Ojala, K. Valkealahti, E. Oja, M. Pietikäinen, Texture discrimination with multidimensional distributions of signed gray-level differences. Pattern Recognit. **34**(3), 727–739 (2001)
35. T. Ojala, M. Pietikäinen, D.A. Harwood, Comparative study of texture measures with classification based on feature distributions. Pattern Recognit. **29**, 51–59 (1996)
36. T. Ojala, M. Pietikainen, T.T. Maenhaa, Multiresolution gray-scale and rotation invariant texture classification with local binary pattern. IEEE Trans. Pattern Anal. Mach. Intell. **24**(7), 971–987 (2002)
37. F. Yuan, A double mapping framework for extraction of shape-invariant features based on multi-scale partitions with Adaboost for video smoke detection. Pattern Recognit. **45**, 4326–4336 (2012)
38. F. Yuan, Z. Fang, S. Wu, Y. Yang, Y. Fang, A real-time video smoke detection using staircase searching based dual threshold Adaboost and dynamic analysis. IET Image Process **9**, 849–856 (2015)
39. G. Zhao, M. Pietikäinen, Dynamic texture recognition using volume local binary patterns, in *Dynamical Vision*, ed. By R. Vidal, A. Heyden, Y. Ma. LNCS vol. 4358 (Springer, Berlin, 2007), pp. 165–177
40. M. Favorskaya, A. Pyataeva, A. Popov, Verification of smoke detection in video sequences based on spatio-temporal local binary patterns. Procedia Comput. Sci. **60**, 671–680 (2015)
41. Y. Xu, Y. Quan, H. Ling, H. Ji, Dynamic texture classification using dynamic fractal analysis, in *IEEE International Conference on Computer Vision* (2011), pp. 1219–1226
42. T. Ojala, M. Pietikainen, D. Harwood, Performance evaluation of texture measures with classification based on Kullback discrimination of distributions, in *12th IAPR International Conference on Pattern Recognition*, vol. 1 (1994), pp. 582–585
43. F. Yuan, J. Shi, X. Xia, Y. Fang, Z. Fang, T. Mei, High-order local ternary patterns with locality preserving projection for smoke detection and image classification. Inform. Sci. **372**, 225–240 (2016)
44. X. Tan, B. Triggs, Enhanced local texture feature sets for face recognition under difficult lighting conditions. IEEE Trans. Image Process. **19**(6), 1635–1650 (2010)
45. F. Yuan, X. Xia, J. Shi, L. Zhang, J. Huang, Learning multi-scale and multi-order features from 3D local differences for visual smoke recognition. Inf. Sci. **468**, 193–212 (2018)
46. T. Chan, K. Jia, S. Gao, J. Lu, Z. Zeng, Y. Ma, PCANet: a simple deep learning baseline for image classification. IEEE Trans. Image Process. **24**, 5017–5032 (2015)
47. A. Filonenko, D.C. Hernández, K. Jo, Fast smoke detection for video surveillance using CUDA. IEEE Trans Ind. Inform. **14**(2), 725–733 (2018)
48. C. Tao, J. Zhang, P. Wang, Smoke detection based on deep convolutional neural networks, in *International Conference on Industrial Informatics - Computing Technology, Intelligent Technology, Industrial Information Integration*, vol. 1 (2016), pp. 150–153
49. R.C. Gonzalez, R.E. Woods, *Digital Image Processing*, 3rd edn. (Prentice-Hall, Englewood Cliffs, NJ, USA, 2006)
50. A. Filonenko, L. Kurnianggoro, K.-H. Jo, Smoke detection on video sequences using convolutional and recurrent neural networks, in *Computational Collective Intelligence, Part II, LNCS*, vol. 10449, ed. by N.T. Nguyen, G.A. Papadopoulos, P. Jędrzejowicz, B. Trawiński, G. Vossen (Springer International Publishing AG, 2017), pp. 558–566

51. C. Szegedy, W. Liu, Y. Jia, P. Sermanet, S. Reed, D. Anguelov, D. Erhan, V. Vanhoucke, A. Rabinovich, Going deeper with convolutions. CoRR (2014), arXiv:1409.4842v1
52. F. Chollet, Xception: deep learning with depthwise separable convolutions. CoRR (2017), arXiv:1610.02357v3
53. F. Yuan, L. Zhang, X. Xia, B. Wan, Q. Huang, X. Li, Deep smoke segmentation. Neurocomputing **357**, 248–260 (2019)
54. J. Long, E. Shelhame, T. Darrell, Fully convolutional networks for semantic segmentation, in *IEEE International Conference on Computer Vision and Pattern Recognition* (2015), pp. 3431–3440
55. K. Muhammad, S. Khan, V. Palade, I. Mehmood, V.H.C. De Albuquerque, Edge intelligence-assisted smoke detection in foggy surveillance environments. IEEE Trans. Ind. Inform. **16**(2), 1067–1075 (2020)
56. S. Khan, K. Muhammad, S. Mumtaz, S.W. Baik, V.H.C. De Albuquerque, Energy-efficient deep CNN for smoke detection in foggy IoT environment. IEEE Internet Things J. **6**(6), 9237–9245 (2019)
57. L. He, X. Gong, S. Zhang, L. Wang, F. Li, Efficient attention based deep fusion CNN for smoke detection in fog environment. Neurocomputing **434**, 224–238 (2021)
58. G. Xu, Y. Zhang, Q. Zhang, G. Lin, Z. Wang, Y. Jia, J. Wang, Video smoke detection based on deep saliency network. Fire Safety J. **105**, 277–285 (2019)
59. S. Ren, K. He, R. Girshick, J. Sun, Faster R-CNN: towards real-time object detection with region proposal networks, in *Advances in Neural Information Processing Systems* (2015), pp. 91–99
60. Y. Jia, M. Han, Category-independent object-level saliency detection, in *IEEE International Conference on Computer Vision* (2013), pp. 1761–1768
61. M. Liu, M. Zhu, Mobile video object detection with temporally-aware feature maps, in *IEEE International Conference on Computer Vision and Pattern Recognition* (2018), pp. 5686–5695
62. A.G. Howard, M. Zhu, B. Chen, D. Kalenichenko, W. Wang, T. Weyand, M. Andreetto, H. Adam, MobileNets: efficient convolutional neural networks for mobile vision applications. CoRR (2017), arXiv:1704.04861
63. W. Liu, D. Anguelov, D. Erhan, C. Szegedy, S. Reed, C.-Y. Fu, A.C. Berg, SSD: Single shot multibox detector, in *Computer Vision – ECCV 2016, LNCS*, vol. 9905, ed. by B. Leibe, J. Matas, N. Sebe, M. Welling (Springer, Cham, 2016), pp. 21–37
64. Videos for Smoke detection (2020), http://imagelab.ing.unimore.it/visor/video_videosInCategory.asp?idcategory=8, Accessed 15 Jan 2020
65. Database of Bilkent University (2020) http://signal.ee.bilkent.edu.tr/VisiFire/Demo/SmokeClips/. Accessed 15 Jan 2020
66. DynTex. (2020), http://projects.cwi.nl/dyntex/, Accessed 15 Jan 2020
67. Video smoke detection (2020), http://staff.ustc.edu.cn/~yfn/vsd.html, Accessed 15 Jan 2020
68. Smoke Detection Dataset (2020), https://mivia.unisa.it/datasets/video-analysis-datasets/smoke-detection-dataset/, Accessed 15 Jan 2020
69. V-MOTE Database (2020), http://www2.imse-cnm.csic.es/vmote/english_version/index.php, Accessed 15 Jan 2020
70. Wildfire Observers and Smoke Recognition (2020), http://wildfire.fesb.hr/index.php?option=com_content&view=article&id=62&Itemid=71, Accessed 15 Jan 2020
71. X. Wu, X. Lu, H. Leung, An adaptive threshold deep learning method for fire and smoke detection, in *2017 IEEE International Conference on Systems, Man, and Cybernetics* (2017), pp. 1954–1959
72. G. Xu, Y. Zhang, Q. Zhang, G. Lin, J. Wang, Deep domain adaptation based video smoke detection using synthetic smoke images. Fire Safety J. **93**, 53–59 (2017)
73. Y. Ganin, V. Lempitsky, Unsupervised domain adaptation by backpropagation. CORR (2014), arXiv:1409.7495
74. B. Sun, K. Saenko, Deep coral: correlation alignment for deep domain adaptation, in *European Conference on Computer Vision*, LNCS, vol. 9915 (. Springer, Cham, 2016), pp 443–450
75. M. Favorskaya, A. Pakhirka, Animal species recognition in the wildlife based on muzzle and shape features using joint CNN. Procedia Comput. Sci. **159**, 933–942 (2019)

76. I.J. Goodfellow, J. Pouget-Abadie, M. Mirza, B. Xu, D. Warde-Farley, S. Ozair, A. Courville, Y. Bengio, Generative adversarial nets, in *27th International Conference on Neural Information Processing Systems*, vol. 2 (2014), pp. 2672–2680
77. Y. Xie, E. Franz, M. Chu, N. Thuerey, TempoGAN: a temporally coherent, volumetric GAN for super-resolution fluid flow. J ACM Trans. Graph. **37**(4), 95.1–95.15 (2018)
78. F. Yuan, J. Shi, X. Xia, Y. Yang, Y. Fang, R. Wang, Sub oriented histograms of local binary patterns for smoke detection and texture classification. KSII Trans. Internet Inf. Syst. **10**(4), 1807–1823 (2016)
79. Z. Guo, L. Zhang, D. Zhang, A completed modeling of local binary pattern operator for texture classification. IEEE Trans. Image Process. **19**(6), 1657–1663 (2010)
80. S.R. Dubey, S.K. Singh, R.K. Singh, Multichannel decoded local binary patterns for content-based image retrieval. IEEE Trans. Image Process. **25**(9), 4018–4032 (2016)
81. X. Qi, R. Xiao, C.-G. Li, Y. Qiao, J. Guo, X. Tang, Pairwise rotation invariant co-occurrence local binary pattern. IEEE Trans. Pattern Anal. Mach. Intell. **36**(11), 2199–2213 (2014)
82. Z. Lei, M. Pietikainen, S.Z. Li, Learning discriminant face descriptor. IEEE Trans. Pattern Anal. Mach. Intell. **36**(2), 289–302 (2014)
83. R. Mehta, K. Egiazarian, Texture classification using dense micro-block difference. IEEE Trans. Image Process **25**(4), 1604–1616 (2016)

Margarita N. Favorskaya is a full Professor, Head of Department at the Reshetnev Siberian State University of Science and Technology, Krasnoyarsk, Russian Federation. She received her Dr. Sci. in Theoretical Informatics from Siberian Federal University, Krasnoyarsk, Ph.D. in Control of Technical Systems from St. Petersburg State University of Aerospace Instrumentation, St. Petersburg and engineering diploma from Rybinsk State Aviation Technological University, Rybinsk. Professor Favorskaya is a member of KES organization since 2010, the IPC member and Chair of invited sessions of around 40 international conferences. She serves as the Reviewer/Associate Editor/Honorary Editor in Neurocomputing, Pattern Recognition Letters, Computer Methods and Programs in Biomedicine, Engineering Applications of Artificial Intelligence, Journal Intelligent Decision Technologies, International Journal of Knowledge-Based and Intelligent Engineering Systems, International Journal of Reasoning-based Intelligent Systems, International Journal of Knowledge Engineering and Soft Data Paradigms, also Guest Editor/Book Editor (Springer). She is the author or the co-author of more than 200 publications and 20 educational manuals in computer science. She co-edited twenty books for Springer recently. She supervised nine Ph.D. and presently supervising four PhD students. Her current research interests are deep learning, digital image and video processing, pattern recognition, decision support systems, artificial intelligence, and information technologies.

Chapter 9
Machine Learning for Identifying Abusive Content in Text Data

Richi Nayak and Hee Sook Baek

Abstract The proliferation of social media has created new norms in society. Incidents of abuse, hate, harassment and misogyny are widely spread across social media platforms. With the advancements in machine learning techniques, advanced text mining methods have been developed to analyse text data. Social media data poses additional challenges to these methods due to their nature of short content and the presence of ambiguity, errors and noises in content. In the past decade, machine learning researchers have focused on finding solutions dealing with these challenges. Outcomes of these methods boost the social media monitoring capability and can assist policymakers and governments to focus on key issues. This chapter will review various types of machine learning techniques including the currently popular deep learning methods that can be used in the analysis of social media data for identifying abusive content.

Keywords Abusive content detection · Hate speech detection · Deep learning · Natural language processing · Transformer model · Attention model · Generative model

9.1 Introduction

Around the globe, social media consumers utilise their favourite online social media services to exchange ideas and opinions, interact and collaborate on diverse topics. For decades, these services have offered innovative and constructive environments for users to enhance their productivity and well-being. With changes in social media trends over time, these services have become a double-edged sword. They do not just facilitate communication and information exchange, they also have started to replicate societal issues including bullying and crimes [1]. In 2020, the United Kingdom

R. Nayak (✉) · H. S. Baek
Faculty of Science, School of Computer Science & Center for Data science, Queensland University of Technology, Brisbane, Australia
e-mail: r.nayak@qut.edu.au

M. Virvou et al. (eds.), *Advances in Selected Artificial Intelligence Areas*, Learning and Analytics in Intelligent Systems 24,
https://doi.org/10.1007/978-3-030-93052-3_9

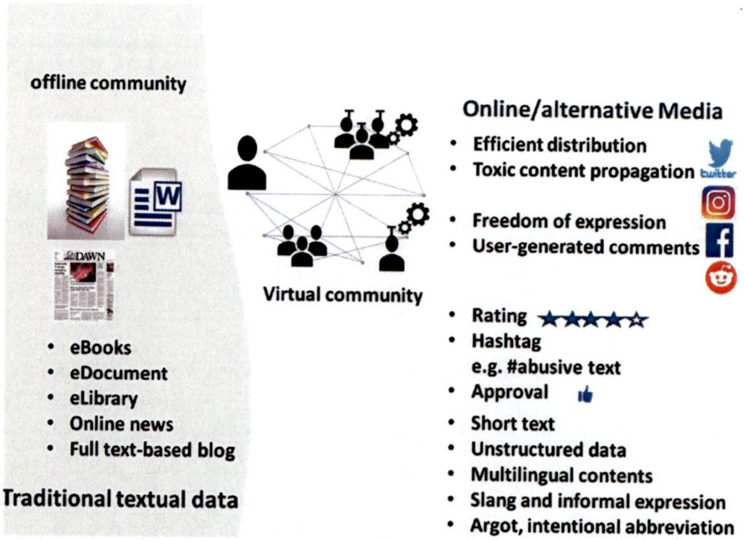

Fig. 9.1 Online textual data that appear in social media

Home Office reported the upward trends of hate crimes from 2012 to 2019 on social media.[1] The European Union[2] and the United Nations[3] have also linked abusive and hate speech on social media to critical crimes.

Associated with negative sides of text interactions, violent verbal expression, propaganda and harassment have threatened vulnerable people. Due to the lack of standards and interventions, hate expressions can cause serious societal issues [2]. With increasing public attention to issues associated with hate expression, community standards have been set for users to consume online media services in the majority of the providers such as Facebook,[4] Twitter,[5] Reddit[6] and YouTube.[7] Social scientists and lawmakers have also made efforts to build appropriate rules [3].

These regulations play a vital role to prevent users from expressing hate in cyberspaces. A critical factor for applying these rules and policy is to understand incidents of hate expressions and comprehend their characteristics to developing mitigated solutions.

With the advancements in machine learning techniques, text mining methods have been developed to analyse the abusive content posted as text data. Figure 9.1 depicts

[1] https://www.gov.uk/government/statistics/hate-crime-england-and-wales-2019-to-2020.

[2] https://www.europarl.europa.eu/thinktank/en/document.html?reference=IPOL_STU(2020)655135.

[3] https://www.un.org/en/genocideprevention/hate-speech-strategy.shtml.

[4] https://www.facebook.com/communitystandards#hate-speech.

[5] https://help.twitter.com/en/rules-and-policies/hateful-conduct-policy.

[6] https://www.redditinc.com/policies/content-policy.

[7] https://support.google.com/youtube/answer/2801939?hl=en.

types of textual data generated by social media users. These users can be professional journalists and writers or ordinary participants who can enjoy consuming and directing content, as well as interacting with other users by using online functions such as rating, approvals, comments and feedback in real-time [4]. Consequently, social media data poses additional challenges to text mining methods due to their nature of short content and the presence of ambiguity, errors and noises in content. Text mining methods need to deal with additional challenges such as the lack of labelled data, the sparseness of data representation, the ambiguity of latent meaning, use of foreign language and other symbols within the text and many others. Understanding latent meaning in conversation becomes especially challenging because of disagreement between annotators and experts in different fields and personal opinions [5].

This chapter will review various types of machine learning techniques including the currently popular deep learning methods for identifying abusive content. It will first explain the different types of abusive content that appear in social media text data. It will then present the classical and modern machine learning techniques developed to deal with the abusive content. Each family of these methods will be presented in detail. The chapter will end with summarising the recent trends and identifying future research.

9.2 Abusive Content on Social Media and Their Identification

In dictionaries, the word 'abusive' means properties of a speech or a person being 'rude and offensive' or 'criticizing rudely and unfairly'. Relating the meaning of relevant terms to virtual space, types of abuse can be cyberbullying [6, 7], racial [6, 8, 9] or sexual abuse [10, 11]. Figure 9.2 illustrates different types of textual content from video, audio and original text formats used on social media that can express abuse. It also shows the arbitrarily forming online harassment such as insults and bullying via social groups. Moreover, users are not just humans, they can also be system controlled bots, e.g., webbots, chatbots [12–14] and automated QA systems [15, 16].

Researchers detect toxicity in hate speech in online social media by the aid of technology. Researchers from diverse fields including Law [17, 18], Policy [19, 20], Psychology [21] and Healthcare [22, 23] have investigated the problem of understanding offenders and helping potential victims [8, 24]. In general, text mining methods are used to facilitate three tasks of abusive content identification, as shown in Fig. 9.3. They can be categorized as (1) feature extraction [25], (2) content detection [10] and (3) content prediction [5, 26].

Figure 9.4 shows various types of abusive content that appear on social media and different families of methods that have been developed to identify them. Target-based detection of hate speech focuses on predicting target users and groups based on the abusive content they use [20, 27]. For example, a study examined spouse

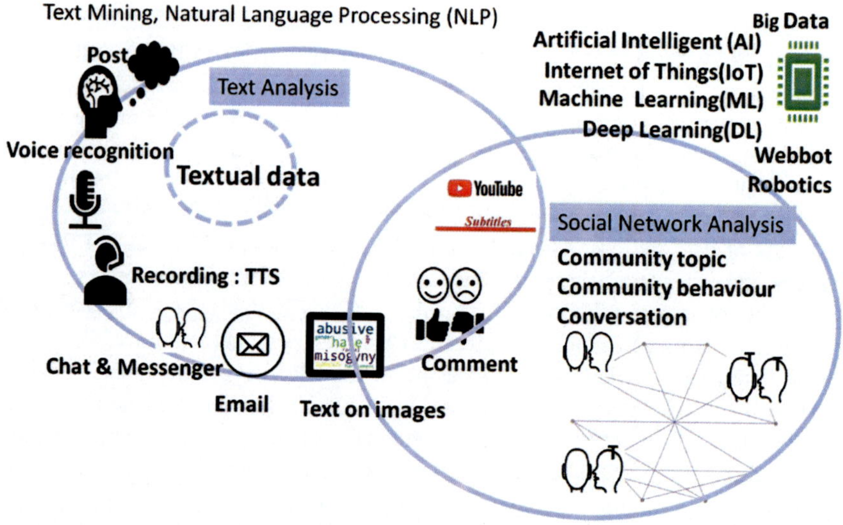

Fig. 9.2 Potential abusive data appearing in online social media

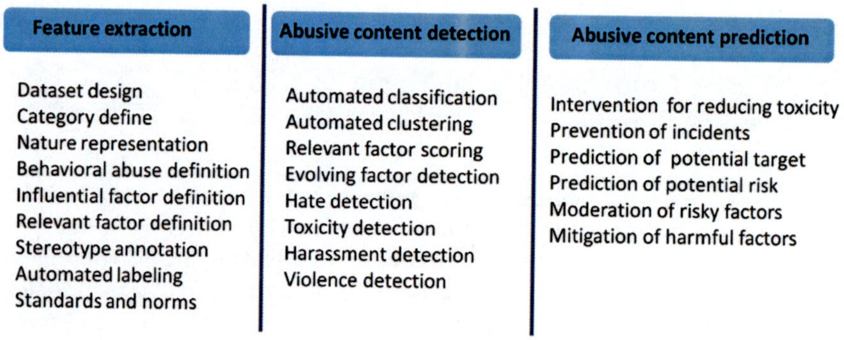

Fig. 9.3 Abusive content identification: three main objectives

abuse especially amidst the COVID-19 pandemic [28]. To understand relationships between users behaviours and contents preference, techniques of community feature mining based on Social Network Analysis (SNA) have been developed [6].

For both misogyny and misandry, gender bias issues are common in hate speech. Stereotypes can be reflective of violent actions not only in the real world but also around virtual spaces [13]. For instance, misunderstanding feminism can lead to anti-feminism in gender issues [29]. Issues connected to toxic languages such as social issues-based, age-based, incident-based features can be referred to as problems of racial or gender abuse. When researchers examine hate speech, the outstanding issues possess a long history in culture, religion and ethnicity [27]. In community-based features, simple token and specific word corpus as toxicity have been collated in terms of explicit abusive behaviours [2]. However, implicit manners and meanings

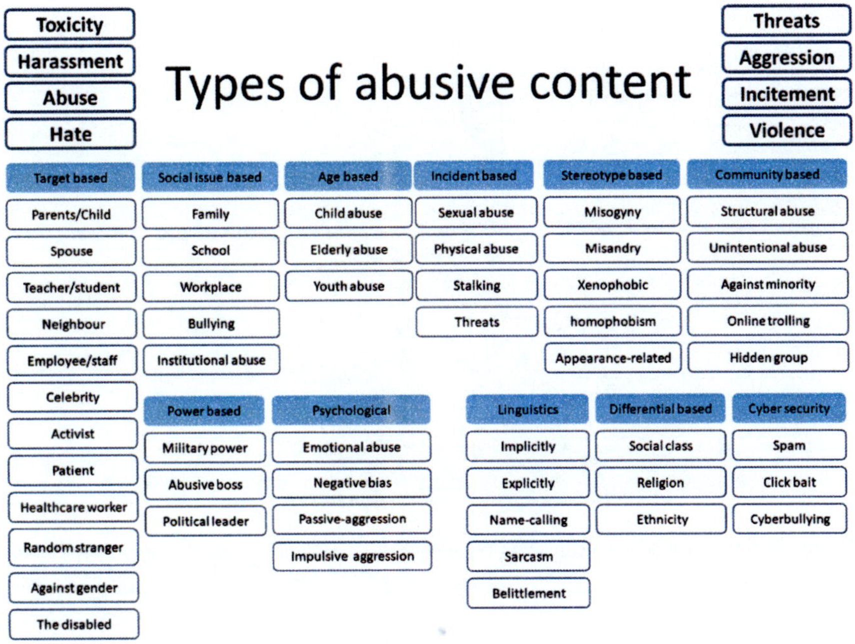

Fig. 9.4 Types of abusive content and their detection

on social networks should be monitored in different ways such as user behaviour and community feature mining [30] because offenders usually intend to hide their detrimental behaviours [8].

The study of Hardaker demonstrates social media intervention in political science [1] and power-based abuse can be connected [17]. Toxic comment detection has been examined for irony and sarcasm [31]. From the psychological perspective, online harassment is categorised into six types of themes such as physical violence, sexual violence, controlling behaviour, manipulation, domination and verbal abuse [21].

Linguistic approaches have been widely emerging into hate speech detection in crime investigation and social media intervention [32]. Detailed language models are developed to understand and predict sequences in Natural Language Processing (NLP). For example, Generative Pre-trained Transformer (GPT) [15], Bidirectional Encoder Representations from Transformers (BERT) [33] and Embeddings from Language Model (EMLo) [34] have been combined with prediction in linguistics. Argument mining has been adopted in research for hate speech [6].

Text mining methods help to discern abusive content by understanding its properties and patterns. These methods use dynamic features to define abusive textual data in designed models and frameworks. Interestingly, despite steep improvement in machine learning, researchers still face ongoing issues related to context analysis in abusive content detection.

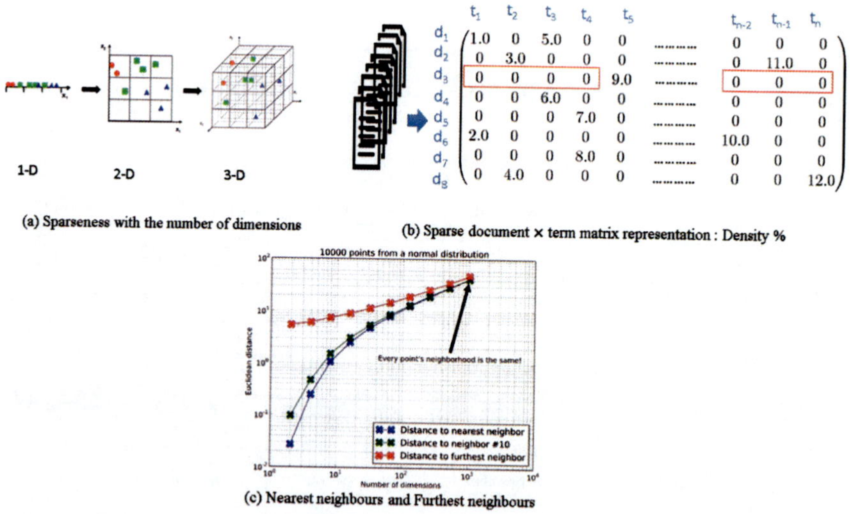

Fig. 9.5 Sparseness in social media text [38] and the distance concentration problem [39]

A common challenge faced by these methods is how to identify the similarity and dissimilarity between text instances. Accurately identifying the similarity among documents is challenged by the sparseness of text representation [35]. A popular data model for text representation is the vector space model (VSM) that records (weighted) presence/count of a term within the document [36]. The short text, which appears on social media platforms, forms short text vectors that are usually extremely sparse compared to other text [35].

This sparse VSM representation results in fewer word co-occurrences and identifying near and far points becomes problematic using the distance measurements [37], as shown in Fig. 9.5.

9.3 Identification of Abusive Content with Classic Machine Learning Methods

Abusive content detection falls into the general research area of text classification. Figure 9.6 shows basic steps to identify abusive content. Data preprocessing and feature representation are critical for machine learning models' success. Depending on the details of incidents, features can be identified by abusive content. The term-based word representations such as n-grams and bag-of-words (BoW) are traditional feature representation methods. Feature representation with the BoW approach is straightforward that can achieve high recall. However, it results in a higher number of false positives by misclassifying text as abusive due to the presence of abusive words [40]. However, abusive words are commonly expressed in sarcastic and joke

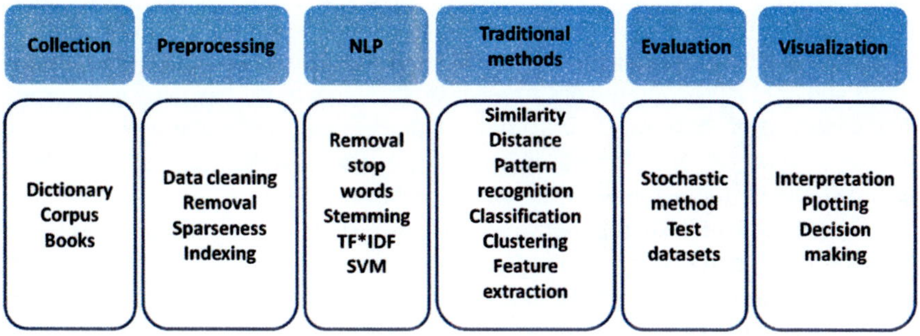

Fig. 9.6 A general process of abusive content detection

manners in social media posts. The use of biagrams with unigrams [41] have shown marginal improvement. Using higher grams may have an adverse effect due to their low frequency [42]. Researchers have leveraged syntactic features and the intensity of hate speech to identify good features for use in classification [43].

There are general methods in text mining for text classification/clustering based on similarity and distance measures. Classifiers such as Decision Tree (DT), Support-Vector Machines (SVM), Naive Bayes (NB), different types of regression functions, and K-Nearest Neighbour (K-NN) have been commonly used to identifying abusive content in text data [20, 44]. Advanced models such as Gradient Boosted Decision Trees (GBDT) [12], Random Forest (RF) [45], Bayesian Logistic Regression (BLR) [45] and Multinomial Naïve Bayes (MNB) [46] have also been used. No one model performs best in all datasets. For instance, the Linear Discriminant (LD) model has shown the best performance on datasets such as OLID[8] (Offensive Language Identification Dataset) and HASOC (Hate Speech and Offensive Content Identification in Indo-European Languages); however, Convolutional Neural Network-Long Short-Term Memory (CNN-LSTM) is shown to outperform the LD model for the dataset combined with HASOC, OLID and HateBase [44].

The last steps in the process of abusive content detection are evaluation and visualization. Evaluation is a significant step to determine whether designed models have the required efficiency and effectiveness for decision making. Visualization assists in understanding patterns in textual data and interpreting severity within scoring.

9.3.1 Use of Word Embedding in Data Representation

Lack of unique and discriminative linguistic characteristics in abusive content makes it difficult to separate them from non-abusive speech [47]. Social media users commonly use offensive words or expletives in their online dialogue to express satire and jokes [48]. Improving the training dataset such as increasing the size, reducing

[8] https://scholar.harvard.edu/malmasi/olid.

the class imbalance, performing over-sampling and under-sampling may not always guarantee a separation between two classes for machine learners. Machine learning models can not just rely on the occurrences of abusive words, they need to incorporate the context of these words in learning. For example, individual words such as 'Muslim', 'Refugee', and 'Troublemakers' are not always indicative features of hate speech but their combinations can be more indicative [47].

Understanding different types of word embedding in feature representation assists in generating a good quality model. Context-based embedding such as Continuous BoW (cBoW) and Skip-Grams embrace different annotation tactics. Compared to the term-based words representations, sentence-based words presentations are more challenging for abusive content representation [49]. Researchers have technically examined the improved word embedding methods such as Word2Vec [50], ELMo [34], BERT [51] and FastText [50]. To analyse context to interpret hidden meanings, Mikolov's word embedding, fastText subword embedding and BERT WordPiece model [52] have shown to be useful.

9.3.2 Ensemble Model

Ensemble models are designed to obtain better performance by combining multiple machine learning models learned on different sets of datasets or trained with different sets of methods [49]. An ensemble model of SVM, logistic regression and naive Bayes with character n-grams, sentiment and lexicons has shown to perform well for misogyny detection [54, 55]. An ensemble classifier of the FastText model combined the RoBERTa model trained on the HASOC[9] data has shown improved performance on hate speech detection [44]. An ensemble model was obtained by combining the decisions of ten convolutional neural networks with different weights [56]. As illustrated in Fig. 9.7, an ensemble model was build to classify different features of hate speech with aggressive (OAG), covertly aggressive (CAG) and non-aggressive (NAG) components [53]. Similarly, the ensemble model that was compromised of BERT combined with multitask learning models achieved better performance [57].

9.4 Identification of Abusive Content with Deep Learning Models

Performance of classical algorithms highly depends on feature engineering and feature representation [58, 59]. On the other hand, neural networks-based deep learning models have become popular due to their ability to automatically learn abstract features from the given input feature representation and reduced dependency on manual feature engineering [12]. For example, a Deep Neural Network (DNN) model has

[9] https://hasocfire.github.io/hasoc/2020.

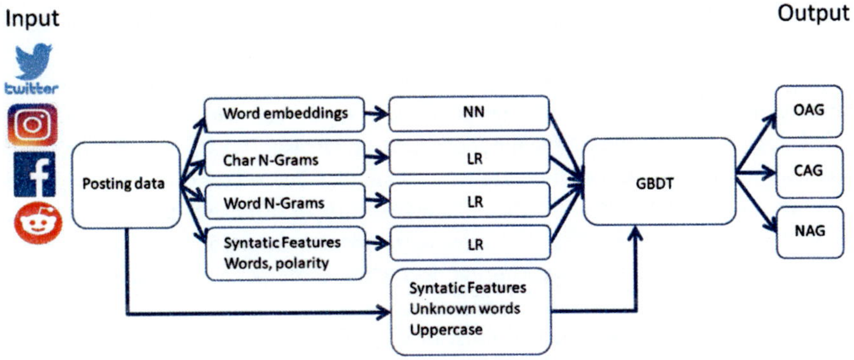

Fig. 9.7 Example of an ensemble model for abusive content detection [53]

been shown to extract discriminative features that can capture the semantics of hate speech [47]. Input to deep learning models can be various forms of feature encoding and embedding, including those used in the classic methods.

Algorithm design in this family focuses on the selection of the network topology to automatically extract useful abstract features. Popular network architectures are Convolutional Neural network (CNN), Recurrent Neural Network (RNN) and Long Short-Term Memory (LSTM). CNN is well known for extracting patterns similar to phrases and *n*Grams [12, 60]. On the other hand, RNN and LSTM have been found effective for sequence learning such as order information in text [12] and have been successful in text classification and prediction [61]. Considering dynamic features of hate speech patterns and related properties, advanced deep learning models such as the transformer model, generative model and attention model have recently been applied in abusive content identification.

9.4.1 *Taxonomy of Deep Learning Models*

Deep learning approaches have experienced a surge with unprecedented success in the fields of computer vision and natural language processing. There exist four categories of deep learning approaches namely supervised, unsupervised, semi-supervised, self-supervised learning [26, 62].

Supervised learning (i.e. prediction) stands for a machine learning approach adopting labelled data [10, 20]. In abusive content identification, supervised learning is known as a robust method [10]. These models include an output layer that can predict the output values based on the knowledge representation learned by the architecture using the training data. However, the unbalanced labelled data in abuse-based datasets (i.e. lack of labelled abuse instances) has been linked to relatively poor model performance [7]. Unsupervised learning (i.e. clustering) relies on the task of analyzing patterns and features of (unlabelled) data collection [63, 64]. However, clustering methods lack the predictive nature as well as pose problems due to their subjective interpretation.

Aimed at dealing with the drawbacks of supervised and unsupervised learning, semi-supervised learning approaches utilise training mixed with labelled and unlabelled data. An Offensive Language Identification Dataset (OLID) was divided into three sets according to three annotations: (1) a labelled dataset for offensive language detection; (2) an unlabelled dataset for categorization of offensive language; and (3) a labelled dataset with offensive language target identification [57]. The authors proposed ensemble-based multi-tasking models using three annotated datasets on the pre-trained language model BERT [57].

A supervised learning task is framed in a special form to predict only a subset of information using the rest. In this way, all the information needed, both inputs and labels, has been provided. This is known as self-supervised learning. This idea has been widely used in language modelling to refer to techniques that do not use human-annotated datasets to learn (textual) representations of the data (i.e. representation learning). A self-supervised learning model with StructBERT was found to perform better on sentence-based prediction than the baseline model using the BERT [62]. Similarly, the pre-trained model BERT was utilised using self-supervised loss to effectively apply the Next Sentence Prediction (NSP) [26]. Based on multitask learning defined as self-learning [20], self-attention learning [65] is introduced.

9.4.2 Natural Language Processing with Advanced Deep Learning Models

While identifying abusive content on social media, user interaction and relationships play a significant role. Users may use expletives or offensive words in their close circle. Also, abusive language can be used in their online dialogue to express satire and jokes. Identifying the context of content can reveal useful information. Pre-trained models in supervised and unsupervised learning are utilised to detect given targeted concepts including abusive content [66]. In general, advanced models have a complex architecture compromised of different layers that are flexible to have connections according to the underlying problem. Next, we present some of the advanced deep learning models that have been used in abusive content identification.

Deep Transformer Model (DTM) Transformer models are originally designed for the task of cross-linguistic translation using the encoder-decoder framework [67]. Figure 9.8 shows a simplified transformer model. Bidirectional encoder representations from Transformers (BERT) [33] and Generative Pre-trained Transformer (GPT) [16] are two popular transformer models. The BERT was introduced with two tasks: (1) understanding the common language patterns with a Market Language Model (MLM); and performing downstream tasks such as Question Answering (QA) and Natural Language Inference (NLI) [33]. GPT is a language model, released by Ope-

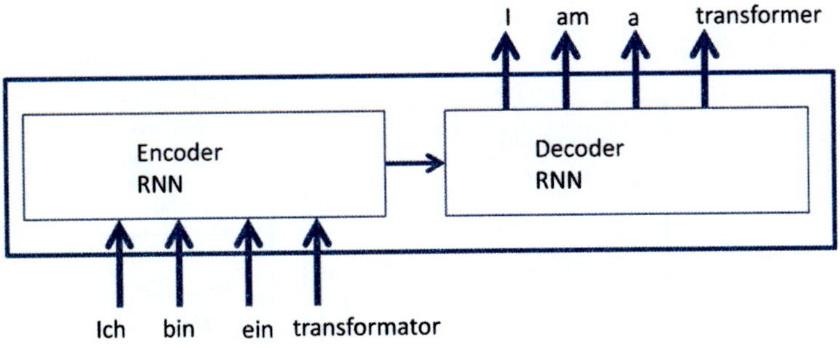

Fig. 9.8 A general (simple) transformer model

nAI [16], that can generate sequences of words in a deep learning method [15]. The BERT and GPT models can enable visualisation aided by an attention model [16].

Overcoming the lack of sufficient labelling datasets and existing biases in hate speech, the transformer models based on the BERT show an improved performance [68]. An increased accuracy was achieved with a BERT + CNN model as compared to BERT + Nonlinear layer or BERT + LSTM [68]. Recently, a transformer-based model aimed at misogyny detection showed a good performance [54, 55]. The proposed model was designed with two subtasks in different categories (e.g. misogyny and non- misogyny, five behaviour categories and classification of target) and compared the performance with the models of SVM, LSTM and Gated Recurrent Units (GRU) in different language datasets. Another proposed BERT model was also found to outperform Linear SVM, Embedding from Language Models (ELMo) and LSTM in an abusive language detection task [69].

Deep Attention Model (DAM) General transformer consists of the encoder for input sequence and the decoder for output sequence as shown in Fig. 9.8. To deal with computational problems of long sentence sequences, an attention model is adopted in language models [70].

Figure 9.9 illustrates a transformer model based on an attention mechanism. The attention mechanism is added to provide importance to the part of the data, such as a pixel in an image or a word in a sentence, (by giving it a higher weight) out of a large collection. For example, it can highlight the relationship between words in one sentence or close context. In a sentence, when the model receives "dressing", it expects to encounter a cloth word soon. If it receives "green", it gives it lesser weight as compared to "shirt".

A hate speech classifier was proposed using attention-based CNN and RNN [71]. For the sequence modelling, an attention model based on encoder-decoder has been adopted in abusive content detection [64, 70]. A BERT based attention model with multi-heads was used [72]. In a model based on ELMo [34], the deep contextualised word embedding representation is designed with self-attention.

Deep Generative Model (DGM) For semantic-based analysis of sentences, generative methods are common concepts in deep learning [74]. A Generative model

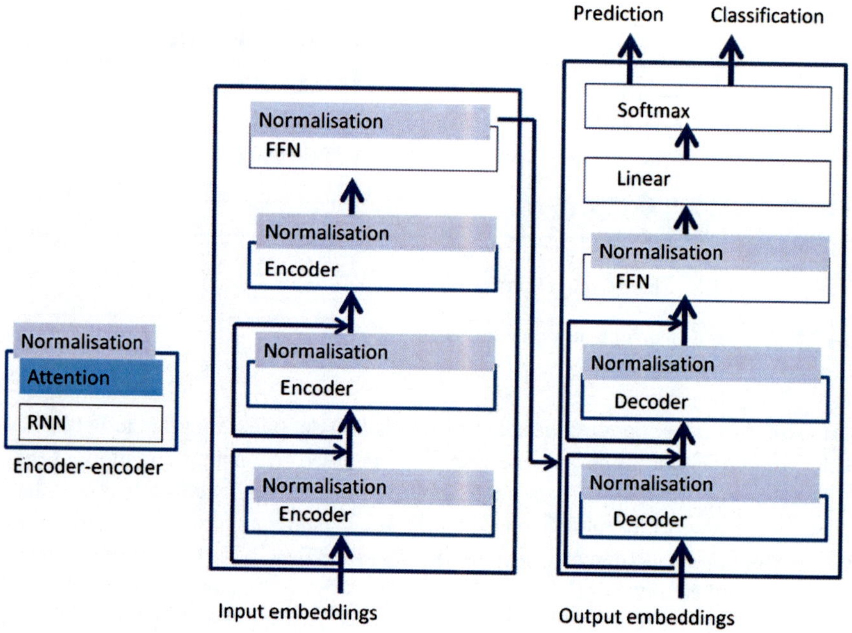

Fig. 9.9 An attention model

learns the true data distribution of a dataset using unsupervised learning and, then, it can generate new data points with some variations [15, 75]. A deep generative model learns a function that can approximate the model distribution to the true data distribution. Variational Autoencoders (VAEs) [74] and Generative Adversarial Networks (GANs) [73] are two of the most commonly used and efficient generative learning approaches. Recently, Generative Pre-trained Transformer (GPT) model was designed in the deep generative models to generate, summarise and answer text as an unsupervised learning approach [15, 68].

To deal with data scarcity in hate speech datasets, a generative model can provide a function to generate textual data closed to the conditional probability of relevant synthetic data [73]. Using the adversarial generative-based model, a method was developed to generate text concerning hate speech detection [73], as illustrated in Fig. 9.10. In another research, the Sparse Additive Generative Model (SAGE) was adopted aimed at understanding individual exposures to hate speech in online college communities [23]. Investigating psychological endurance to hate expressions, the model identifies discriminating n-grams (n=1,2) between the comments of low and high endurance users.

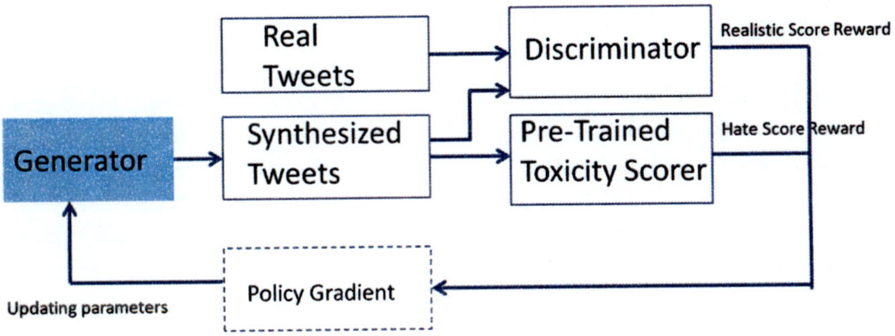

Fig. 9.10 An architecture in hate speech detection: HateGAN [73]

9.5 Applications

Due to the complex nature of abusive content [74], collaborative efforts in different fields have contributed and emerged as an outcome of abusive content identification. The abusive content identification on social media can lead to several applications including detection, prediction and moderation of online toxicity.

The abusive content identification helps investigating and detecting cyberbullying [7, 76], gender abuse [10] and threats [77, 78]. Researchers have extended this to analyse the online game community [79] by understanding relevant patterns and features of gender hate issues. Issues focusing on the youth groups (students 14 to 19 years old) were examined for hate identification based on online behaviours, risk perception, parental supervision [80]. Based on identified abusive content, a method was developed to investigate psychological endurance to hate expression in online college communities [23]. The study discussed the hate speech category. For argument mining [81], a conceptual architecture combined with sentiment and opinion mining was proposed to find a deeper understanding and reasoning of contextualisation information. With the core method based on network analysis integrated with argument and stance detection, one of the applications includes trolling detection [81]. Analysis of textual content expressing harassment is a common approach in online security [82] such as spam detection [31], social bots detection [14] and misinformation detection [13].

Based on the abusive content detection, researchers in social science, policing and law see the opportunity in the prediction of malicious actions, victims and offenders [19, 27]. For example, understanding harassment targets at an early phase, standards and norms can be analysed and built as a preemptive intervention to defer violent actions [83]. Prediction of potential victims and offenders can help in the invention of related hate-crime and social events [78].

The abusive content detection also leads to intervention measures in healthcare [22, 80] and education [22]. A detection method offers a tool to monitor inappropriate situations which can be collated as evidential data to generate new norms in community and society [17, 82].

9.6 Future Direction

A myriad of machine learning methods has been developed for abusive content identification riding on the success of deep learning with natural language processing. However, these methods share the same difficulties faced by the deep learning methods such as the lack of labelled data, lack of quality data, poor feature representation, presence of multi-modality in social media data and poor efficiency. Moreover, the machine learning models should be aimed at the prediction of hidden incidents, especially in harassment with implicit expression [6, 82] or without hate expression [31].

Data scarcity and comprehensive annotation are challenges in abusive content detection [14, 44]. Researchers have designed datasets covering multilingual features to improve the data quality [6]. Deep learning models using progressive transfer learning and active learning are promising directions. Transfer learning helps to first learn the general linguistic features from the commonly available datasets such as Wikipedia and movie reviews and then use a language model to transfer the linguistic knowledge to a machine learning model that can be trained on a small labelled dataset [84]. Active learning helps to grow a small labelled dataset by interactively query a user (or a machine learning model) to label new data points with the desired outputs [85]. Our prior works using transfer and active learning for abusive content identification [84, 85] are just a beginning and many more methods can be developed. Reducing auto-labelling errors in semi-supervised learning will be a feasible approach for future studies [5, 52].

An improved feature representation learning latent features within the data can improve the quality of abuse identification [86]. Figure 9.11 shows different states of feature representation. The left-hand side box shows the most common methods. The middle box shows the methods that are in infancy. For example, as for the analysis of conversation, sentence-based prediction can improve the quality of abuse identification [62]. However, not much work has been done in this area. Understanding context-based conversation can be improved by using the combinations of advanced models such as transformer, attention and generative models.

The right-hand side box in Fig. 9.11 shows the future directions. Dynamic feature extraction by understanding complex contexts such as polysemy and objectivity in linguistics is an ongoing issue [87]. There are evolving features in ecosystems of social networking platforms since participants freely share ideas and interact with each other using informal forms of expression. Because a cause is not only one factor, multi-domain detection and analysis of interaction can play an important role [86]. Multi-model deep learning comprising of diverse datasets such as posts content as well as images can understand latent relationships in deep learning models [5]. Similarly, the complex context in textual data such as aggression and sarcasm can also learn to improve the result of hate speech detection in deep learning models. For example, examination of the relationship between mental health and psychological experience in online social networks can improve the model quality [23].

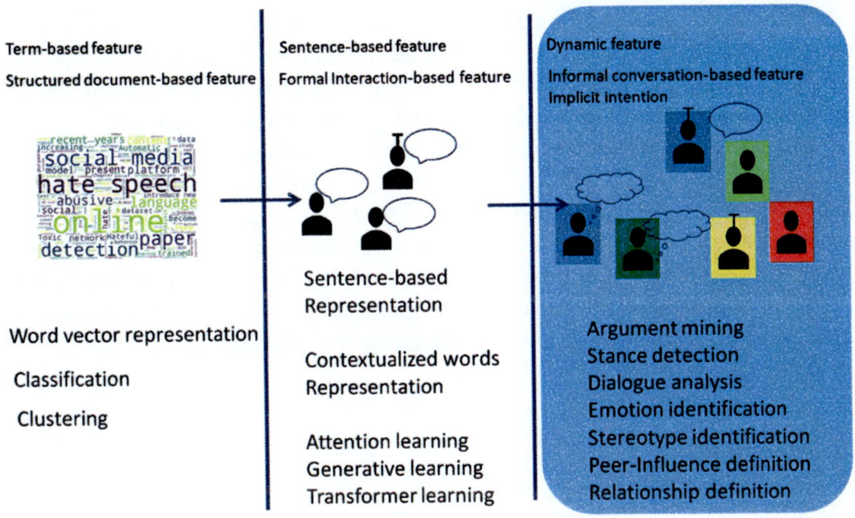

Fig. 9.11 Feature representation: now to future

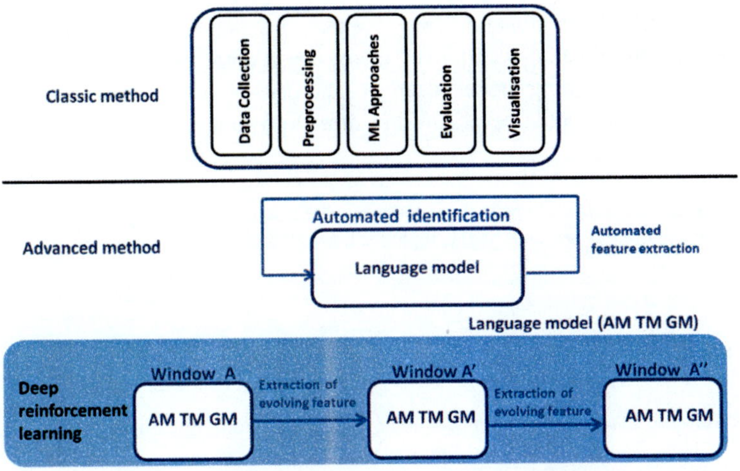

Fig. 9.12 Status of deep learning models in abusive content identification

There exist many challenges for machine learning models. Figure 9.12 shows advanced and classic methods that have been developed to identify abusive content. These methods usually work on static datasets [17, 32]. However, for the ever-changing features of textual data in social media, a model can be defined using a temporal window to analyse the current state [88]. Eventually, the output can extract new features to understand timely abusive content property [4, 73]. Deep reinforcement learning is an emerging direction in this field.

In addition to obtaining high accuracy of abusive content identification, system performance is also a key factor [70, 74]. The system complexity with the time and memory efficiency should be considered to build a real-world solution and should be a fundamental requirement of architectural systems in automation [2].

Finally, peer-to-peer systems and security measures should be combined with abusive content identification systems. Individuals' feeling of abuse, hate and harassment is a personal factor [22, 77]. To monitor and detect abusive content, a tool based on a peer attention method can provide a solution. For example, a decentralised detection system was found useful for identifying cyberbullying in school communities [22]. With security concerns and a 'user-centric approach' [22], the authors proposed an infrastructure based on Blockchain and the Internet of Medical Things (IoMT).

9.7 Conclusion

With the proliferation of social media, incidents of abuse, hate, harassment and misogyny are widely spread across social media platforms. With the advancements in machine learning and natural language processing techniques, it has become feasible to identify abusive content. This chapter presented a comprehensive survey of machine learning methods employed in abusive content identification. It focused on advanced deep learning models used as state-of-the-art in abusive content identification. It also presented several future directions that research can be focused on.

Acknowledgements I would like to acknowledge my research team, especially Dr Md Abul Bashar, who has been conducting research on this topic for a few years.

References

1. J.W. Howard, Free speech and hate speech. Annu. Rev. Polit. Sci. Annu. Rev. **22**, 93–109 (2019). https://doi.org/10.1146/annurev-polisci-051517-012343
2. A. D'Sa, I. Illina, D. Fohr, BERT and fastText embeddings for automatic detection of toxic speech, in *2020 International Multi-Conference on: "Organization of Knowledge and Advanced Technologies" (OCTA), 1–5* (2020), https://doi.org/10.1109/OCTA49274.2020.9151853
3. M. Sap, D. Card, S. Gabriel, Y. Choi N. Smith, The risk of racial bias in hate speech detection, in *Proceedings of the 57th Annual Meeting of the Association for Computational Linguistics, ACL* (2019), pp. 1668–1678, https://doi.org/10.18653/v1/P19-1163
4. T. Balasubramaniam, R. Nayak, M.A. Bashar, Understanding the spatio-temporal topic dynamics of covid-19 using nonnegative tensor factorization: a case study, in *Proceedings of the 2020 IEEE Symposium Series on Computational Intelligence (SSCI). Institute of Electrical and Electronics Engineers Inc., United States of America*, pp. 1218-1225, https://doi.org/10.1109/SSCI47803.2020.9308265
5. A. Obadimu, E. Mead, N. Mead, Identifying latent toxic features on youtube using non-negative matrix factorization, in *The Ninth International Conference on Social Media Technologies,*

Communication, and Informatics: Valencia, Spain, International Academy, Research, and Industry Association (2019), pp. 25–31
6. Z. Ashktorab, "The continuum of harm" taxonomy of cyberbullying mitigation and prevention, in *Online Harassment*. Human–Computer Interaction Series, ed. by J. Golbeck (Springer, Cham, 2018), https://doi.org/10.1007/978-3-319-78583-7_9
7. E. Raisi, B. Huang, Weakly supervised cyberbullying detection with participant-vocabulary consistency. Soc. Netw. Anal. Min. **8**(1), 1–17 (2018). https://doi.org/10.1007/s13278-018-0517-y
8. A. Al-Hassan, H. Al-Dossari, Detection of hate speech in Arabic tweets using deep learning. Multimedia Syst. (2021). https://doi.org/10.1007/s00530-020-00742-w
9. M. Mozafari, R. Farahbakhsh, N. Crespi, Hate speech detection and racial bias mitigation in social media based on BERT model. PloS One **15**(8), e0237861–e0237861 (2020), https://doi.org/10.1371/journal.pone.0237861
10. M. Anzovino, E. Fersini, P. Rosso, Automatic identification and classification of misogynistic language on twitter, in *Natural Language Processing and Information Systems. NLDB 2018*. Lecture Notes in Computer Science, ed. by M. Silberztein, F. Atigui, E. Kornyshova, Métais, E., F. Meziane, vol. 10859 (Springer, Cham, 2018), https://doi.org/10.1007/978-3-319-91947-8_6
11. J. Sekeres, O. Ormandjieva, C. Suen, J. Hamel, Advanced data preprocessing for detecting cybercrime in text-based online interactions, in *Pattern Recognition and Artificial Intelligence. ICPRAI 2020*, ed. by Y. Lu, N. Vincent, P.C. Yuen, W.S. Zheng, F. Cheriet, C.Y. Suen. Lecture Notes in Computer Science, vol. 12068. (Springer, Cham, 2020), https://doi.org/10.1007/978-3-030-59830-3_36
12. P. Badjatiya, S. Gupta, M. Gupta, V. Varma, Deep learning for hate speech detection in tweets (2017). https://doi.org/10.1145/3041021.3054223
13. S. Boberg, L. Frischlich, T. Schatto-Eckrodt, F. Wintterlin, T. Quandt, Between overload and indifference: detection of fake accounts and social bots by community managers, in *Disinformation in Open Online Media. MISDOOM 2019*, ed. by C. Grimme, M. Preuss, F. Takes, A. Waldherr. Lecture Notes in Computer Science, vol. 12021. (Springer, Cham, 2020), https://doi.org/10.1007/978-3-030-39627-5_2
14. S. Cresci, Detecting malicious social bots: story of a never-ending clash, in *Disinformation in Open Online Media. MISDOOM 2019*, ed. by C. Grimme, M. Preuss, F. Takes, A. Waldherr. Lecture Notes in Computer Science, vol. 12021. (Springer, Cham, 2020), https://doi.org/10.1007/978-3-030-39627-5_7 (
15. L. Floridi, M. Chiriatti, GPT-3: its nature, scope, limits, and consequences. Minds Mach. **30**, 681–694 (2020). https://doi.org/10.1007/s11023-020-09548-1
16. J. Vig, Visualizing attention in transformer-based language representation models (2019)
17. C. Hardaker, Social media interventions and the language of political campaigns: from online petitions to platform policy changes, in *Professional Communication. Communicating in Professions and Organizations*, ed. by L. Mullany (Palgrave Macmillan, Cham, 2020), pp. 227–247, https://doi.org/10.1007/978-3-030-41668-3_12
18. M. Naldi, A conversation analysis of interactions in personal finance forums, in *Text Analytics. JADT 2018. Studies in Classification, Data Analysis, and Knowledge Organization*, ed. by D.F. Iezzi, D. Mayaffre, M. Misuraca (Springer, Cham, 2020), https://doi.org/10.1007/978-3-030-52680-1_6
19. L. Mullany, L. Trickett, The language of 'misogyny hate crime': politics, policy and policing, in *Professional Communication. Communicating in Professions and Organizations*, ed. by L. Mullany (Palgrave Macmillan, Cham, 2020), https://doi.org/10.1007/978-3-030-41668-3_13
20. J. Pereira-Kohatsu, L. Quijano-Sánchez, F. Liberatore, M. Camacho-Collados, Detecting and monitoring hate speech in twitter. Sensors (Basel, Switzerland) **19**(21), 4654 (2019). https://doi.org/10.3390/s19214654
21. A. Walker, K. Lyall, D. Silva, G. Craigie, R. Mayshak, B. Costa, S. Hyder, A. Bentley, Male victims of female-perpetrated intimate partner violence, help-seeking, and reporting behaviors: a qualitative study. Psychol. Men Masculinity **21**(2), 213–223 (2020). https://doi.org/10.1037/men0000222

22. N. Ersotelos, M. Bottarelli, H. Al-Khateeb, G. Epiphaniou, Z. Alhaboby, P. Pillai, A. Aggoun, Blockchain and IoMT against Physical Abuse: bullying in schools as a case study. J. Sens. Actuator Netw. **10**(1), 1 (2021). https://doi.org/10.3390/jsan10010001
23. K. Saha, E. Chandrasekharan, M. De Choudhury, Prevalence and psychological effects of hateful speech in online college communities, in *Proceedings of the 10th ACM Conference on Web Science* (2019), pp. 255–264, https://doi.org/10.1145/3292522.3326032
24. B. Haddad, Z. Orabe, A. Al-Abood, N. Ghneim, Arabic offensive language detection with attention-based deep neural networks, in *Language Resources and Evaluation Conference, European Language Resources* (2020), pp. 76–81. https://www.aclweb.org/anthology/2020.osact-1.12.pdf
25. M. Wiegand, M. Siegel, J. Ruppenhofer, Overview of the GermEval 2018 shared task on the identification of offensive language. in *Proceedings of GermEval 2018, 14th Conference on Natural Language Processing (KONVENS 2018), Vienna, Austria September 21, 2018. Vienna, Austria: Austrian Academy of Sciences, 2018* (2018), pp. 1–10
26. Z. Lan, M. Chen, S. Goodman, K. Gimpel, P. Sharma, R. Soricut, ALBERT: a lite BERT for self-supervised learning of language representations (2019), https://arxiv.org/abs/1909.11942v6
27. J. Salminen, S. Sengän, J. Corporan, S. Jung, B. Jansen, Topic-driven toxicity: exploring the relationship between online toxicity and news topics. PloS One **15**(2), e0228723 (2020). https://doi.org/10.1371/journal.pone.0228723
28. A. Workman, E. Kruger, T. Dune, Policing victims of partner violence during COVID-19: a qualitative content study on Australian grey literature. Polic. Soc. 1–21 (2021), https://doi.org/10.1080/10439463.2021.1888951
29. D. Ging, E. Siapera, *Gender Hate Online Understanding the New Anti-Feminism*, 1st edn. (Springer International Publishing, 2019), https://doi.org/10.1007/978-3-319-96226-9
30. F. Ye, C. Chen, Z. Zheng, Deep autoencoder-like nonnegative matrix factorization for community detection, in *Proceedings of the 27th ACM International Conference on Information and Knowledge Management* (2018), pp. 1393–1402, https://doi.org/10.1145/3269206.3271697
31. J. Risch, R. Krestel, Toxic comment detection in online discussions, in *Deep Learning-Based Approaches for Sentiment Analysis. Algorithms for Intelligent Systems*, ed. by B. Agarwal, R. Nayak, N. Mittal, S. Patnaik (Springer, Singapore, 2020), https://doi.org/10.1007/978-981-15-1216-2_4
32. E. Dixon, Automation and harassment detection, in *Online Harassment*. Human–Computer Interaction Series, ed. by J. Golbeck (Springer, Cham, 2018), https://doi.org/10.1007/978-3-319-78583-7_5
33. J. Devlin, M. Chang, K. Lee, K. Toutanova, BERT: pre-training of deep bidirectional transformers for language understanding (2018), https://arxiv.org/pdf/1810.04805.pdf
34. M. Peters, M. Neumann, M. Iyyer, M. Gardner, C. Clark, K. Lee, L. Zettlemoyer, Deep contextualized word representations, in *Proceedings of the 2018 Conference of the North American Chapter of the Association for Computational Linguistics: Human Language Technologies. ACL*, vol. 1 (2018), pp. 2227–2237, https://doi.org/10.18653/v1/N18-1202
35. C. Aggarwal, C. Zhai, *Mining Text Data*, 1st edn. (Springer, New York, 2012). https://doi.org/10.1007/978-1-4614-3223-4
36. I. El-Khair, Term weighting, in *Encyclopedia of Database Systems*, ed. by L. LIU, M. ÖZSU (Springer, Boston, MA, 2009), https://doi.org/10.1007/978-0-387-39940-9_943
37. A. Zimek (ed.), *Clustering High-Dimensional Data in Data Clustering* (Chapman and Hall/CRC, 2019), pp. 201–230
38. Purude University, Predictive modeling & machine learning laboratory (2016)
39. A. Egg, Locality-sensitive hashing (LSH) (2017)
40. I. Kwok, Y. Wang, Locate the hate: detecting tweets against blacks, in *Twenty-Seventh AAAI Conference on Artificial Intelligence* (2013), pp. 1621–1622. https://dl.acm.org/doi/10.5555/2891460.2891697
41. M. Molina-González, F. Plaza-del Arco, M. Martïn-Valdivia, L. Ureña López, Ensemble learning to detect aggressiveness in mexican spanish tweets, in *Proceedings of the First Workshop for Iberian Languages Evaluation Forum (IberLEF 2019), CEUR WS Proceedings* (2019), pp. 495–501. http://ceur-ws.org/Vol-2421/MEX-A3T_paper_1

42. Y. Li, A. Algarni, N. Zhong, Mining positive and negative patterns for relevance feature discovery, in *Proceedings of the 16th ACM SIGKDD International Conference on Knowledge Discovery and Data Mining* (ACM, Washington, 2010), pp.753–762, https://doi.org/10.1145/1835804.1835900
43. L. Silva, M. Mondal, D. Correa, F. Benevenuto, I. Weber, Analyzing the targets of hate in online social media, in *Tenth International AAAI Conference on Web and Social Media* (2016), https://arxiv.org/pdf/1603.07709.pdf
44. G. Kovács, P. Alonso, R. Saini Challenges of hate speech detection in social media: data scarcity, and leveraging external resources. SN Comput. Sci. **2**(2), (2021), https://doi.org/10.1007/s42979-021-00457-3
45. W. Mohotti, R. Nayak, Efficient outlier detection in text corpus using rare frequency and ranking. ACM Trans. Knowl. Discov. Data **14**(6) (2020), https://doi.org/10.1145/3399712
46. D. Schabus, M. Skowron, M. Trapp, One million posts: a data set of german online discussions, in *Proceedings of SIGIR '17, August 07-11* (2017), pp. 1241–1244, https://doi.org/10.1145/3077136.3080711
47. Z. Zhang, L. Luo, Hate speech detection: a solved problem? *The Challenging Case of Long Tail on Twitter* (2018)
48. W. Wang, L. Chen, K. Thirunarayan, A. Sheth, Cursing in english on twitter, in *Proceedings of the 17th ACM Conference on Computer Supported Cooperative Work & Social Computing* (ACM, 2014), pp. 415–425
49. S. MacAvaney, H. Yao, E. Yang, K. Russell, N. Goharian, O. Frieder, Hate speech detection: challenges and solutions. PloS One **14**(8), e0221152–e0221152 (2019). https://doi.org/10.1371/journal.pone.0221152
50. O. Makhnytkina, A. Matveev, D. Bogoradnikova, I. Lizunova, A. Maltseva, N. Shilkina, Detection of toxic language in short text messages, in *Speech and Computer SPECOM 2020*, ed. by A. Karpov, R. Potapova. Lecture Notes in Computer Science, vol. 12335. (Springer, Cham, 2020), https://doi.org/10.1007/978-3-030-60276-5_31
51. L. Xie, X. Zhang, Gate-fusion transformer for multimodal sentiment analysis, in *Pattern Recognition and Artificial Intelligence. ICPRAI 2020*, ed. by Y. Lu, N. Vincent, P.C. Yuen, W.S. Zheng, F. Cheriet, C.Y Suen. Lecture Notes in Computer Science, 12068. Springer, Cham, 2020), https://doi.org/10.1007/978-3-030-59830-3_3
52. A. D'Sa, I. Illina, D. Fohr, Towards non-toxic landscapes: automatic toxic comment detection using DNN (2019), pp. 21–25, https://arxiv.org/ftp/arxiv/papers/1911/1911.08395.pdf
53. J. Risch R. Krestel, *Aggression Identification Using Deep Learning and Data Augmentation, ACL* (2018), pp. 150–158, https://www.aclweb.org/anthology/W18-4418
54. M.A. Bashar, R. Nayak, N. Suzor, Regularising LSTM classifier by transfer learning for detecting misogynistic tweets with small training set. Knowl. Inf. Syst. **62**(10), 4029–4054 (2020). https://doi.org/10.1007/s10115-020-01481-0
55. E. Pamungkas, V. Basile, V. Patti, Misogyny detection in twitter: a multilingual and cross-domain study. Inf. Process. Manag. **57**(6), 102360 (2020). https://doi.org/10.1016/j.ipm.2020.102360
56. S. Zimmerman, C. Fox, U. Krushwitz, Improving hate speech detection with deep learning ensembles (2018)
57. W. Dai, T. Yu, Z. Liu, P. Fung, Kungfupanda at SemEval-2020 Task 12: BERT-based multi-task, learning for offensive language detection, https://arxiv.org/abs/2004.13432
58. T. Davidson, D. Warmsley, M. Macy, I. Weber, Automated hate speech detection and the problem of offensive language (2017), https://arxiv.org/abs/1703.04009
59. G. Xiang, B. Fan, L. Wang, J. Hong, C. Rose, Detecting offensive tweets via topical feature discovery over a large scale twitter corpus, in *Proceedings of the 21st ACM International Conference on Information and Knowledge Management* (ACM, 2012), pp. 1980–1984
60. M.A. Bashar, R. Nayak, QutNocturnal@HASOC'19: CNN for hate speech and offensive content identification in Hindi language, in *Working Notes of FIRE 2019 - Forum for Information Retrieval Evaluation*, vol. 2517, ed. by P. Mehta, P. Rosso, P. Majumder, M. Mitra (Sun SITE Central Europe, Germany, 2019), pp. 237–245

61. Y. Kim, Convolutional neural networks for sentence classification, in *Proceedings of the 2014 Conference on Empirical Methods in Natural Language Processing (EMNLP)* (2014), pp. 1746–1751. https://arxiv.org/pdf/1408.5882.pdf
62. W. Wang, B. Bi, M. Yan, C. Wu, Z. Bao, J. Xia, L. Peng, L. Si, StructBERT: incorporating language structures into pre-training for deep language understanding (2019)
63. D. Gordeev, V. Potapov, Toxicity in texts and images on the internet, in *Speech and Computer. SPECOM 2020*, ed. by A. Karpov, R. Potapova. Lecture Notes in Computer Science, vol. 12335 (Springer, Cham, 2020), pp. 156–165, https://doi.org/10.1007/978-3-030-60276-5_16
64. N. Reimers, I. Gurevych, Sentence-BERT: sentence embeddings using siamese BERT-networks (2019), https://arxiv.org/pdf/1908.10084.pdf
65. V. Sinh, N. Minh, A study on self-attention mechanism for AMR-to-text generation, in *Natural Language Processing and Information Systems. NLDB 2019*, ed. by E. Métais, F. Meziane, S. Vadera, V. Sugumaran, M. Saraee. Lecture Notes in Computer Science, vol. 11608. (Springer, Cham, 2019), https://doi.org/10.1007/978-3-030-23281-8_27
66. T. Wullach, A. Adler, E. Minkov, Towards hate speech detection at large via deep generative modeling. IEEE Int. Comput. (2020). https://doi.org/10.1109/MIC.2020.3033161
67. T. Wolf, V. Sanh, J. Chaumond, C. Delangue, TransferTransfo: a transfer learning approach for neural network based conversational agents (2019)
68. M. Mozafari, R. Farahbakhsh, N. Crespi, A BERT-based transfer learning approach for hate speech detection in online social media (2019), https://arxiv.org/pdf/1910.12574.pdf
69. S. Swamy, A. Jamatia, B. Gambäck, Studying generalisability across abusive language detection datasets, in *Proceedings of the 23rd Conference on Computational Natural Language Learning (CoNLL),Association for Computational Linguistics* (2019), pp 940–950, https://doi.org/10.18653/v1/K19-1088
70. A. Vaswani, N. Shazeer, N. Parmar, J. Uszkoreit, L. Jones, A. Gomez, L. Kaiser, I. Polosukhin, Attention is all you need (2017)
71. A. Koratana, K. Hu, Toxic speech detection, in *32nd Conference on Neural Information Processing Systems* (2018)
72. K. Clark, U. Khandelwal, O. Levy, C. Manning, What does BERT look at? An analysis of BERT's attention (2019), https://arxiv.org/abs/1906.04341
73. R. Cao, R. Lee, HateGAN: adversarial generative-based data augmentation for hate speech detection, in *Proceedings of the 28th International Conference on Computational Linguistics* (2020), pp. 6327–6338. https://doi.org/10.18653/v1/2020.coling-main.557
74. S. Minaee, N. Kalchbrenner, E. Cambria, N. Nikzad, M. Chenaghlu, J. Gao, Deep learning based text classification: a comprehensive review (2020), https://arxiv.org/pdf/2004.03705.pdf
75. M.A. Bashar, R. Nayak, TAnoGAN: time series anomaly detection with generative adversarial networks, in *Proceedings of the 2020 IEEE Symposium Series on Computational Intelligence (SSCI). Institute of Electrical and Electronics Engineers Inc., United States of America* (2020), pp. 1778–1785, https://doi.org/10.1109/SSCI47803.2020.9308512
76. J. Chen, S. Yan, K.C. Wong, Verbal aggression detection on Twitter comments: convolutional neural network for short-text sentiment analysis. Neural Comput. Appl. **32**, 10809–10818 (2020). https://doi.org/10.1007/s00521-018-3442-0
77. M.A. Bashar, R. Nayak, K. Luong, T. Balasubramaniam, Progressive domain adaptation for detecting hate speech on social media with small training set and its application to COVID-19 concerned posts. Soc. Netw. Anal. Min. **11**, 69 (2021). https://doi.org/10.1007/s13278-021-00780-w
78. S. Ghosh, A. Mondal, K. Singh, J. Maiti, P. Mitra, Potential threat detection from industrial accident reports using text mining, in *Intelligent Computing and Communication. ICICC 2019. Advances in Intelligent Systems and Computing*, vol. 1034 (Springer, Singapore, 2020), pp. 109–123, https://doi.org/10.1007/978-981-15-1084-7_12
79. S. Aghazadeh, A. Burns, J. Chu, H. Feigenblatt, E. Laribee, L. Maynard, A. Meyers, J. O'Brien, L. Rufus, GamerGate: a case study in online harassment, in *Online Harassment. Human–Computer Interaction Series*, ed. by J. Golbeck (Springer, Cham. 2018), https://doi.org/10.1007/978-3-319-78583-7_8

80. N. Harriman, N. Shortland, M. Su, T. Cote, M. Testa, E. Savoia, Youth exposure to hate in the online space: an exploratory analysis. Int. J. Environ. Res. Public Health **17**(22), 1–14 (2020). https://doi.org/10.3390/ijerph17228531
81. A. Lytos, T. Lagkas, P. Sarigiannidis, K. Bontcheva, The evolution of argumentation mining: from models to social media and emerging tools. Inf. Process. Manag. **56**(6), 102055 (2019). https://doi.org/10.1016/j.ipm.2019.10205
82. C. Blaya, Cyberhate: a review and content analysis of intervention strategies. Aggress. Violent Behav. **45**, 163–172 (2019). https://doi.org/10.1016/j.avb.2018.05.006
83. S. Dowlagar, R. Mamidi, HASOCOne@FIRE-HASOC2020: Using BERT and multilingual BERT models for hate speech detection (2021), https://arxiv.org/pdf/2101.09007.pdf
84. M. Bashar, R. Nayak, N. Suzor, B. Weir, Misogynistic tweet detection: modelling cnn with small datasets (2020). https://doi.org/10.1007/978-981-13-6661-1_1
85. M. Bashar, R. Nayak, Active learning for effectively fine-tuning transfer learning to downstream task. ACM Trans. Intell. Syst. Technol. **12**(2), 1–24 (2021), https://doi.org/10.1145/3446343
86. A. de los Riscos, L. D'Haro, ToxicBot: a conversational agent to fight online hate speech, in *Conversational dialogue systems for the next decade*, ed. by L.F. D'Haro, Z. Callejas, S. Nakamura. Lecture Notes in Electrical Engineering, vol. 704. (Springer, Singapore, 2021), https://doi.org/10.1007/978-981-15-8395-7_2
87. J. Salminen, M. Hopf, S. Chowdhury, S. Jung, H. Almerekhi, B. Jansen, Developing an online hate classifier for multiple social media platforms. Hum.-Centric Comput. Inf. Sci. **10**(1), 1–34 (2020), https://doi.org/10.1186/s13673-019-0205-6
88. T. Balasubramaniam, R. Nayak, K. Luong, M.A. Bashar, Identifying covid-19 misinformation tweets and learning their spatio-temporal topic dynamics using nonnegative coupled matrix tensor factorization. Soc. Netw. Anal. Min. **11**(1), 57 (2021). https://doi.org/10.1007/s13278-021-00767-7

Richi Nayak is Leader of the Applied Data Science Program at the Centre of Data Science and Professor at Queensland University of Technology, Brisbane Australia. She has a driving passion to address pressing societal problems by innovating Artificial Intelligence field underpinned by fundamental research in machine learning. Her research has resulted in the development of novel solutions to address industry-specific problems in Marketing, K-12 Education, Agriculture, Digital humanities, and Mining. She has made multiple advances in social media mining, deep neural networks, multi-view learning, matrix/tensor factorization, clustering and recommender systems. She has authored over 180 high-quality refereed publications and has h-index of 30 currently. Her research leadership is recognised by multiple best paper awards and nominations at international conferences, QUT Postgraduate Research Supervision awards, and the 2016 Women in Technology (WiT) Infotech Outstanding Achievement Award in Australia. She holds a PhD in Computer Science from the Queensland University of Technology and a Masters in Engineering from IIT Roorkee.

Chapter 10
Toward Artifical Intelligence Tools for Solving the Real World Problems: Effective Hybrid Genetic Algorithms Proposal

Jouhaina Chaouachi and Olfa Harrabi

Abstract In this chapter, we encounter the resolution of real world problems using evolutionary computing methods. This chapter demonstrates how the design of an Interactive Decision Support System (IDSS) integrating a hybrid genetic algorithm during the optimization process is a performing approach to provide optimal solutions. This approach is applied to two NP-Hard problems: the University Course Timetabling problem (UCT) and the Solid Waste collection Problem (SWP). To prove the practical interest of our work, the case of Business School of Carthage (IHEC) with a huge number of courses and teachers was treated along to solve the University Timetabling problem. Moreover, the case of the city of Sidi Bousaid is treated along during our experimental study to solve the Solid Waste collection Problem. The experimental results demonstrate the performance of our proposed system compared to the manual process on both solution quality and computation time effort.

10.1 Introduction

Artificial Intelligence (AI) is used to implement innovative solutions that cannot be achieved using traditional approaches. Many approaches are used in Artificial Intelligence, including mathematical optimization, evolutionary computing methods, artificial neural networks, and methods based on statistics, probability and economics. Genetic algorithms are categorized as evolutionary computing methods that rapidly attract a growing interest in Artificial Intelligence area. A key motivating this

J. Chaouachi (✉)
Institute of Advanced Business Studies of Carthage, Carthage University,
IHEC Carthage Presidency-2017, Tunis, Tunisia

O. Harrabi
Higher Institute of Management of Tunis, Tunis University,
41, Liberty Street-Bouchoucha - 2000, Bardo, Tunisia

M. Virvou et al. (eds.), *Advances in Selected Artificial Intelligence Areas*,
Learning and Analytics in Intelligent Systems 24,
https://doi.org/10.1007/978-3-030-93052-3_10

trend is not only its theoretical interest, but also its practical interest to encounter the resolution of many real world applications in both scientific and engineering areas. In this chapter, we encounter the resolution of some real world problems using Genetic Algorithms. We demonstrate how a Genetic Algorithm integrated in an Interactive Decision Support System (IDSS) is a performing approach to help the Decision-Maker in his decision-making task. Our approach is applied to two difficult problems: the University Course Timetabling problem (UCT) and the Solid Waste collection Problem (SWP). In a first section, this chapter tackles some of the recent challenge using hybrid evolutionary approach proposal to solve the "University Course Timetabling problem" (UCT). UCT is an ongoing challenge that most of educational institutions face when scheduling courses. The problem assigns lectures to specific numbers of time slots and rooms while including several conflicting constraints into account. Hence, effective decision making is strongly required to provide the timetablers a useful toolkit. From a practical aspect, we proposed the design of an Interactive Decision Support System (IDSS) that employs a new hybrid genetic algorithm during the optimization process to provide a best solution satisfying at most teachers' preferences. The elaborated evolutionary approach combines a genetic algorithm with an iterated simulated annealing local search technique. One important feature of our simulated annealing is the use of the destroy/repair method as neighborhood structure. Moreover, our proposed hybrid genetic algorithm relies on the use of a dedicated crossover operators to generate offspring solutions. To prove the practical interest of our work, the case of Business School of Carthage (IHEC) with a huge number of courses and teachers was treated along.

The second section of this chapter addresses the methodological aspects of evolutionary methods by the design of a parameter tuning procedure to cleverly detect the most promising genetic operators and parameter values. Indeed, genetic algorithms are categorized as efficient global search heuristic in solving high computational problems. However, their performance is heavily dependent on the choice of the appropriate operator especially crossover. Moreover, the crossover operator efficiency could considerably be influenced by the problem definition, the fitness function and the instance structure. For the practical demonstration of these phenomena, we tackle a real world application "the Solid Waste collection Problem" (SWP). We focus therefore on determining the most promising set of parameters by varying the performance measures. The case of the city of Sidi Bou Said is solved in our experimental study.

The remainder of this chapter is organized as follows. Section 10.2 briefly presents the university course timetabling problem with the corresponding literature review. In the same section, we introduce the problem model and the proposed mathematical formulation. The architecture of the proposed computer-based system with the different modules is therefore described in detail. We finally expose the computational results and the capabilities of the developed system in handling real case studies. In Sect. 10.3, we define the SWM, then, we describe in details our proposed solution approach. In the same section, we show the computational results when handling a real case study: the municipality of Sidi Bou Said. Finally, some conclusions are reported in Sect. 10.4.

10.2 University Course Timetabling UCT

The University Course Timetabling problem deals with the design of timetables that satisfy educational organization needs. It is an important task but an ongoing challenge that the involved staff face.

10.2.1 Problem Statement and Preliminary Definitions

There are some preliminary definitions that must be outlined to be familiarized with the university environment. The term *Events* means a set of scheduled activities (exams, lab sessions, lectures, etc.). The *time slots* are designated to schedule the events. The *resource* are mainly used to provide all the events' requirements. They are classified into human resources and technical resources. We mean by *constraints* the rules to be respected during the events scheduling in order deliver feasible timetables. The *agents* are the persons who attend the events such as faculty members, students, invigilators, etc. The *conflict* must be avoided with respect to some important constraints in order to deliver feasible timetables. Finally, the *class* are a set of students belonging to the same curricula program.

We rationally propose therefore a general definition that captures the context of the university course timetabling problem as follows. The UCT is the scheduling of a given events over a weekly time slots in accordance with a strictly respected constraints.

10.2.2 Related Works

The university course timetabling problem has been approached using several operational research methods given its complexity [1]. In this context, several surveys detailed descriptions of the proposed approaches to solve the university course timetabling problem starting from [2] to the recent [3].

In this chapter, we are interested by pointing out research works proposing graph theory optimization techniques, integer programming models and computer-based systems to solve the UCT problem.

In the early decade of sixties, the first reference considering the graph coloring methods for the resolution of the UCT referred back to the work of [4] where the authors proved that the problems could be analogized. More than 15 years later, the research study of [5, 6] maintain the interest of automating the university course timetabling process using the graph tool with a least number of colors. Slightly different, to construct a feasible timetable and reduce the chromatic number of graph, [7] has invoked the split vertices idea. In the same context, [8] observed that a two part graph coloring could be an effective tool for the resolution of university timetabling problem.

Besides graph tool approaches, a part of researches focused on proposing Integer Programming (IP) models. Firstly, in [9] authors constructed desirable timetables using the IP as effective resolution methods. Later, a linear programming model is proposed in [10] to automate the design of the university timetables. Then, in [11], authors attempted a solution proposal using mathematical programming. From a different point of view, the work in [12, 13] focused on decomposing the UCT into two sub-problems: the classroom assignments problem and the faculty-course assignments and proposed IP formulations as a strategy for solving these sub-problems. Slightly different, authors in [14] have proposed a relaxation for their IP model using a heuristic approach. Then, in the same context, in [15], the process of assigning the classes to professors was modeled using a mixed integer program and an optimal solution was generated with he help of a branch and bound algorithm. Later, [16] focus on developing a two-stage relaxation procedure to solve an integer programming formulation and guaranty the quality of the timetables. Last but not least, to automate the problem resolution, [17] proposed a 0-1 IP model whilst implementing constraints related to classroom consistency.

A recent practice to deliver feasible course timetables uses computer-based systems in order to support course-scheduling process. These research works employed usually linear programming techniques during the solution optimization. In this regard, let us introduce the most representative research work including mainly: [18–24].

10.2.3 Problem Modelization and Mathematical Formulation

In this section, we model the UCT problem using the graph tool where each lecture of courses is represented by a graph vertice and the occurred conflicts by edges. In other words, adjacent nodes are lectures sharing a number of common students or instructors. A proper coloring of the graph G using a minimal number of colors k corresponds therefore to a feasible course schedule in k time slots: each node receives exactly one color such as all the adjacent nodes are colored differently. Adversely, the course schedule is non feasible and the involved events are conflicting. An illustrative example is provided in Fig. 10.1 to show how can the graph model solve the course timetabling problem.

In what follows, we formulate the UCT problem given an graph coloring-based integer linear programming formulation. The proposed model adapted the notion of weights for the different colors that express the preferences of the teachers. Accordingly, a color k has two different labels: 0 if the section schedule is preferred; 1 otherwise. We list bellow the necessary notation to explain our proposed model:

	Physics	Systems Programming	Database Systems	Information Management Systems	Web Tools
Physics		✓	✓	✓	
Systems Programming	✓		✓	✓	✓
Database Systems	✓	✓		✓	
Information Management Systems	✓	✓	✓		✓
Web Tools		✓		✓	

(a) Courses distributions

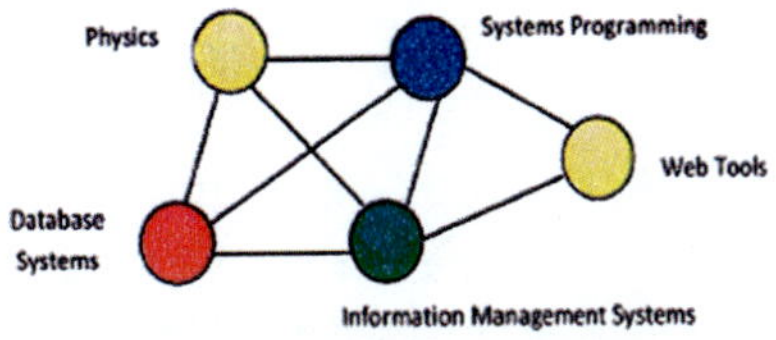

(b) A 4-coloring of the course conflict graph

Fig. 10.1 Relationship between the graph coloring problem and the course timetabling problem

Parameters

Let

- A is a set of all the students' groups
- S is a set sections (lecture of courses attended by only a group of students)
- E_m is a set of sections in conflict (sharing common students of instructors); the mth row of the conflict matrix
- M is a set of values of m
- F is a set of instructors
- T is a set of time slots
- P_f is a set of time slots preferred by the instructor f
- C_{it} is the cost of time slot t if it is assigned to the section i
- C_{it} $\begin{cases} 0 \text{ if the time slot } t \text{ belongs to the set of preferred schedules for section } i \\ (i.e. t \in P), \forall t \in T, \forall i \in S \\ 1 \text{ otherwise} \end{cases}$

Decision variables

- x_{it} $\begin{cases} 1 \text{ if the section } i \text{ is assigned to a time slot } t, \\ \forall t \in T, \forall i \in S \\ 0 \text{ otherwise} \end{cases}$

The model

Using the foregoing notations, the final specification of the proposed ILP model takes the following form:

$$Minimize \sum_{i \in S} \sum_{t \in T} c_{it} x_{it} \tag{10.1}$$

$$subject\ to: \sum_{t \in T} x_{it} = 1 \quad \forall i \in S \tag{10.2}$$

$$x_{it} + x_{jt} \leq 1 \quad \forall (i, j) \in E, \forall t \in T \tag{10.3}$$

$$x_{it} \in \{0, 1\} \quad \forall t \in T, \forall i \in S \tag{10.4}$$

The objective function (10.1) is a cost function that needs to be minimized. It defines the preferences of scheduling a section i in a time slot t in term of cost C_{it}. In our context, we assign each section to some time slots in $\{1, \ldots, t\}$ by affecting a cost value 0 if the scheduling is preferred; 1 otherwise. Constraints (10.2) ensure that each section is taught once (i.e. it must be assigned to only one time slot). Constraints (10.3) force a conflicting sections pair (i, j) to be assigned to different time slots. Finally, constraints (10.4) state that the decision variables x_{it} is binary-valued.

10.2.4 An Interactive Decision Support System (IDSS) for the UCT Problem

In this section, we propose an Interactive Decision Support System (IDSS) to solve the UCT problem. The proposed IDSS provides the timetablers a useful toolkit to result not only feasible timetables, but also treats the preferences of teachers as a priority. The main architecture of our IDSS consists on:

- **Input Information Module**: It organizes the manipulated data into structured database files and tables. Throughout the Schedule table, the user has the opportunity to easily handle and modify all the necessary data related to a resulting timetable.
- **Control System Module**: This module offers the user an interface to check and adjust the system for information managing. Throughout this interface, the IDSS system collects the needed data in a form readily amenable to analysis: questionnaire.
- **Optimization Module**: The UCT problem was treated along during the optimization process using a hybrid evolutionary approaches that combines the use of a Genetic Algorithm (GA) and an Iterated Simulated Annealing approach (ISA). The main reason for the interest of GA techniques is their ability to generate different solutions in order to better explore the search solution space. The general routine of the proposed hybrid evolutionary approach is summerized in Algorithm 1. This evolutionary search aims to minimize the sum of colors in a set of k-colorings. For this purpose, GA starts with an initial population randomly generated. Then, until a maximum number of generations is reached, GA performs different evolutionary steps. Firstly, two solutions are selected using the binary tournament selection. Then, the proposed genetic algorithm relies on the Sum Partition Crossover operator (SPX) [25] to improve the solution by exchanging information contained in the current selected parents. Mutation operator is thereafter used to improve the obtained offspring resulted by the crossover step. In our context, we adapt The ISA approach to improve the obtained solution since it acquires a high capability

Algorithm 1 Pseudo-code of the Hybrid Genetic Algorithm for the UCT

```
Input:A graph G, population of solutions (Pop)
Output: The best found solution (Best-Indiv)
Begin
While a maximum number of iterations is not reached do
  For each Individual in the population do
    (parent1, parent2) <— Selection (Pop);
    Offspring <— Crossover (parent1, parent2);
     Offspring' <— Mutation ISA (Offspring);
     Pop <— Replacement (Offspring', Pop);
  End for
End while
Best-Indiv <— (Best-Individual (Pop));
Return Best-Indiv
End
```

on intensifying the search process. As neighboring operators, the elaborated ISA performed two neighborhood structures the [Destruction /construction] method and a Swap operator. Finally, GA inserts the new solution during the population updating mechanism.

- **Report Module**: This module receives the solution results and produces the course timetables for each class and the teachers' timetables scheduling.

In Fig. 10.2, we display the designed IDSS architecture.

10.2.5 Empirical Testing

The system was implemented on a i3 processor with 4 Gb of available memory. The evolutionary approach was coded using C++.

10.2.5.1 Benchmark Instance

The collected data are extracted from the winter semester of the *Advanced Business Studies of Carthage (IHEC)* institute. IHEC of Carthage offers about 20 undergraduate programs (16 masters degree and 6 bachelor degrees). An undergraduate study involves at least six semesters. The institute includes mainly six departments with different educational streams coached by several instructors. When refering to the curriculum program, one notes that there are about 365 courses which must be scheduled in the winter semester of the IHEC Business School (it may comprise of lectures, lectures with tutorials, lectures with lab works and/or integrated courses).

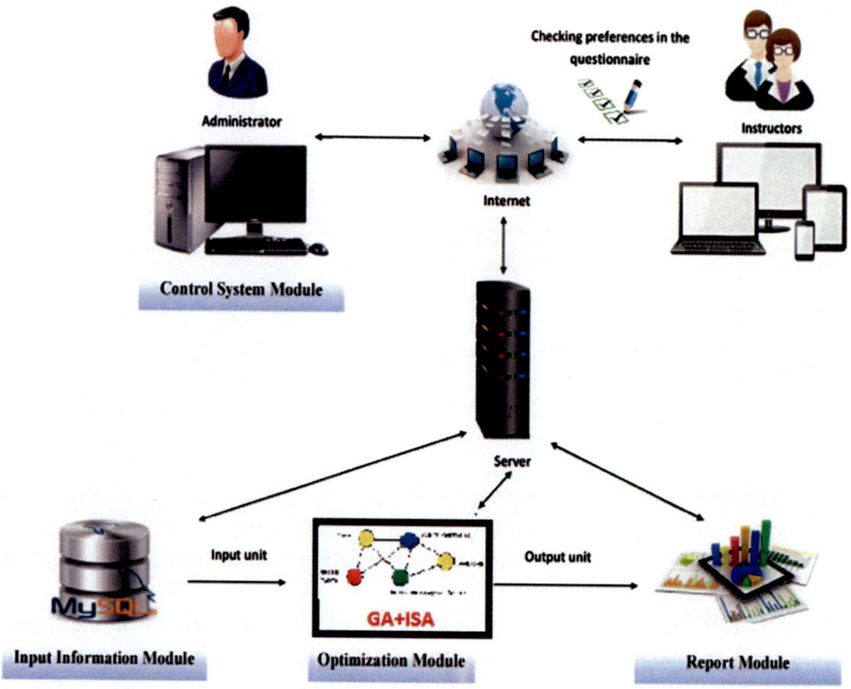

Fig. 10.2 Architecture of the proposed IDSS for the UCT problem

10.2.6 Evaluation and Results

To evaluate the efficiency of the *IDSS* system, we provide a comparison with a manual process method considered to solve the same instances. In order to appraise the solution quality, we refer to the following performance indicators (see Table 10.1).

Where the satisfaction rate SR calculated as follows:

$$SR = \sum_{i=1}^{N} \left(\frac{R_i}{P_i} \right) / N * 100 \tag{10.5}$$

With:

- N: Total number of instructors.
- P_i: Number of time slots preferred by the instructor i.
- R_i: Number of assigned sections satisfying the preference of the instructor i.

Table 10.2 reports the comparison of computational performance relative to the IDSS system and the manual process

Table 10.1 Timetabling performance indicators

Indicator	Description
Time	The total time for the generation of the course timetables
Cost	The cost of assigning a section to a non-preferred time slot
#Success	The number of times when a section meets the corresponding preferences
SR	The satisfaction rate

Table 10.2 Comparison of computational performances: manual process versus *IDSS system*

	Semester I	
	Manual process	*IDSS system*
Time	6 weeks	1 s
Cost	1119	40
#Success	105	1192
SR	5.94%	98.03%

The experiments show that the IDSS system clearly outperforms the manual process and provides remarkable improvements which lead us to draw these observations:

- The total time required to deliver the optimal timetable was dramatically cut from 6 weeks to one second. At this level, it worthy to note that, besides boosting in satisfying the instructors' preferences, the *IDSS* system successfully delivers feasible solutions.
- Regarding the *cost* indicator, IDSS system leads to a reduction in failing to meet the instructor's preferences. Indeed, it results only 40 non-preferred section schedules compared to 1119 for the manual process.
- More improvements are noted for the *#success* indicator where the IDSS system exhibits a good performance since it delivers 1192 preferred assignments compared to 105 for the manual process.
- As a result, the satisfaction rate *SR* has been improved when integrating the IDSS system and was ranging from 5.94% to 98.03%.

Overall, the experimental results show that supporting the timetable process through a computer-based system turned out to be an efficient strategy since it achieves both feasible and satisfactory course schedules.

10.3 Solid Waste Management Problem

In the most cites, the Solid Waste Management problem (SWM) shows a wealth of pertinence due to its increasing impact on the environmental issues. In municipalities, the solid waste system deals with managing the waste from its source of generation until its final disposal. In other words, the system includes the following necessary operations during the transformation steps [26]:

(1) Involving generation,
(2) Source-separation,
(3) Storage,
(4) Collection,
(5) Transfer and transport,
(6) Processing and recovery,
(7) Disposal

This research work deals with the collection step and focuses on minimizing the distance traveled by the municipality vehicles as a fitness function.

10.3.1 Related Works

Due to the interest of the SWM problem, many resolution approaches including mainly *approximate methods* [27–30], *exact approaches* [31–33], *data communication technologies* [34–37], *data acquisition technologies* [38–40] and *spatial technologies* [41–44] have been proposed. For an excellent overview of the solid waste management problem, interested readers could refer to [45–47]. Figure 10.1 shows a general-view of the aforementioned methods (Fig. 10.3).

We briefly detail in what follows some of techniques used to solve the SWM problem: exact methods and approximate approaches. Due to the hardness of the

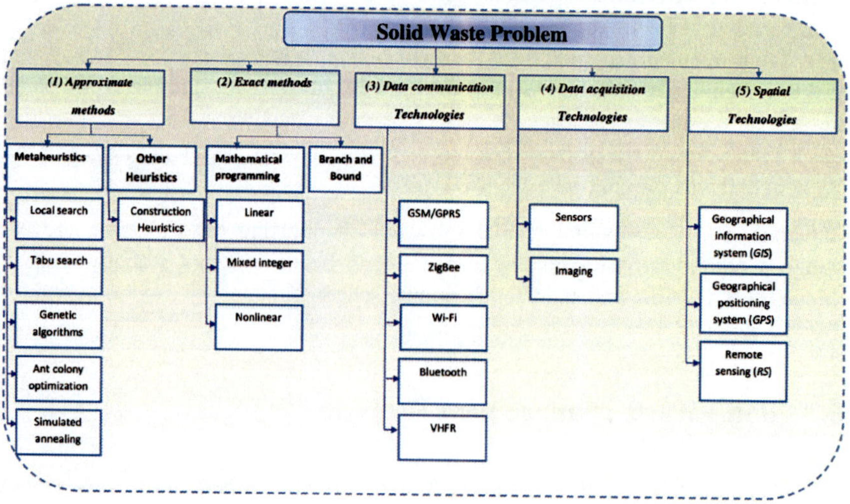

Fig. 10.3 Different methods to solve SWM problem

problem [48], there are restricted works attempting to solve the SWM exactly. These works include mainly an integer linear programming method, a mixed integer linear programming method, an integer non linear method and the branch and bound approach. An early mixed integer linear programming model was proposed in [49] to solve the facility location problem while minimizing the cost flows. In the same context, an additional mixed integer model combined with an improvement procedure was proposed in [50] in order to tackle large problem instance with up to 2,000 arcs and 40 vehicles. Concerning nonlinear programming models, authors attempt in [33] to reduce the handling cost of processing garbage using a nonlinear objective function. Next, authors in [51] developed a heuristic to strengthen their proposed linear programming model that attempts mainly on minimizing the cost routing. Then, a new linear programming model was proposed in [52] to fulfill the orders with a minimal cost possible. An alternative of exact method consists on proposing the branch and bound algorithm. In this context, one notes the research work in [53] that mainly focuses on solving the collection routing problem. To experimentally assess the efficiency of this approach, authors have treated along a real case of study with up to 160 customers.

At this level, it is worth to note that computational performance of these method was not competitive on rather easily instance since additional heuristic are usually performed to handle harder benchmarks. For the practical resolution of the solid waste management problem, many approximate approaches were developed. These approaches belong mainly to two classes: construction heuristics and metaheuristics. Construction heuristics try to build iteratively the feasible solutions from scratch by making, in each step, the most favorable choice for a decision variable. Each choice depends on the decisions made in the previous steps. Such choices can thus be viewed as local decision rules that generally lead to sub-optimal solutions. There are few references citing this class of heuristics in the context of the solid waste management problem [30, 54, 55]. Concerning metaheuristic approaches, one notes a local search algorithm [56] that mainly performs a neighborhood operator consisting on modifying two selected customers in a tour. In the same paper, authors investigate a tabu search approach that explores the search space of the collection vehicles problem by means of neighborhood moves. The approach uses a temporary memorization structure "tabu list" to escape local optima during a number of iterations equal to the "tabu list" size. The tabu search method outperforms recent local search algorithms in term of solution cost. To tackle the SWM, researchers investigate an additional category of metaheuristic approaches "genetic algorithms" that were frequently used as powerful evolutionary methods. By the use of different variation operators, genetic algorithm focus on incrementally improving the solution quality. Indeed, given a population of solutions, this approach employs the selection, crossover, mutation and replacement operators during the search process. In the context of the SWM, different objective function were treated along using genetic algorithms: minimizing the distance traveled by the collection vehicles [57], minimizing the cost travels of the collection vehicles [58] and dealing with the uncertainty encountered during waste planning process [59]. Additionally, the SWM problem was tackled with the help of ant colony optimization. It is a swarm intelligent based meta-heuristic in

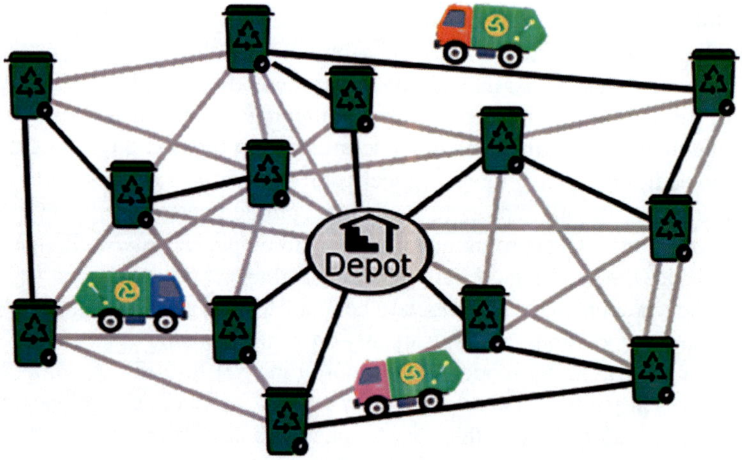

Fig. 10.4 Modelling the SWP using the CVRP

which researches study the behavior patterns of social insects (bees, ants, etc.) in order to simulate processes. Generally, two optimization objectives of SWM were addressed when adapting this resolution method: minimizing the collection time [60] and minimizing the collection traveled distance [28]. Slightly different, researchers addressed another practical metaheuristic method "simulated annealing" to automate the bins location [61] and to minimize the collection routing [30]. It is a local search technique that operates with modified random moves to intensify the search process and escape from getting trapped into local minima.

10.3.2 The Mathematical Formulation Model

Formally, the SWM problem could be modeled using the CVRP where the potential aim focuses on collecting the waste from a set of collection nodes (bins) as illustrated in Fig. 10.4. In this context, a homogeneous fleet of vehicles with fixed capacity should start from a depot and retain to it. During this process, the being challenging tasks consist on reducing the distance traveled by these vehicles.

We list bellow same necessary sets and notions to formulate the SWM problem given an integer linear programming model of the CVRP.

- V : is a set of vertices,
- E : is a set of edges,
- N : is a set of bins,
- K : is a set of vehicles,
- Q : is the maximum capacity of vehicles,
- c_i: is the waste quantity inside each bin i,

- $d_{i,j}$: is the traveled distance from bin i to bin j ,where $i \neq j$
- $q_{i,j}$: is the load of vehicle k while traversing the arc (i, j),
- $x_{i,j}^k = \begin{cases} 1 \text{ if vehicle } k \text{ traverses arc } (i, j) \\ 0 \text{ otherwise} \end{cases}$
- $y_i^k = \begin{cases} 1 \text{ if the bin } i \text{ belongs to the route of the vehicle } k \\ 0 \text{ otherwise} \end{cases}$

Hence, the specification of the mathematical formulation is as follows:

$$Minimize \sum_{i=0}^{n}\sum_{j=0}^{n}\sum_{k=1}^{m} X_{i,j}^k d_{i,j} \tag{10.6}$$

To make the CVRP model realistic, the following constraints must be respected:

$$\sum_{j=0}^{n} X_{0,j} = 1 \quad \forall k = \{1, 2, \ldots, m\} \tag{10.7}$$

$$\sum_{j=1}^{n} q_{0,j}^k = 0 \quad \forall k = \{1, 2, \ldots, m\} \tag{10.8}$$

$$\sum_{j=0}^{n}\sum_{k=1}^{m} X_{i,j}^k = 1 \quad \forall i = \{0, 1, \ldots, n\} \tag{10.9}$$

$$\sum_{i=0}^{n} q_{i,j} - \sum_{i=0}^{n} q_{j,i} = c_j \quad \forall j = \{1, 2, \ldots, n\} \tag{10.10}$$

$$\sum_{i=0}^{n}\sum_{k=1}^{m} c_j X_{i,j}^k \leq Q \quad \forall j = \{1, 2, \ldots, n\}, \forall k = \{1, 2, \ldots, m\} \tag{10.11}$$

$$\sum_{i=0}^{n}\sum_{k=1}^{m} X_{i,0}^k = 1 \quad \forall k = \{1, 2, \ldots, m\} \tag{10.12}$$

$$d_{i,j} = d_{j,i} \quad \forall i = \{0, 1, \ldots, n\}, \forall j = \{0, 1, \ldots, n\} \tag{10.13}$$

$$\sum_{j=0}^{n} X_{i,j}^k = \sum_{j=0}^{n} X_{j,i}^k = Y_i^k \quad \forall i = \{1, 2, \ldots, n\}, \forall k = \{1, 2, \ldots, m\} \tag{10.14}$$

$$X_{i,j}^k \in \{0, 1\} \tag{10.15}$$

$$Y_i^k \in \{0, 1\} \tag{10.16}$$

The objective function of this problem in Eq. (10.6) focuses on reducing the distance traveled by the vehicle. Constraints (10.7) and (10.8) show that each vehicle must leave the depot 0 empty. Whereas constraints (10.9) specify that bin i is visited by not more than one vehicle. Constraints (10.10) show that the vehicle empties the bins visited. Constraints (10.11) indicate that the total collected waste from the visited bin in a tour must not exceed the vehicle capacity. Constraints (10.12) ensure that all vehicles must arrive at the depot 0. Constraints (10.13) show that the distance between two nodes is the same in both directions. Constraints (10.14) ensure the continuity condition: a vehicle arriving at a bin must leave for another destination. Constraints (10.15) and (10.16) state that the decision variables are binary-valued.

10.3.3 A Genetic Algorithm Proposal for the SWM

The Genetic Algorithms (GA) are among the evolutionary search methods that interestingly improve the quality of the generated solution by means of many probabilistic operators. Moreover, these methods are more likely to converge towards global solutions especially when tacking high computational complexity problem with a large solution space. The generic routine of the developed approach is summarized in Algorithm 2.

Algorithm 2 Pseudo-code of Genetic Algorithm for SWM

Input: Population Pop of size N
Output: The best collection circuit with a minimal traveled distance f^*
Begin
Generate randomly a population Pop
While Stopping condition is not met **do**
  Random population initialization (Pop, N);
  Evaluate Pop.
  For $i \in \{1, \cdots, P_N\}$ **do**
    $(I_1, I_2) \longleftarrow$ Selection (Pop);
    $I_3 \longleftarrow$ Crossover (I_1, I_2);
    Mutation (I_3);
    $Pop \longleftarrow$ Replacement (I_3, Pop)
  End for
End While
Return The best found solution with f^*
End

During the evolutionary process, we aim to minimize the total collection distance. This means that, given a space of many solutions, the various genetic operators attempt to reach feasible solutions with a minimum traversed distance. To that aim, we vary the following operator to detect the best arrangement of bins that vehicle must visit:

- One-point crossover: it randomly selects one crossover point. Then, the tails of the individuals are swapped to get new offsprings.
- Two-points crossover: it randomly selects two crossover points within a chromosome. Then interchanges the two parents chromosomes between these points to produce new offsprings.
- New-randomly crossover: it is a new developed crossover which ensures the validity of a resulted solution (i.e. no genes can be duplicated). In this context,the operator begins by getting a copy of both parents. Then, it randomly chooses a numbers of bins to remove from the offspring starting from a position P. Finally, it reconstructs the offspring by adding from its end the removed bins while respecting the predefined order.
- Order crossover: it randomly selects a substring from the parent. Then, it produces a proto-child by copying the substring into the corresponding position. Therefore, it deletes the gene already existing in the substring from the 2nd parent. The resulted sequence of gens contains the gens that the proto-child needs. Finally, the operator places the gens into the unfixed positions of the proto-child according to the order of the sequence to produce an offspring.

10.3.4 Experimental Study and Results

The objective of this section is threefold. Firstly, it is noteworthy to estimate the performance of integer linear programming model previously. The mathematical formulation was coded using C++ and solved by Cplex 12.6. Due to memory limitations and the large tested instance, experimentation terminated abnormally. Therefore, we focus on the other hand on enhancing the performance of the proposed genetic algorithm through determining the well adapted crossover operator. Finally, to fully appraise the efficiency of the genetic approach, we compare the output collection circuit delivered by the municipality of Sidi Bou Said against our method. During implementation, we performed the experimentation on a server with an Intel (R) Pentium (R) processor CPU B960 @2.20 Ghz and 6 GO memory (RAM). The genetic algorithm was programmed in Java. To visualize data, we opted for the software ArcGis 10.6.

10.3.4.1 Benchmark Instances

The collected data correspond to the collection circuit traveled by the vehicle in the municipality of Sidi Bou Said. The city of Sidi Bou Said is located in the north east of Tunis. It is among the tourist cities that has an exceptional cultural character, with population of 7164 inhabitants and producing about 3000 tons of household solid waste per year. "Sidi Bou Said" city is divided into 16 routes, with an average length of the collection of 7283m. The collection circuit is twice a day (day/night). The city is divided into two zones: For the high zone, the collection is performed

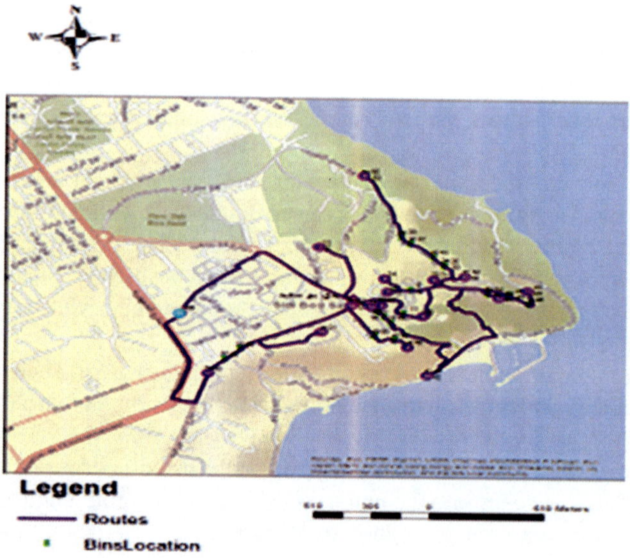

Fig. 10.5 Description of the municipality instance: location map and collection points

from house-to-house due to the narrow streets. For the lower zone, the municipality collects the waste from the public bins. There are two types of collection bins:

- 25 bins of 770L as a capacity.
- 45 bins of 120L as a capacity.

The collection is distributed on eventually 16 routes. In Fig. 10.5, we visualize all the relative data of the tested instance.

10.3.4.2 Used Metrics

To assert the efficiency of our genetic algorithm, we use the average relative percentage deviation GAP(%) measured as follows:

- $Gap = (\frac{Solution - Solution^*}{Solution^*}) * 100$

  Where $Solution^*$ is the best obtained value of solutions.

10.3.4.3 Analysis of the Genetic Algorithm

The challenging task focuses on enhancing the being existing solution of the municipality which means minimizing the distance traveled by the municipality vehicle during its collection circuit. To achieve this, on one hand, we show the variation of many crossover operators to detect the most promising one. On the other hand, we

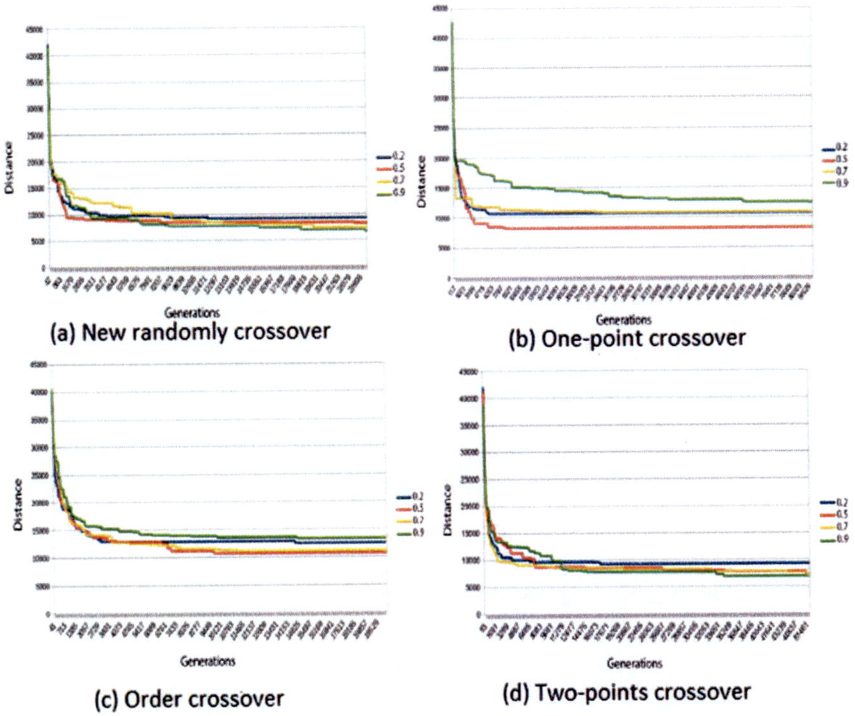

Fig. 10.6 The variation of circuit collection for the Sidi Bou Said instance with various crossover operators

tune the crossover rate parameter in order to improve the efficiency of the developed method. Experimental study show that all the crossover operators iteratively perform better results as the number of generations increases. These observations are clearly illustrated in Fig. 10.6.

The same figure clearly shows that it is instructive to favor the use of 0.5 as crossover rate for the one-point crossover, 0.9 for the two-point crossover, [0.5,0.7] for the order crossover and 0.5 for the new-randomly crossover. Pushing our analysis a step further, we adopt different crossover operator in order to detect the best adapted operator for our genetic algorithm. In Fig. 10.7, we provide the comparison between the tested operator in term of fitness value.

The same figure shows the ability of the new-randomly crossover in guiding the search process and delivering a best solution with a minimal traveled distance value. In this context, it is worth mentioning that the best adapted configuration to the new-randomly crossover is 0.9 as a crossover rate and 30 as a population size.

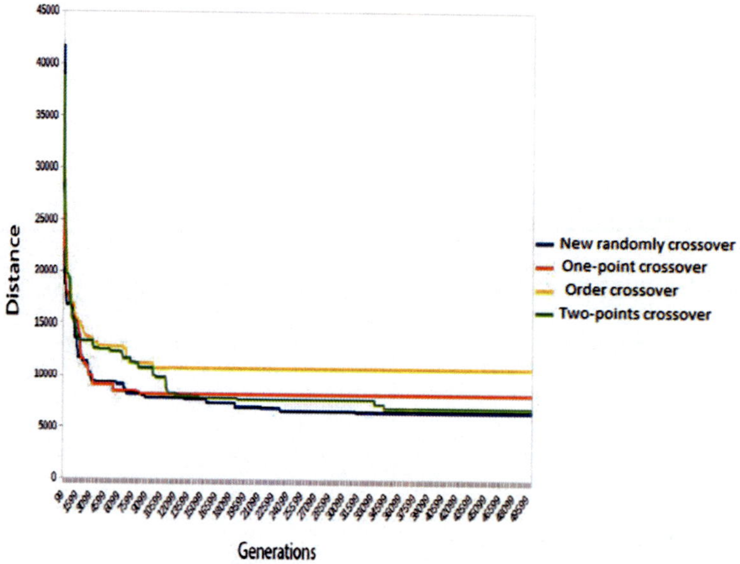

Fig. 10.7 Comparison between the different crossover operators

10.3.4.4 Comparison of GA Solution and Municipality Solution

To assess the practical performance of our GA, we provide a comparison with existing solution in the municipality of Sidi Bou Said in term of total traveled distance, CPU time, GAP (%) and the arrangement of bins. The results are displayed in Table 10.3.

Looking at Table 10.3, the performance of the GA was made evident since it requires only 53.072s to achieve a minimal distance equals to 6590m compared to an actual traveled distance 7283m in the municipality of Sidi Bou Said. The latter is achieved within approximately 10 h of extremely hard task work. This observation is consistent when looking at the $GAP(\%)$ which is 0% for GA vs a 10.5% for the solution of municipality. A global view of the experimental analysis is illustrated in Fig. 10.8.

In Fig. 10.9, we illustrate the solution of the GA thanks to the ArcGIS software.

Moreover, Fig. 10.10 provides evidence that using the GA approach has lead to a collection circuit with a minimal traveled distance.

Table 10.3 Comparison of the genetic algorithm versus municipality solution

	Municipality solution	GA solution
Bins arrangement	0-1-2-3-4-5-6-7-8-9-10-11-12-13-14-15-16-17-18-19-20-21-22-23-24-25-26-27-28-29-30-31-32-33-34-35-36-37-38-39-40-41-42-43-45-46-44-46-47-48-49-50-51-52-53-54-55-56-57-58-59-60-61-62-63-64-65-66-6-68-69-70-0	0-62-61-51-50-52-7-6-11-10-8-9-58-57-55-60-53-59-56-54-4-5-22-21-15-16-20-19-17-18-47-46-48-49-39-41-38-42-45-43-44-40-34-28-35-30-31-26-27-36-29-37-32-33-24-25-23-3-2-1-63-64-14-13-12-66-68-69-70-65-67-0
Distance(m)	7283	6590
$GAP(\%)$	10.5 %	0 %
Time(s)	10 h	53.072

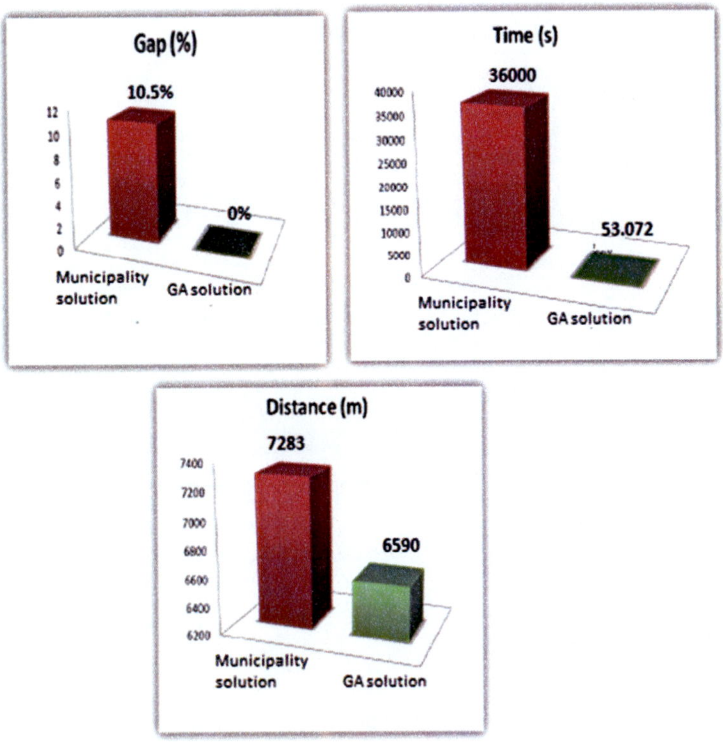

Fig. 10.8 Detailed comparison of the competitor approaches

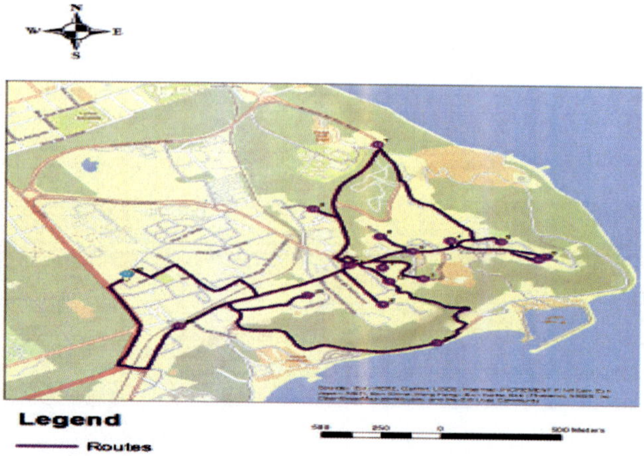

Fig. 10.9 New collection circuit resulted by the Genetic Algorithm

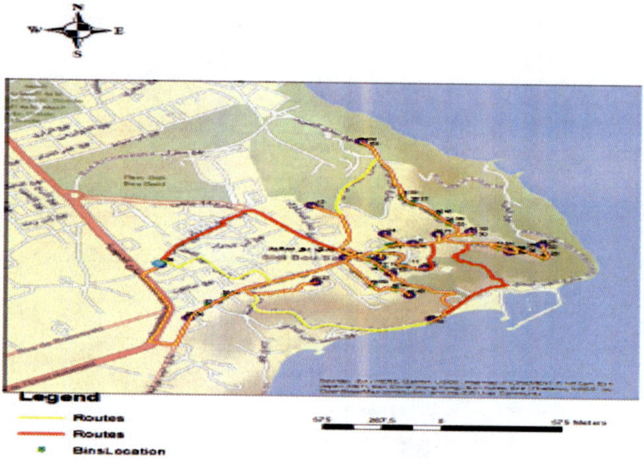

Fig. 10.10 New collection circuit resulted by the Genetic Algorithm

10.4 Conclusion

Throughout this chapter, we have proved that artificial intelligence tools and specifically the application of evolutionary approaches could perform a straightforward alternative to solve hard realistic problems.

Firstly, we have proposed an interactive decision support system that potentially considers the instructors' preferences regarding time slots scheduling. Our IDSS is consolidated using a new hybrid evolutionary approach to solve the UCT problem. The proposed system results in fulfilling most of the qualitative policy goals such as respecting each instructor's due and avoiding teaching staff's conflicts.

Computational experiments conducted on a real data collection exhibit strong evidence that the our timetabling system is robust. Moreover, it significantly improves upon manual process approach in term of many indicators performance. We would expect that this research effort proves its efficiency in the complete solution of the university course timetabling process. In a second section of our conducted research, we potentially carry an adaptive genetic algorithm to solve the solid waste collection problem. The indisputable significance of our method considers an attempt to predict the best configuration in term of crossover operators, crossover rates and the population size. Computational experiments conducted our real data provided by the municipality of Sidi Bou Said confirms a strong evidence of the new-randomly crossover. Results are plotted to show the efficiency of all the tested operators. Future research could focus on designing a more elaborated method to calibrate the parameters of the evolutionary approaches.

References

1. A.S. Asratian, D. de Werra, A generalized class-teacher model for some timetabling problems. Eur. J. Oper. Res. **143**(3), 531–542 (2002)
2. M.W. Carter, G. Laporte, Recent developments in practical course timetabling, in *International Conference on the Practice and Theory of Automated Timetabling* (1997), pp. 3–19
3. H. Babaei, J. Karimpour, A. Hadidi, A survey of approaches for university course timetabling problem. Comput. Ind. Eng. **86**, 43–59 (2015)
4. D.J. Welsh, M.B. Powell, An upper bound for the chromatic number of a graph and its application to timetabling problems. Comput. J. **10**(1), 85–86 (1967)
5. D. De Werra, An introduction to timetabling. Eur. J. Oper. Res. **19**(2), 151–162 (1985)
6. T.A. Redl, A study of university timetabling that blends graph coloring with the satisfaction of various essential and preferential conditions (Unpublished doctoral dissertation). Rice University (2004)
7. S.M. Selim, Split vertices in vertex colouring and their application in developing a solution to the faculty timetable problem. Comput. J. **31**(1), 76–82 (1988)
8. H.A. Razak, Z. Ibrahim, N.M. Hussin, Bipartite graph edge coloring approach to course timetabling, in *2010 International Conference on Information Retrieval & Knowledge Management, (CAMP)* (2010), pp. 229–234
9. E.A. Akkoyunlu, A linear algorithm for computing the optimum university timetable. Comput. J. **16**(4), 347–350 (1973)
10. J.A. Breslaw, A linear programming solution to the faculty assignment problem. Socio-Econ. Plan. Sci. **10**(6), 227–230 (1976)
11. R.H. McClure, C.E. Wells, A mathematical programming model for faculty course assignments. Decis. Sci. **15**(3), 409–420 (1984)
12. J.A. Ferland, S. Roy, Timetabling problem for university as assignment of activities to resources. Comput. Oper. Res. **12**(2), 207–218 (1985)
13. J. Aubin, J.A. Ferland, A large scale timetabling problem. Comput. Oper. Res. **16**(1), 67–77 (1989)
14. A. Tripathy, Computerised decision aid for timetabling–a case analysis. Discret. Appl. Math. **35**(3), 313–323 (1992)
15. T.H. Hultberg, D.M. Cardoso, The teacher assignment problem: a special case of the fixed charge transportation problem. Eur. J. Oper. Res. **101**(3), 463–473 (1997)
16. S. Daskalaki, T. Birbas, Efficient solutions for a university timetabling problem through integer programming. Eur. J. Oper. Res. **160**(1), 106–120 (2005)

17. M.A. Bakir, C. Aksop, A 0-1 integer programming approach to a university timetabling problem. Hacet. J. Math. Stat. **37**(1)
18. J.A. Ferland, C. Fleurent, Saphir: a decision support system for course scheduling. Interfaces **24**(2), 105–115 (1994)
19. L.R. Foulds, D.G. Johnson, Slotmanager: a microcomputer-based decision support system for university timetabling. Decis. Support Syst. **27**(4), 367–381 (2000)
20. T.R. Hinkin, G.M. Thompson, Schedulexpert: Scheduling courses in the cornell university school of hotel administration. Interfaces **32**(6), 45–57 (2002)
21. D. Johnson, A database approach to course timetabling. J. Oper. Res. Soc. **44**(5), 425–433 (1993)
22. T.-P. Liang, C.-C. Lee, E. Turban, Model management and solvers for decision support, in *Handbook on Decision Support Systems*, vol. 1 (Springer, Berlin, 2008), pp. 231–258
23. J. Miranda, eClasSkeduler: a course scheduling system for the executive education unit at the Universidad de Chile. Interfaces **40**(3), 196–207 (2010)
24. J. Stallaert, Automated timetabling improves course scheduling at UCLA. Interfaces **27**(4), 67–81 (1997)
25. H. Bouziri, O. Harrabi, Behavior study of genetic operators for the minimum sum coloring problem, in *, 2013 5th international conference on Modeling, Simulation and Applied Optimization (ICMSAO)* (2013), pp. 1–6
26. E. Rada, M. Grigoriu, M. Ragazzi, P. Fedrizzi, Web oriented technologies and equipments for MSW collection, in *Proceedings of the International Conference on Risk Management, Assessment and Mitigation-Rima*, vol. 10, (2010), pp. 150–153
27. C.A. Arribas, C.A. Blazquez, A. Lamas, Urban solid waste collection system using mathematical modelling and tools of geographic information systems. Waste Manag. Res. **28**(4), 355–363 (2010)
28. J. Bautista, J. Pereira, Ant algorithms for urban waste collection routing, in *International Workshop on Ant Colony Optimization and Swarm Intelligence* (2004), pp. 302–309
29. A.M. Benjamin, J. Beasley, Metaheuristics for the waste collection vehicle routing problem with time windows, driver rest period and multiple disposal facilities. Comput. Oper. Res. **37**(12), 2270–2280 (2010)
30. S. Sahoo, S. Kim, B.-I. Kim, B. Kraas, A. Popov Jr., Routing optimization for waste management. Interfaces **35**(1), 24–36 (2005)
31. G. Huang, N. Sae-Lim, L. Liu, Z. Chen, An interval-parameter fuzzy-stochastic programming approach for municipal solid waste management and planning. Environ. Model. Assess. **6**(4), 271–283 (2001)
32. A. Ustundag, E. Cevikcan, Vehicle route optimization for RFID integrated waste collection system. Int. J. Inf. Technol. Decis. Mak. **7**(04), 611–625 (2008)
33. X. Wu, G.H. Huang, L. Liu, J. Li, An interval nonlinear program for the planning of waste management systems with economies-of-scale effects–a case study for the region of hamilton, ontario, canada. Eur. J. Oper. Res. **171**(2), 349–372 (2006)
34. M.L. Ali, M. Alam, M.A.N.R. Rahaman, RFID based e-monitoring system for municipal solid waste management, in *2012 7th International Conference on Electrical and Computer Engineering* (2012), pp. 474–477
35. I. Hong, S. Park, B. Lee, J. Lee, D. Jeong, S. Park, Iot-based smart garbage system for efficient food waste management. Sci. World J. (2014)
36. A. Malakahmad, P.M. Bakri, M.R.M. Mokhtar, N. Khalil, Solid waste collection routes optimization via GIS techniques in Ipoh city, Malaysia. Procedia Eng. **77**, 20–27 (2014)
37. F. McLeod, G. Erdogan, T. Cherrett, T. Bektas, N. Davies, C. Speed, S. Norgate, Dynamic collection scheduling using remote asset monitoring: case study in the UK charity sector. Transp. Res. Rec. **2378**(1), 65–72 (2013)
38. J.M. Gutierrez, M. Jensen, M. Henius, T. Riaz, Smart waste collection system based on location intelligence. Procedia Comput. Sci. **61**, 120–127 (2015)
39. O.M. Johansson, The effect of dynamic scheduling and routing in a solid waste management system. Waste Manag. **26**(8), 875–885 (2006)

40. A. Rovetta, F. Xiumin, F. Vicentini, Z. Minghua, A. Giusti, H. Qichang, Early detection and evaluation of waste through sensorized containers for a collection monitoring application. Waste Manag. **29**(12), 2939–2949 (2009)
41. C. Adeofun, H. Achi, G. Ufoegbune, A. Gbadebo, J. Oyedepo, Application of remote sensing and geographic information system for selecting dumpsites and transport routes in Abeokuta, Nigeria, in *COLERM Proceedings* (2012), pp. 1264–278
42. M. Arebey, M. Hannan, H. Basri, R.A. Begum, H. Abdullah, Integrated technologies for solid waste bin monitoring system. Environ. Monit. Assess. **177**(1–4), 399–408 (2011)
43. J. Senthil, S. Vadivel, J. Murugesan, Optimum location of dust bins using geo-spatial technology: a case study of Kumbakonam town, Tamil Nadu, India. Adv. Appl. Sci. Res. **3**(5), 2997–3003 (2012)
44. G. Tavares, Z. Zsigraiova, V. Semiao, M. Carvalho, d.G., Optimisation of MSW collection routes for minimum fuel consumption using 3d GIS modelling. Waste Manag. **29**(3), 1176–1185 (2009)
45. J. Beliën, L. De Boeck, J. Van Ackere, Municipal solid waste collection and management problems: a literature review. Transp. Sci. **48**(1), 78–102 (2014)
46. H. Han, E. Ponce Cueto, Waste collection vehicle routing problem: literature review. Promet-Traffic Transp. **27**(4), 345–358 (2015)
47. M. Hannan, M.A. Al Mamun, A. Hussain, H. Basri, R.A. Begum, A review on technologies and their usage in solid waste monitoring and management systems: issues and challenges. Waste Manag. **43**, 509–523 (2015)
48. H.Ş Düzgün, S.O. Uşkay, A. Aksoy, Parallel hybrid genetic algorithm and GIS-based optimization for municipal solid waste collection routing. J. Comput. Civ. Eng. **30**(3), 04015037 (2016)
49. N.-B. Chang, Y.-C. Yang, S. Wang, Solid-waste management system analysis with noise control and traffic congestion limitations. J. Environ. Eng. **122**(2), 122–131 (1996)
50. L. Bodin, A. Mingozzi, R. Baldacci, M. Ball, The rollon-rolloff vehicle routing problem. Transp. Sci. **34**(3), 271–288 (2000)
51. E. de Oliveira Simonetto, D. Borenstein, A decision support system for the operational planning of solid waste collection. Waste Manag. **27**(10), 1286–1297 (2007)
52. H. Krikke, I. le Blanc, M. van Krieken, H. Fleuren, Low-frequency collection of materials disassembled from end-of-life vehicles: on the value of on-line monitoring in optimizing route planning. Int. J. Prod. Econ. **111**(2), 209–228 (2008)
53. L. De Meulemeester, G. Laporte, F. Louveaux, F. Semet, Optimal sequencing of skip collections and deliveries. J. Oper. Res. Soc. **48**(1), 57–64 (1997)
54. S.K. Amponsah, S. Salhi, The investigation of a class of capacitated arc routing problems: the collection of garbage in developing countries. Waste Manag. **24**(7), 711–721 (2004)
55. R. Hansmann, U. Zimmermann, Integrated vehicle routing and crew scheduling in waste management (part I), in *Dagstuhl Seminar Proceedings* (2009)
56. N. Bianchessi, G. Righini, Heuristic algorithms for the vehicle routing problem with simultaneous pick-up and delivery. Comput. Oper. Res. **34**(2), 578–594 (2007)
57. N.-B. Chang, Y. Wei, Siting recycling drop-off stations in urban area by genetic algorithm-based fuzzy multiobjective nonlinear integer programming modeling. Fuzzy Sets Syst. **114**(1), 133–149 (2000)
58. V. Maniezzo, *Algorithms for Large Directed Carp Instances: Urban Solid Waste Collection Operational Support* (University of Bolonha, UBLCS Technical Report Series, Bolonha, Italy, 2004), p. 27
59. J.S. Yeomans, Solid waste planning under uncertainty using evolutionary simulation-optimization. Socio-Econ. Plan. Sci. **41**(1), 38–60 (2007)
60. N.V. Karadimas, G. Kouzas, I. Anagnostopoulos, V. Loumos, Urban solid waste collection and routing: the ant colony strategic approach. Int. J. Simul. **6**(12–13), 45–53 (2005)
61. R. Muttiah, B. Engel, D. Jones, Waste disposal site selection using GIS-based simulated annealing. Comput. Geosci. **22**(9), 1013–1017 (1996)

Jouhaina Chaouachi is a full Professor at Business School of Carthage (IHEC), University of Carthage in Quantitative Methods and Computer Science Department. She is the Academic Dean of IHEC Carthage. She is a lead researcher in ECSTRA Laboratory. She received her PhD from University of Tunis in Modeling and Computer Science. She has continuing in the academic streamline since 1998. She has published a number of papers in many prestigious international journals and conferences. Her areas of interests include Artificial Intelligence, Operations Research, Metaheuristics, Transportation problems, Electric Vehicle Routing Problems and Decision Support Systems.

Olfa Harrabi received her PhD from the High Institute of Management (ISG) Tunis, University of Tunis in computer science. She has published a number of papers in international journals and conferences. She is currently an Assistant of computer Science at the Higher School of Economic and Commercial Sciences (ESSECT) Tunis, Tunis University. Her main research interests include Artificial intelligence, operations research, optimization, metaheuristics and decision support systems.

Chapter 11
Artificial Neural Networks for Precision Medicine in Cancer Detection

Smaranda Belciug

Abstract Automated diagnosis has a crucial role in the medical decision-making process. Artificial Intelligence models are ubiquitous in the healthcare system, aiding physicians in providing fast and accurate diagnosis. When it comes to cancer, besides the accuracy, the speed of the diagnosing process is essential. The last decade shed a light onto cancer treatment, through precision medicine and artificial neural networks. Precision medicine uses microarrays of deoxyribonucleic acid and mass spectrometry. Artificial neural networks through their adaptive learning and nonlinear mapping properties can personalize their hyperparameters in order to provide personalized diagnosis, followed by personalized treatment. The goal of this chapter is to present several novel adaptive neural networks that embed genomic knowledge into their architecture, increasing their diagnosis performance and computational speed, while decreasing computational cost.

Keywords Single-hidden layer feedforward neural network · Logistic regression · Statistical assessment · Gene expression · Adaptive hidden nodes initialization

11.1 Introduction

Cancer. The word drops like an ax on our head when we hear it. We all know people who fought it, some won, some lost the battle, and some are in the middle of the fight. Worldwide statistics reveals that cancer is the second leading cause of death, surpassed only by cardiovascular diseases (https://ourworldindata.org/cancer—accessed March 3, 2021). Every sixth death is caused by cancer according to the Institute for Health Metrics and Evaluation (IHME—http://healthdata.org—accessed March 3,2021). It seems that no matter what we do (eat well, sleep well, no smoking, no drinking, exercise, etc.) the danger is still out there due to genetics and environmental issues. Perhaps, cancer is a game of luck. Bad luck. It appears that there are more and more cases of cancer discovered daily, which leads to general

---

S. Belciug (✉)
Department of Computer Science, Faculty of Sciences, University of Craiova, Craiova, Romania
e-mail: sbelciug@inf.ucv.ro

M. Virvou et al. (eds.), *Advances in Selected Artificial Intelligence Areas*,
Learning and Analytics in Intelligent Systems 24,
https://doi.org/10.1007/978-3-030-93052-3_11

paranoia. We believe that the increase is due to national screening programs and Artificial Intelligence (AI) that detect more cases. Some good news come from a 2019 report of the CDC (Central of Disease Control) which shows a decline by 19% in death rates caused by cancer [1], due to early, rapid, and accurate diagnosis obtained through the use of AI.

Historically speaking, we have come a long way in our fight against cancer. Progress has been made, but unfortunately, it's not enough.

To fight cancer, we first need to understand how it works. Cancer can start in any part of the body, whether it is an organ, skin, or blood. Cancer is unique, just like the person who has it. Even if there are different types of cancer that start with a certain type of cell, the speed of its growth and spread makes it one of a kind. So, cancer tailors itself to the carrier. Thus, to have a chance at fighting and defeating cancer we need tailored diagnosis and tailored treatment [2].

Classifying cancer is done by examining the morphological features of biopsy tissues. Even if this is the standard technique, it offers incomplete evidence, failing in providing crucial information regarding the tumor, such as the capacity for invasion, metastases, or rate of proliferation, etc. [3]. The last decade brought a new perspective when it comes to differentiating cancer: precision medicine. Precision medicine takes into account the genetic variability, environment, and lifestyle when it comes to diagnosing and treating cancer.

Precision medicine produces massive amounts of data. Huge datasets that contain RNA ribonucleic acid sequencing expression data, or mass spectrometry (MS), need to be processed fast and accurate. These types of datasets need machine learning (ML) techniques to be processed due to their size. The results obtained by applying ML on genetic data are outstanding in terms of differentiating cancer [4–8]. Artificial neural networks (NNs) are the most used machine learning (ML) techniques.

Treating cancer is all about personalized medicine: tailored diagnosis, tailored treatment, etc. Naturally, some data scientists shifted their research towards tailored personalized NNs. Developing new artificial intelligence algorithms is not enough. The statistical analysis of the obtained results is crucial. We believe that the statistical evaluation of the novel algorithms' performance is the most important part. Sadly, this part in often overlook.

Therefore, in this chapter, we shall present some novel NNs models that have been designed, implemented, and tested successfully on precision medicine cancer datasets, followed by a brief description of the statistical tests that have been used to validate their results.

11.2 The fLogSLFN Model

A novel personalized single hidden layer feedforward neural network (SLFN) named the *fLogSLFN,* which can be applied on a two-class decision problem was developed for differentiating breast and lung cancer in [9]. The model uses logistic regression for computing the weights between the input and the hidden layer in a SLFN. In this manner, knowledge from the dataset is embedded in the network. The network bias is the computed intercept, and the weights are the regression coefficients.

Besides initializing the weights in this way, the author also developed a filtering module for speeding up the computational time. Only the statistically significant input features are kept in the network's architecture. The filtering module uses the p-value which is computed together with the regression coefficients and intercept. If an attribute has its p-level below 0.05, it is kept as input feature. This embedded filter helps avoiding the '*curse of dimensionality*'.

This method is fast because its training is performed in just one step using linear algebra. Mathematically speaking, the weights between the hidden and the output layer are computed using the Moore–Penrose pseudo-inverse matrix.

We can measure the correlation between the input variables and the output variable, if we assume that the features are independent of each other. In this manner we can say that if we have an object $\mathbf{x}_k = (x_1^k, \ldots, x_m^k; y_j)$, where x_i^k is the i-th attribute, y_j, $j = 1, 2$, the class label, from a statistical point of view, the features x_i^k are governed by a random variable X_i. This presumption implies that the whole training set is a random sample of X_i. The same principle is applied for the class label also, that represents a random sample of variable Y. Y is a categorical random variable.

Thus, we can highlight the correlation between the predictors (X_i), and the dependent variable (Y).

Recall the logistic regression equation:

$$logit(p) = a + X \cdot b$$

where:

$$X = \begin{pmatrix} 1x_1^1 x_2^1 \ldots x_m^1 \\ \ldots \\ \ldots \\ 1x_1^N x_2^N \ldots x_m^N \end{pmatrix}$$

N being the number of training samples.

The $logit(p)$ transformation is defined as:

$$logit(p) = \ln\left(\frac{p}{1-p}\right),$$

p being the proportion of subjects that have malignant tumors, whereas $1 - p$ being the proportion of subjects that have benign tumors.

We shall briefly present the *fLogSLFN* algorithm.

Input:

Let us denote:

- the training dataset with *Train;*
- the number of hidden neurons with *nH;*
- the weights between the input and hidden layer with w_{ih}, where i is an input neuron, and h a hidden neuron;
- the hidden-output layer matrix with *M;*
- the Moore–Penrose inverse of M with $M^+ = (M^T \cdot M)^{-1} \cdot M^T$;
- the class label y_j is *one-hot-encoded* as it follows: $y_1 \sim (0, 1)$, $y_2 \sim (1, 0)$;
- the ground truth output with *o*;
- the weight matrix between the hidden layer and the output layer with β;
- the non-linear activation function is the modified hyperbolic tangent $f(u) = 1.7159 \cdot \tanh\left(\frac{2u}{3}\right)$.

The hyperbolic function was chosen due to its fast convergence, [10].

Method:

1. Apply the logistic regression and obtain the regression coefficients $b_i, i = 1, 2, \ldots, m$ together with the intercept, b_0, and the corresponding *p*-value.
2. Filter the input features, keeping only the attributes that have the p-value < 0.05.
3. Assign the regression coefficients to the input hidden layer weights, $w_{ih} \leftarrow b_i, i = 1, 2, \ldots, m, h = 1, 2, \ldots, nH$.
4. Compute the hidden layer output matrix *M*, taking into account the training set, *Train*;
5. Compute $\beta = M^+ \cdot o$.

The *fLogSLFN* was applied successfully on three datasets regarding breast and lung cancer, the accuracy ranging between 64.70 and 98.66% (mldata.org). The results are competitive with other state-of-the-art techniques such as the multilayer perceptron (obtained accuracy between 50.81 and 89.91%), radial basis functions (obtained accuracy between 46.75 and 80.95%), logistic regression with LASSO penalty (obtained accuracy between 49.94 and 86.34%). To prove that the results are correct, the authors performed a thorough statistical analysis that included Kolomogorov-Smirnov and Lilliefors, Shapiro–Wilk, Levene, Brown Forsythe, *t*-test, and Mann Whitney U tests, along with one-way ANOVA.

Even if the novel method enhances differentiating between malignant and benign tumors using DNA arrays, it still has a major drawback: the fact that it cannot be applied when dealing with multiple decision-classes. This issue concerned the author also, and luckily enough they developed a new study which mended it. In another study they extrapolated the two-case problem to a multiple decision classes through two novel models. In the next subchapter we shall discuss the two methods.

11.3 Parallel Versus Cascaded LogSLFN

In the previous section we have discussed a special NN, that uses logistic regression to initialize and filter the synaptic weights between the input and the hidden layer in a SLFN. As you recall, the fLogSLFN can be used only if we are dealing with a two-class decision problem. The aim of one study was to extend the logSLFN to the case of multiple classes [11]. Two approaches have been considered: a parallel logSLFN, and a cascaded logSLFN.

The parallel approach transforms the multi-class case into a binary one, by applying the one-against-all method. Technically, the classical logSLFN is applied in parallel for all the classes, using for each class the corresponding logistic regression coefficients. The other approach, the cascaded logSLFN transforms the multi-class case into a binary one in cascade by computing the regression coefficients for the first class, after which it eliminates them. The process continues until all the classes are covered.

Recall, that we mentioned the fact that we are going to presume that the training set is a random sample X_i, and that all the attributes are independent of each other. The same thing goes for the output label set, that can be thought of as a categorical random variable Y.

The training of both methods is done also in just one step, by using linear algebra. β, the matrix that contains the weights between the hidden and the output layer, is computed through the Moore–Penrose inverse matrix. Again, the activation function is the modified hyperbolic tangent $f(u) = 1.7159 \cdot \tanh\left(\frac{2u}{3}\right)$.

Another important issue must be discussed: the number of hidden neurons, *nH*. In some cases, if we take into account different universal approximation theorems, we can choose *nH* using different methods. For example, in [12] it is proven through some approximation theorems that NNs can approximate any continuous function well if we have a large enough number of hidden units. Another example is in [13], where the authors state that a single hidden layer NN that has any nonpolynomial activation function can approximate any continuous function if the number of hidden neurons is large. Other two studies show that NN that have a given number of hidden units can approximate univariate and multivariate functions [14, 15].

The parallel logSLFN also known as the *pLogSLFN* converts the multi-decision class problem into multiple binary problems and applies the logSLFN algorithm in parallel for each of these problems. For instance, let us presume that we have 4 decision classes, then by applying the one-against-all method, the *pLogSLFN* will divide this problem into the first binary problem that is 1 class against 2, 3, and 4 classes. The second binary problem is class 2 against 1, 3, and 4, the third is class 3 against 1, 2, and 4, and the last binary class is class 4 against 1, 2, and 3.

After dividing the initial problem as such, we compute for each binary case the logistic regression coefficients and intercept and assign them to the weights between the input and hidden layer. After this step is over, we apply the logSLFN for each binary problem. We can summarize the *pLogSLFN* as follows:

Method:

1. For each corresponding class label $y_i, i = 1, 2, \ldots, q$, where q is the number of classes, we transform it using the *one-hot-encoding* rule for categorical data, as follows: $o_1 \sim (0, 1)$ for class Ω_1, and $o_2 \sim (1, 0)$ for the rest of the classes $\Omega_2, \Omega_3, \ldots, \Omega_q$, for the first binary problem; $o_1 \sim (0, 1)$ for class Ω_2, and $o_2 \sim (1, 0)$ for the rest of the classes $\Omega_1, \Omega_3, \ldots, \Omega_q$, for the second binary problem; ...; $o_1 \sim (0, 1)$ for class Ω_q, and $o_2 \sim (1, 0)$ for the rest of the classes $\Omega_1, \Omega_2, \ldots, \Omega_{q-1}$, for the last binary problem.
2. Apply the logistic regression and compute the intercept $b_0^1, b_0^2, \ldots, b_0^q$, and regression coefficients $b_i^1, b_i^2, \ldots, b_i^q$ for each binary case.
3. Compute the input hidden matrices $W^1, W^2, \ldots, W^q$, and hidden output matrices $\beta^1, \beta^2, \ldots, \beta^q$ using the logSLFN.
4. Classify the testing data with the matrices obtained at step 3.

The cascaded logSLFN, known as the *cLogSLFN*, transforms the multiple decision problem into binary problems using a cascaded approach. Let us return to our 4-class decision problem. By converting the multiple decision problem into binary cases, we obtain the following problems: the first binary case is class 1 against 2,3, and 4. After this step is over, we apply the logSLFN on it, we classify the data accordingly, and exclude all the items in the dataset that have the class label 1. The newly obtained dataset has fewer items, and also fewer classes: 2, 3 and 4. We transform this 3-class problem into a binary one and repeat the above presented steps, until we remain with the last binary problem that contains classes 4 and 5.

The *cLogSLFN* can be summarized as follows:

Method:

1. Use the *one-hot-encoding* rule to transform the class labels $y_i, i = 1, 2, ..., q$, where q is the number of classes into: $o_1 \sim (0, 1)$ for class Ω_i, and $o_2 \sim (1, 0)$ for the rest of the classes $\Omega_j, j \neq i$.
2. Apply the logistic regression and compute the intercept b_0^i, and regression coefficients $b_k^i, k = 1, 2, \ldots, m$, m being the number of features.
3. Compute the input hidden matrix W^i and hidden output matrix β^i by using the logSLFN algorithm.
4. Remove the items that have been classified as class i from the training dataset.
5. Repeat steps 1 through 4 until we reach a binary problem with the classes q and $q - 1$.
6. Use the obtained input hidden matrices and hidden output matrices to classify the testing data.

The two approaches have been applied on two datasets for differentiating breast, kidney, colon, lung, prostate and liver cancer (https://archive.ics.uci.edu/ml/datasets/gene+expression+cancer+RNA-Seq). The statistical analysis revealed that the *pLogSLFN* outperforms the *cLogSLFN* on datasets that have not been previously preprocessed. The explanation is that if *cLogSLFN* misclassified data at a certain step, and afterwards deletes the misclassified items for the training set, then that error propagates throughout the whole algorithm. Indeed, the error propagates in both cases, but in the *cLogSLFN* case the training set is getting smaller and smaller at each step, making it easier for the NN to make classification mistakes. The author states that during the tests, no item was classified as belonging to two different classes.

Another result is that whatever the *pLogSLFN* gains in accuracy, it loses in running time: the computational cost in more expensive. The explanation is that the *pLogSLFN* runs the algorithm on the whole dataset each time, whereas the *cLogSLFN* runs it on smaller and smaller datasets.

The *pLogSLFN* obtained 99.91% accuracy versus *cLogSLFN* which obtained 99.94% when applied on the dataset regarding breast, kidney, colon, prostate, and lung cancer. On the liver cancer dataset, the *pLogSLFN* obtained 80.77% accuracy, whereas the *cLogSLFN* obtained only 52.00%. The two models were compared with state-of-the-art NNs, such as 3-multilayer perceptron, radial basis functions, and extreme learning machines. The results were statistically validated using the following tests: Kolmogorov–Smirnov & Lilliefors, Shapiro–Wilk W, Levene, Brown-Forsythe, *t*-test, one-way ANOVA and Tukey's post-hoc test.

11.4 Adaptive SLFN

A statistical strategy that embeds knowledge into hidden nodes and also filters features based on their significance is presented in [16]. Instead of the random initialization of the weights between the input and the hidden layer, the authors estimated the statistical relationship between the features and the class label using the non-parametric Goodman–Kruskal Gamma rank correlation. In this manner, the synaptic weights are assigned a value which is a consistent quantification of the knowledge inserted within the data. The filtering mechanism takes into account the corresponding *p*-value.

Mathematically speaking, if we again consider that the sample data is governed by the random variable X_i, and that the labels are governed by the categorical random variable Y, then we can quantify the relationship between them with a non-parametric rank. In real-world applications we are dealing with non-linear monotonic relationships between variable and the presence of tied observations within the data, so this is why the non-parametric Goodman Kruskal Gamma rank correlation had been chosen.

The Goodman Kruskal Gamma Γ is computed using the concordant (C) and discordant pairs (D) as it follows: $\Gamma = \frac{C-D}{C+D}$. Concordant and discordant pairs mean comparing two pairs of data points and see whether they match or not. They are computed from ordered variables and tell us whether the pairs agree (concordant) or disagree (discordant).

The 'adaptive' part of the algorithm means two different things. First it lets us know that the SLFN adapts itself to the problem at hand, by tuning the synaptic weights taking into account the Γ rank, the natural relationship between input and output. The second meaning refers to the fact that the SLFN adapts itself by taking into account the relevance of each feature through the significance level.

In this study, the authors took into account only the strength of the relationship, not its direction. Each synaptic weight between the input and the hidden layer is computed using the following formula:

$$w_{ih} = \Gamma\left(\left(x_i^k - mean_i^j\right), y_k\right),$$

where $i = 1, 2, \ldots, m$; $j = 1, 2, \ldots, q, k = 1, 2, \ldots, N, h = 1, 2, \ldots, nH$. For each decision class Ω_j, $j = 1, 2, \ldots, q$, and for each attribute A_i, we compute $mean_i^j$ as the mean attribute per class.

The filtering module is inspired by the *backward stepwise regression*, were in the beginning all the features are considered important explanatory variables for the class label. Thus, the adaptive SLFN considers at first all the variables being important, and then it removes the unimportant features one at a time. The cut-off value of the significance level was set at 0.05. The null and alternative hypothesis in this case state: H_0 : there is no association between the features and the class label, in other words zero correlation; H_1: there is an association between the features and the class label.

The computation of the significance level p was inspired by the p-level of the Kendall Tau independence test [17]. Hence, at first, the rank correlation gamma was computed, followed by computing the statistical significance z as such:

$$z = \frac{3 \cdot \Gamma \cdot \sqrt{n(n-1)}}{\sqrt{2(2n+5)}}.$$

The z score used the Z (Normal) distribution, two-tailed, 1 – cumulative p. The attributes that did not surpass $|z| \leq 1.96$ were removed from the network.

We shall briefly present the adaptive SLFN algorithm, where the activation function is given by the hyperbolic tangent, $f(u) = 1.7159 \cdot \tanh\left(\frac{2u}{3}\right)$:

Method:

1. For each decision class Ω_j, $j = 1, 2, \ldots, q$, the corresponding class label y_j is transformed using the *one-hot-encoding rule*, as follows: $y_1 \sim (0, 0, \ldots, 1)$, $y_2 \sim (0, 0, \ldots, 1, 0)$, $\ldots$, $y_q \sim (1, 0, \ldots, 0)$.
2. Compute the mean attribute per class $mean_i^j$, for each attributeA_i, $i = 1, 2, \ldots, m$, and each class Ω_j, $j = 1, 2, \ldots, q$
3. Compute the Goodman Kruskal Gamma rank correlation, Γ:

$$\Gamma = \frac{C - D}{C + D}$$

4. Apply the filtering module and remove the unimportant features from the network.

5. Assign the Goodman Kruskal Gamma rank correlation $|\Gamma|$ corresponding to $|z| > 1.96$ to each synaptic weight$w_{ih}, i = 1, 2, .., m, h = 1, 2, \ldots, nH$:

$$w_{ih} = \Gamma\left(\left(x_i^k - mean_i^j\right), y_k\right).$$

6. Compute the hidden layer output matrix *M*.
7. Compute the output weight $\beta = M^+ \cdot o$.
8. Use the obtained input hidden matrices and hidden output matrices to classify the testing data.

The adaptive SLFN was comparable in terms of results with other state-of-the-art NNs when applied on DNA microarray dataset concerning breast, colon, lung, and ovarian cancer. The adaptive SLFN obtained 54.62% and 83.81% accuracies on two different breast cancer datasets, 84.53% accuracy on the colon cancer dataset, 92.84% on the lung cancer dataset, and 72.15% on the ovarian cancer dataset. The model was statistically benchmarked with other state-of-the-art NNs such as extreme learning machine, radial basis functions, 3-multilayer perceptron, and support vector machines.

11.5 Statistical Assessment

It is obvious that in the last decade or so we have entered a new era, the era where we are using artificial NNs to fight cancer. NNs are algorithms of stochastic nature, so to trust the results we need a measure of their performance. Whether we are using NNs to diagnose cancer, or to determine the right dosage of the chemotherapy drugs, or radiotherapy, or we are developing new drugs and vaccines that help cure or prevent cancer, we need to prove beyond a doubt that the results obtained are true, trustworthy and robust, and that they can be replicated. For this, we must perform a statistical analysis. We have seen in the previous sections of this chapter, that all the four novel NNs have been benchmarked using certain statistical tests. Because, statistical analysis is an important feature of artificial intelligence, and yet many researchers fail to understand and use it, we end our chapter by briefly explaining the statistical tests that have been used for validating the novel NNs. We encourage the reader, to use these tests whenever she/he is developing new machine learning algorithms.

The first statistical tool that we are going to discuss is the *p*-value. We have seen throughout this chapter, that two models the *fLogSLFN* and the adaptive SLFN have used the *p*-level to filter the attributes that weren't important and eliminate them from the network. The *p*-value is a powerful statistical tool that has its value between 0 and 1. The cut-off value that indicates the level of significance is 0.05. A *p*-value which is less or equal to 0.05 implies the fact that we can reject the hypothesis, because there exists enough evidence to support the alternative hypothesis. In this case the null hypothesis states that "the results are no significant enough", whereas the alternative hypothesis states that "the results are statistically significant". If the p-level is greater than 0.05, then we accept the null hypothesis. The authors used the p-level to determine which attributes were indeed strongly correlated to the outcome.

For each novel model some statistical tests were applied. In what follows we shall describe shortly each of these tests, and more especially we shall explain why it is necessary to apply them.

The first tests are the *Kolmogorov–Smirnov Goodness of Fit* and *Lilliefors test*, and the *Shapiro Wilk W test*. These tests are used for verifying if the data distribution is normal or not. Here, we need to mention the fact that when we are speaking about the data, we are referring to the data sample which contains the computer runs, not the actual data that the models have been applied on. We want and need to verify the normality, as well as the equality of variances, in order to check whether we can or cannot apply comparison test such as the t-test or one-way ANOVA. If the data sample is not normally distributed, then we need to find alternatives to these comparison tests. All three tests have the following working hypothesis:

- H_0: the data is governed by the Normal distribution.
- H_1: the data is not governed by the Normal distribution.

We can apply the tests by hand, which is not recommended, or by using different statistical software on the market.

In many cases, after applying these tests we find ourselves in the situation where the data sample containing the computer runs is not normally distributed, which might lead to more complicated tests that need to be applied. In this circumstance, we instruct the reader to use a large enough sample, over 30 computer runs, so that the *Central Limit Theorem* could be applied. If the sample size is large enough, more than 30, the distribution is nearly Gaussian [18].

The next two tests that have been used are *Levene* and *Brown-Forsythe* tests. Both tests verify the equality of variances of two data samples. The difference between the two is that while Levene test uses the mean to compute the statistic, Brown-Forsythe uses the median. Both tests can be used even if the data samples are not governed by the normal distribution. If the samples are indeed normally distributed, we can use Bartlett test instead. In the case where we are not satisfied with the results, meaning that the samples do not have equal variances, we can bypass this in practice by having the exact number of observations in the samples.

After verifying these presumptions, the normality and the equality of variances, we can proceed in applying comparison tests such as the t-test and one-way ANOVA. If the samples are not governed by the Normal distribution, we can use the alternative test Mann–Whitney U test.

If we need to compare multiple algorithms, we can use either the t-test (for comparing two independent samples) or the one-way ANOVA (for comparing more than two independent samples). Whatever test we may choose, at first, we need to verify two assumptions: the sample distribution and equality of variances. If the sample is governed by the normal distribution and the samples' variances are equal, then we can proceed with performing the tests.

Understanding the t-test results is rather simple. If the p-level associated with the obtained t-value is less than 0.05, then we reject the null hypothesis and accept the alternative one. When it comes to one-way ANOVA things tend to become a little more complicated. When using on-way ANOVA we compute the F distribution, after which we determine whether we accept or not the null hypothesis. The only problem is that the null hypothesis only states whether there are or not difference between all the data samples but does not specify between which data samples there are differences. In order to find out the answer to this question one more test must be applied: an ad-hoc or a post-hoc test.

The ad-hoc analysis is done during the ANOVA test (e.g. least significant difference). The post-hoc is done after the ANOVA test (e.g. Tukey Honest Significant difference). One-way ANOVA analyses the residuals variances, which are computed as the difference between the mean of each group and every object in the sample.

By performing these tests, we add value to our research. In many medical research papers statistical analysis is missing, and this issue leads to questioning the reported results.

11.6 Conclusions

Little by little, day by day, the way we tackle the fight against cancer changes. From setting the diagnosis to determine the treatment plan, we use artificial intelligence in order to improve our chances. Even if medicine progresses every day, we still do not know precisely how and why cancer affects people. Is it genetics, lifestyle, bad luck? Apparently, anything can trigger it.

The healthcare system produces massive amounts of data. This data cannot be processed by humans. Artificial Intelligence lends a helping 'hand' and solves a lot of complex problems. Besides the classical standard ways of diagnosing cancer, medicine took a step forward and evolved into precision medicine. Each person is unique, and so it is cancer. To improve the chances of survival, we must diagnose it fast and accurate. DNA arrays contain important information, information that can be processed by using different artificial intelligence tools such as NNs. The goal of this chapter was to present three novel NNs that use the idea of precision medicine in tailoring the hidden nodes of SLFNs and making them problem dependent. All the new methods proved to be competitive with state-of-the-art NNs when applied on publicly available datasets. The chapter ends with a short presentation of the statistical tests that have been performed in order to establish the true potential of these methods in terms of performance.

References

1. CDC. cdc.gov/nchs/data/nvsr/nvsr68/nvsr68–05–508.pdf (2019)
2. S. Belciug, Artificial Intelligence in Cancer: Diagnostic to tailored treatment, Elsevier (2020).
3. A. Perez-Diez, A. Morgun, N. Shulzhenko, Microarrays for cancer diagnosis and classification, in *Madame Curie Bioscience Database*, Austim (TX): Landes, Bioscience (2013)
4. F. Duan, F. Xu, Applying multivariate adaptive splines to identify genes with expressions varying after diagnosis in microarray experiments. Cancer Inform. **16** (2017). https://doi.org/10.1177/1176935117705381
5. Y. Yamamoto, A. Saito, A. Tateishi, H. Shimojo, H. Kanno, A. Tsuchiya, K.I. Ito, E. Cosatto, H.P. Graf, R.R. Moraleda, N. Eils, N. Grabe, Quantitative diagnosis of breast tumors by morphometric classification of microenvironmental myoepithelial cells using a machine learning approach. Sci. Rep. **25**(2017). https://doi.org/10.1038/rsep46732
6. C.F. Aliferis, D. Hardin, P.P. Massion (2002). Machine learning models for lung cancer classification using array comparative genomic hybridization. Proc. AMIA. Symp.: 7–11.
7. X. Wang, R. Simon, Microarray-based cancer prediction using single genes. BMC Bioinf. **12**, 391 (2011). https://doi.org/10.1186/1471-2105-12-391
8. O. Klein, F. Kanter, H. Kulbe, P. Jank, C. Denkert, G. Nebrich, W.D. Schmitt, Z. Wu, C.A. Kunze, J. Sehouli, S. Darb-Esfahani, I. Braicu, J. Lellmann, H. Thiele, E.T. Taube, MALDI-Imaging for classification of epithelial ovarian cancer histo-types from tissue microarray using machine learning methods. Proteomics Clin. Appl. **13**, 1 (2019). https://doi.org/10.1002/prca.201700181
9. S. Belcuig, Logistic regression paradigm for training a single-hidden layer feedforward neural network. Application to gene expression datasets for cancer research. J. Biomed. Inform. **102**, 103372 (2020).
10. J.Y.F. Yam, T.W.S. Chow, A weight initialization method for improving training speed in feedforward neural network. Neurocomputing **219**, 232 (2000)
11. S. Belciug, Parallel versus cascaded logistic regression trained single-hidden feedforward neural network for medical data. Expert Syst. Appl. **170**(114538), 2021 (2021)
12. A. Pinkus, Approximation theory of the MLP model in neural networks. Acta Numer **8**, 143–195 (1999)
13. M. Leshno, V.Y. Lin, A. Pinkus, S. Schocken, Multilayer feedforward networks with a nonpolynomial activation function can approximate any function. Neural Network **6**(6), 861–867 (1993)
14. N.J. Guliyev, V.E. Ismailov, A single hidden layer feedforward network with only one neuron in the hidden layer cand approximate any univariate function. Neural Comput. **28**(7), 1289–1304 (2016). https://doi.org/10.1162/NECO_a_00849
15. Ismailov, VE (2014) On the approximation by neural networks with bounded number of neurons in hidden layers. Journal of Mathematical Analysis and Applications, 417 (2), 963–969. https://doi.org/10.1016/j.jmaa.2014.03.092.
16. S. Belciug, F. Gorunescu, Learning a single-hidden layer feedforward neural networks using a rank correlation-based strategy with application to high dimensional gene expression and proteomic spectra datasets in cancer detection. J. Biomed. Inform. **83**, 159–166 (2018)
17. National Institute of Standards and Technology – NIST (U.S. Department of Commerce), available at: http://www.itl.nist.gov/div898/software/dataplot/refman1/auxillar/kend_tau.htm
18. D.G. Altman, *Practical statistics for medical research* (Chapman and Hall, New York, 1991)

Smaranda Belciug is an Associate Professor at the Department of Computer Science, University of Craiova, Romania. In the last decade, after her PhD, her professional commitments have allowed her to develop the know-how regarding Artificial Intelligence applied in the healthcare sector. Even after her two children were born, she continued her research and achieved many accomplishments along the way. The year her daughter was born, 2020, her two research monographs, Intelligent Decision Support Systems – A journey to Smarter Healthcare and Artificial Intelligence in Cancer: diagnostic to tailored treatment were published by Springer Nature and Elsevier Academic Press. The books were acquired by top Universities such as MIT, Stanford of John Hopkins, and by the National Health Institute. Her studies have been published in Q1 journals or as book chapters in Springer and Elsevier. She is Associate Editor for BMC Medical Informatics and Decision Making, for the Journal of Medical Artificial Intelligence, and for International Journal of Computers in Healthcare. She is an enthusiastic partisan of the multidisciplinary approach in scientific studies, and all her research is driven by this reason. This has been recognized at multiple levels, from the wide variety of nature of the journals she has published into to the variety of journals and conferences that she reviews for.

Part III
Recent Trends in Artificial Intelligence Areas and Applications

Chapter 12
Towards the Joint Use of Symbolic and Connectionist Approaches for Explainable Artificial Intelligence

Cecilia Zanni-Merk and Anne Jeannin-Girardon

Abstract Artificial Intelligence (AI) applications are increasingly present in the professional and private worlds. This is due to the success of technologies such as deep learning and automatic decision-making, allowing the development of increasingly robust and autonomous AI applications. Most of them analyze historical data and learn models based on the experience recorded in this data to make decisions or predictions. However, automatic decision-making based on AI now raises new challenges in terms of human understanding of processes resulting from learning and of explanations of the decisions made (crucial issue when ethical or legal considerations are involved). To meet these needs, the field of Explainable Artificial Intelligence (XAI) has recently developed. Indeed, according to the literature, the notion of intelligence can be considered under four abilities: (a) to perceive rich, complex and subtle information, (b) to learn in a particular environment or context; (c) to abstract, to create new meanings and (d) to reason, for planning and decision-making. These four skills are implemented by XAI with the goal of building explanatory models, to try and overcome shortcomings of pure statistical learning by providing justifications, understandable by a human, for decisions made. In the last few years, several contributions have been proposed to this fascinating new research field. In this chapter, we will focus on the joint use of symbolic and connectionist artificial intelligence with the aim of improving explainability.

C. Zanni-Merk (✉)
Normandy University, INSA Rouen Normandie, LITIS, Rouen, France
e-mail: cecilia.zanni-merk@insa-rouen.fr

A. Jeannin-Girardon
ICube laboratory, University of Strasbourg, CNRS, Strasbourg, France
e-mail: anne.jeannin@unistra.fr

M. Virvou et al. (eds.), *Advances in Selected Artificial Intelligence Areas*,
Learning and Analytics in Intelligent Systems 24,
https://doi.org/10.1007/978-3-030-93052-3_12

12.1 Introduction

Artificial Intelligence (AI) applications are increasingly present in the professional and private sectors. This is due to the success of technologies such as machine learning and in particular deep learning and automatic decision-making that allow the development of increasingly robust and autonomous AI applications. Most of these applications are based on the analysis of data histories, to learn models based on the experience recorded in them and to make decisions or make predictions.

Automatic decision-making by Artificial Intelligence now raises new challenges in terms of human understanding of processes resulting from learning, explanations of decisions taken (crucial issue when ethical or legal considerations are involved) and human-machine communication.

To meet these needs, the field of Explainable Artificial Intelligence (XAI) has recently developed. The first works come from Darpa [1] in 2016, with their project *CwC & XAI* [1] and more recently the Villani[1] report [2] in its part 5 "Which AI ethics?" also emphasizes the urgent need to support research on explainability (p. 145 of the report).

Indeed, the notion of intelligence can be approached according to four axes (Fig. 12.1):

- the ability to perceive rich, complex and subtle information
- the ability to learn in a particular environment or context
- the ability to abstract, to create new meanings
- the ability to reason, for planning and decision-making.

These four skills arc implemented by the new research field called Explainable AI (XAI) with the objective of building explanatory models, which make it possible to fill gaps in statistical learning, by providing justifications, understandable by a human, for decisions or predictions made.

Summarizing, XAI aims to develop machine learning techniques, associating them closely with symbolic approaches (represented in particular by semantic technologies) with three main goals:

- to produce more understandable learning models from the data, while maintaining high accuracy of predictions;
- to facilitate a "symmetrical" communication between people, experts or not, and machines, including computers;
- to enable human users to understand, trust and effectively manage the new constellations of connected smart machines that are developing.

In this article, we will discuss some current works on XAI published in the literature, with the goal of outlining two research directions that we think, are the most promising to go *Towards and Explainable AI.*

[1] Cédric Villani is a French mathematician and politician.

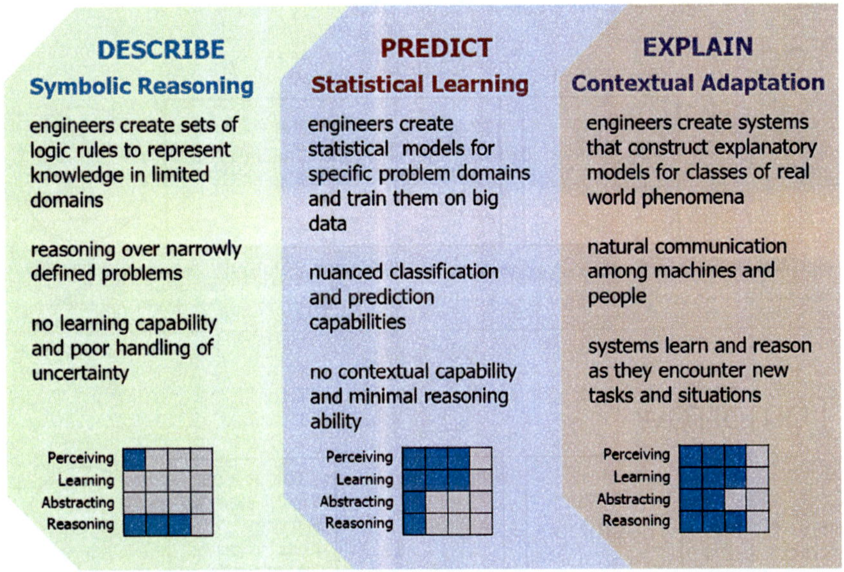

Fig. 12.1 The three waves of Artificial Intelligence [1]

After a literature revie of current works on XAI in Sects. 12.2, 12.3 introduces some works of our teams on the joint use of symbolic and inductive approaches of artificial intelligence for XAI. Finally, Sect. 12.4 presents our conclusions and perspectives of future work.

12.2 Literature Review

New and powerful deep learning techniques (based on deep neural networks or DNN) are extremely efficient to make classifications (see, for example, the impressive results in classification of images of ImageNet [3]). However, when used to make decisions, there are sometimes when they fail [4].

It is important, therefore, that errors made by an artificial intelligence be explained [5]. The reasons why can be classified into four types:

- Improvement: If the cause of the error is understood, this is a source for improvement.
- Control: Explanations can also enable an enhanced control on the behaviour of a system. Explanations can also help prevent things from going wrong. Indeed, understanding more about system behavior provides greater visibility over unknown vulnerabilities and flaws, and helps to rapidly identify and correct errors in low criticality situations (debugging).

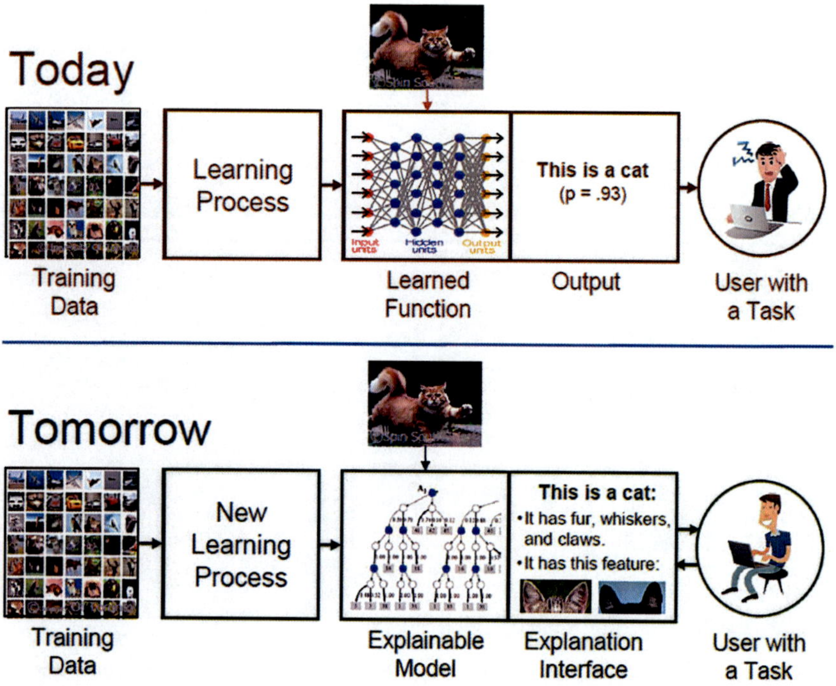

Fig. 12.2 Target architecture for an XAI [1]

- Discovering. Asking for explanations is a helpful tool to learn new facts, to gather information and thus to gain knowledge. It could be expected that, in the future, XAI models will teach us new and hidden laws in different scientific domains.
- Justification: Perhaps the most all-encompassing reason for developing explainability is the General Data Protection Regulation (GDPR) which states

> The data subject should have the right ... to obtain an explanation of the decision reached ... [5].

The researchers from DARPA propose a new architecture for XAI, based on the classical architecture for learning systems (Fig. 12.2 [1]), where a new learning process is associated with two new "modules" that are, to our understanding, the main challenges to focus on:

- The explanation interface, where the notion of "explanation" needs to be formalized,
- The explainable model, where the results of the classification could be associated with a set of rules (or other formalism) to ground the decisions being made, and where the explanation can be extracted from.

12.2.1 The Explainable Interface

An explanation is the ability to answer a how or why question and to answer a follow-up question to clarify a certain situation in a particular context. There are different levels of explanation

1. Interpretable explanations: It is possible to identify why an input goes to an output
2. Explainable explanations: It is possible to explain how inputs are mapped to outputs
3. Reasoned explanations: It is possible to conceptualise how inputs are mapped to outputs

The literature presents two approaches to formalizing explanations. Top-down approaches intend to use theories coming from social sciences and bottom-up approaches study explanation models from different disciplines to abstract a generic explanation model.

Regarding top down approaches in detail, they use results from social science on how humans explain decisions and behaviours to each other [6]. The main conclusions of these works are that explanations are contrastive or counter-factual: people do not ask why event E happened, but rather why event E happened instead of some event F. Also, explanations are selective and focus on one or two possible causes and not all causes for the recommendation. Similarly, one must be selective in the choices of the counterfactual simulations. The use of counterfactual explanation in narrative is a common part of conversation. Automated explanation must be capable of arguing both in favor of an answer as well as against proposed alternative answers. Finally, explanations are social conversation and interaction for transfer of knowledge, implying that the explainer must be able to leverage the mental model of the explainee while engaging in the explanation process.

The authors of [7], in contrast, have studied theories and models of explanation in different disciplines to identify the minimal needed subset of components, including

- An *explanans* (the explanation)
- An *explanandum* (what is being explained)
- A context (or situation) that relates the two
- Some theory (a description that represents a set of assumptions for describing something, usually general. Scientific, philosophical, and commonsense theories can be included here)
- An agent producing the explanation.

These authors produced an Explanation Ontology Design Pattern (Fig. 12.3). However, despite the existence of these works, there is still a lack of a clear and shared definition of explainability.

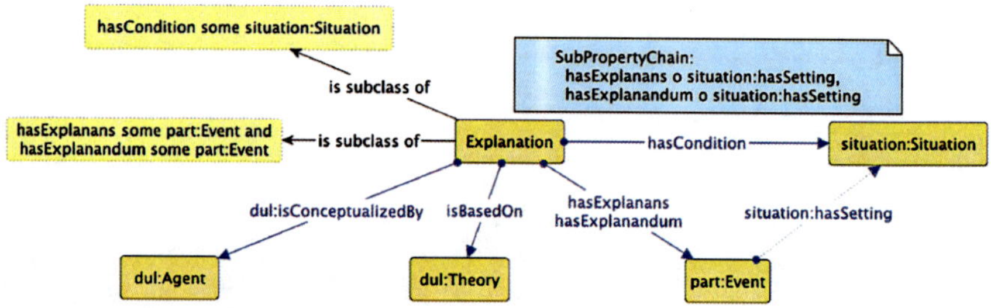

Fig. 12.3 The Explanation Ontology Design Pattern [7]

12.2.2 The Explainable Model

Early work on explanation often focused on producing visualizations of the prediction to assist machine learning experts in evaluating the correctness of the model

Beyond visualisation, research has focused on two broad approaches to explanation [8]

1. *Prediction and Interpretation*: a (usually non-interpretable) model and prediction are given, and a justification for the prediction is produced
2. *Interpretable Models*: design models that are intrinsically interpretable and that can be explained through reasoning.

The *Prediction and Interpretation* approach has focused on interpreting the predictions of complex models, often by proposing to isolate the contributions of individual features to the prediction. There are model-specific approaches, specifically built for different classifiers (Bayesian networks, Multi-Layer perceptrons, RBF networks; SVM classifiers) whose performance heavily depends on the classifier itself. There are the so-called model-agnostic approaches, that measure the effect of an individual feature on an unknown classifier's prediction by checking what the prediction would have been if that feature value was absent and comparing the two using various distance measures. Finally, there are also model approximation approaches whose goal is deriving a simple, interpretable model (such as a shallow decision tree, rule list, or sparse linear model) that approximates a more complex, uninterpretable one (e.g., a neural net).

The disadvantage of these early approaches is that for even moderately complex models, a good global approximation cannot generally be found. The authors of [9] introduce an approach that focuses on local approximations, which behave similarly to the global model only in the vicinity of a particular prediction. Their algorithm is agnostic to the details of the original model.

Concerning the development of *Interpretable Models*, their main idea is to directly produce models that are inherently interpretable, such as shallow rule-based models: decision lists and decision trees, sparse models via features selection or extraction to optimize interpretability, compositional generative models which are constrained to learn hierarchical, semantically meaningful representations of data, techniques to learn more structured, interpretable, causal models. Not much work has been published in the literature yet. However, the disadvantage of creating directly shallow rule-based models or sparse models is that the prediction accuracy and robustness can be compromised. An interesting direction to explore is to try and develop models that are constrained to learn semantically meaningful representations, along with interpretable causal models, where some initial proposals can be found in [10, 11].

12.3 New Approaches to Explainability

For AI systems to be understandable, symbolic systems and reasoning must be integrated to provide operationally effective communications of the internal state and operation of the system.

Meaningful ontologies and knowledge bases, and Symbolic AI methods are fundamental (but currently, mostly ignored) to the most important explainable AI use-case: when in practice, within some context, a final user must understand, trust, and be responsible for the conclusions an AI system draws.

In fact the relationships between symbolic reasoning and machine (deep) learning can be viewed according to two dimensions:

- **Deep learning for semantics**: Ontologies are widely used for knowledge representation and reasoning about semantic content in a structured way. However, manual ontology development is a hard and expensive task that usually requires knowledge of domain experts and skills of ontology engineers. Handcrafted ontologies are often inflexible and complex, which limits their usefulness and makes cross-domain alignment difficult. Recent advances in deep learning [12] have the potential to help mitigate these issues with automatic ontology development and alignment. Hohenecker and Lukasiewicz [13], for example, present a novel deep learning based reasoning approach rather than logic-based formal reasoning to ontology reasoning.
- **Semantics for deep learning**: Conversely, there are quite a few works on semantics for deep learning. Research in semantic data mining [14] has shown that the integration of semantics and data often presents better results on the tasks of data mining and deep learning. However, in most cases of application, formal knowledge representation is usually not well explored. Knowledge is generally encoded in a highly formal and abstract way. In most machine learning and deep learning algorithms, the input is always in the form of a numerical vector. It is often impractical to apply semantics directly to raw data. Previous works have shown that semantics is mostly applied to data pre-processing by enriching input vectors with ontological concepts and properties [15, 16], and post-processing by using distance measures with ontology semantics [17] rather than in the key stages of machine learning approaches which include model design and training. However, knowledge representation approaches such as ontologies have a similar "geometrical" structure to the ones of deep neural networks. Exploring semantics both in the

learning and the interpretation process may enhance effectiveness and flexibility of the machine learning approach. However, few advances have been made from this perspective [18, 19].

For explainability issues, we are more concerned about the second item: semantics for deep learning. And it is needed to be discussed thoroughly where to include knowledge along a learning task to improve it:

- **Before** the learning task: Even if sufficient amounts of data are available, it is often tedious and time consuming to train a DNN from scratch. A common practice is to pre-train such models using data from a relative domain to learn representations that will afterwards be refined. Given the highly curated nature of knowledge bases and ontologies (because they are built by domain experts), such source of information would provide a solid pre-training ground for deep models. However, such knowledge representations cannot be transferred as is: shared and transferable representations of domain knowledge (preserving semantics) must be built in order for DNN to use them.
- **During** the learning task: As it will be discussed in Sect. 12.3.2, there are already some preliminary works on the exploitation of knowledge during a learning task, by using an generic ontology to design the structure of the neural network used. The knowledge in the ontology assists in identifying the most pertinent explanation for the prediction made by the network.
- **After** the learning task: Classification models are compelled to do predictions when given an input, even if these inputs do not belong to the domain used to train the model. It would then be first useful to evaluate the uncertainty of predictions [20] and exploit this (un)certainty to obtain a model that knows and says when it does not know. Unknown inputs could be brought together with ontologies and knowledge bases to determine, at a more or less fine grain, the domain to which they belong. Here again, there is a need for building shared representations to be able to "understand" unknown data exploiting the reasoning mechanisms of ontologies.

Using ontologies before and after learning requires an intermediate model that would learn a shared (and thus, transferable) representation of heterogeneous data and knowledge.

In the following subsections, we will discuss the need of a clear definition of the terminology associated to explanations and explainability and give some preliminary results of our teams showing the joint use of symbolic approaches and machine/deep learning to improve explainability.

12.3.1 Towards a Formal Definition of Explainability

As already evoked, XAI has become quite popular in the past few years, thanks to the publication of new international laws that promote the "right to explanation". Many

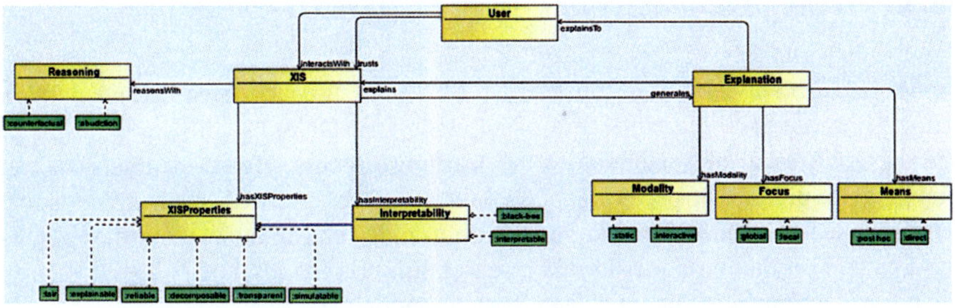

Fig. 12.4 A terminology for a fully contextualized XAI [21]

popular methods have been developed recently to help understand black-box models, but it is not clear yet how an explanation is defined. Furthermore, the community agrees to say that many important terms do not have commonly accepted definitions.

Belluci et al. [21] show that there is a major issue concerning the definitions of terms such as *explainability* or *interpretability*. There is a lack of consensus that slows the development of this field down. To address this problem, they have proposed a terminology that takes into account the full context of an intelligent system and characterizes an explanation.

They have introduced he notion of *eXplainable Intelligent System* (XIS) which corresponds to an explainable model paired with an explanation interface. They have also highlighted the importance of interactivity to provide explanations adapted to a certain kind of user with a certain task.

The main goal of the proposed terminology is the development of different metrics associated to XISs using the clearly defined and unambiguous terms (Fig. 12.4).

The main contribution of this proposal is that an explanation is characterized by three well defined properties: *Focus*, *Means* and *Modality*. *Focus* answers the question "what is being explained?", *Means* answers the question "By what means is it explained?" and finally, *Modality* answers "how is it explained?". These properties assist in understanding what the objective of an explanation is. They also clarify which methods are comparable and why. Additionally, the values associated to each of these properties of an explanation will help associate it implicitly to the context of the XAI task.

Finally, a part of the terminology is dedicated to trust and related concepts in XISs. These concepts do not directly contribute to explainability, but they participate in increasing trust in an XIS. The authors highlight that the design of a Responsible AI is a common goal of the XAI community. Responsible AI will allow XISs to be broadly used because they will be trusted and accountable for their decisions. Explainability is the first milestone to reach this goal.

12.3.2 Using Ontologies to Design the Deep Architecture

Some works have been done on the use of the architecture of a certain domain ontology to design the architecture of an associated DNN [14, 18] as a means to add a certain form of semantics to the learning process. However, this does not immediately mean that perfect explainability will be achieved, even if the native reasoning mechanisms of ontologies can be used to try and understand the reasons for a certain prediction done. In this research line, Huang et al. [22] have explored the use of semantics to enhance deep learning models.

These authors have explored a novel approach to enhance deep learning models by introducing semantics into a deep learning process in the framework of predictive maintenance in Industry 4.0. Specifically, they have proposed an ontology-based model *OntoLSTM* (Fig. 12.5), in which a core manufacturing process ontology is used to design the deep neural networks that will be used for learning, including generic concepts such as lines, machines, sensors.

From a specific production line (Fig. 12.6 proposes an example), the ontology in Fig. 12.5 is instantiated (Fig. 12.7). This instantiation permits the inference of the structure of a deep architecture that will be used to learn high-level cognitive features, including stacked LSTM layers for learning temporal dependencies (Fig. 12.8 shows the resulting deep neural network for the sample line in Fig. 12.6).

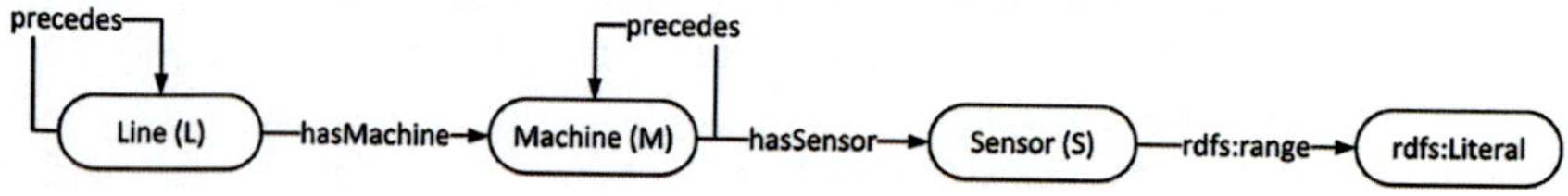

Fig. 12.5 *OntoLSTM* - an abstract ontology for manufacturing time series [22]

Fig. 12.6 A sample production line

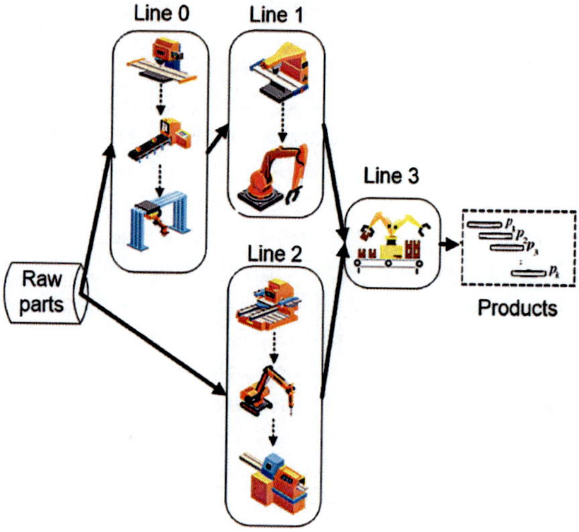

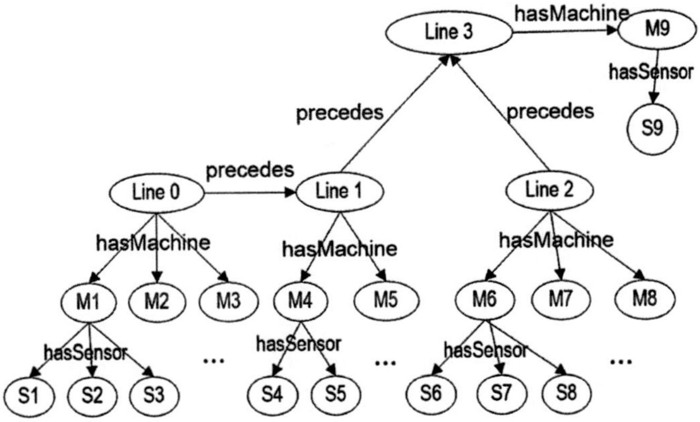

Fig. 12.7 Instantiation of the abstract ontology for the sample production line

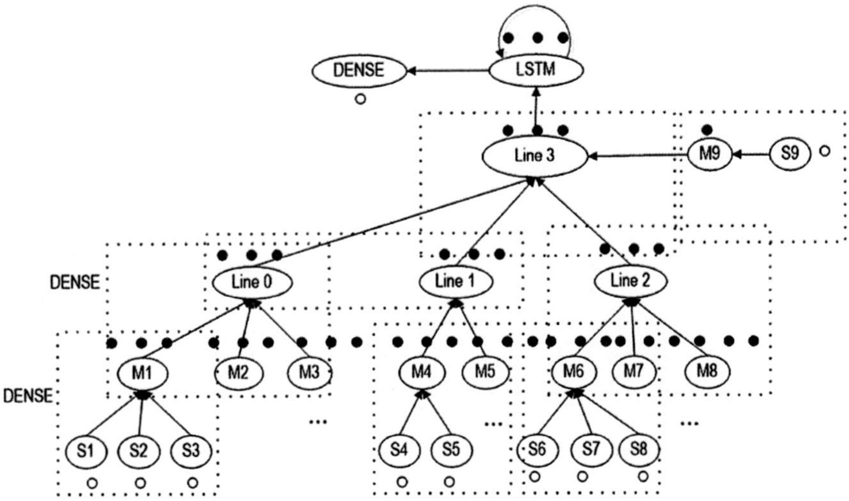

Fig. 12.8 The structure of the resulting network for the sample production line based on *OntoLSTM*

To evaluate the feasibility of the proposed approach, the authors have performed a series of experiments by using a dataset proposed by a real manufacturing company.[2] The goal is to predict which products will fail quality control and to explain the reasons for the failure.

The performance of the OntoLSTM-based deep architecture was compared to DenseLSTM (a deep learning method which stacks several dense layers and a LSTM layer) and AutoencoderLSTM (that stacks Autoencoder and LSTM networks encoding the high dimensional input data to the hidden layers by using relevant activation

[2] https://www.kaggle.com/c/bosch-production-line-performance.

functions and trying to reconstruct the original inputs through the decoder layer to minimize the reconstruction loss). The obtained results are encouraging, as the proposed architecture outperforms the other two methods.

In general, we consider this approach promising to improve explinability. By the use of a generic abstract ontology, the method can be applied to any manufacturing process. Our research groups are currently experimenting this approach on other domains where knowledge can be formally structured to infer the deep architectures to improve the explainability of the learning task.

12.3.3 Coupling DNN and Learning Classifier Systems

Alternative symbolic approaches include Expert Systems, and, in particular, Learning Classifier Systems (LCS) [23]. LCS are rule-based machine learning systems that autonomously build interpretable rules (also called "classifiers") to solve a task in the environment in which they evolve. Such systems can be used to different problems such as classification tasks, function approximations or reinforcement learning tasks. Like many "early-days" machine learning approaches, LCS have fallen into disuse because of the difficulty to scale them up. However, their intrinsic interpretability could very well be a major reason for a come back.

Among LCS, Anticipatory Learning Classifier Systems (ALCS) rely on a cognitive mechanism called Anticipatory Behavioral Control [24]: ALCS build their internal classifiers by discriminating between perceptions of the environment so that they can anticipate possible changes induced by an action in a given situation. In ALCS, the learning process relies on experience only and does not involve stochasticity, thus allowing the system to learn correlations between causes and effects, which provide, to an external eye, explanatory insights regarding the actions performed by the system (which action is performed in what condition). ALCS enhancements and/or variants have been proposed through the years, such as ACS2 [25], YACS [26], XACS [27], MACS [28].

BACS [29] is the first extension of ACS2 to integrate Behavioral Sequences that allow the system to anticipate *sequences of actions*, thus increasing the autonomy of the system and its ability to deal with non deterministic environments. Figure 12.9 illustrates the ability of BACS to deal with uncertain environments thanks to Behavioral Sequences. This model has been further enhanced by Probability Enhanced Predictions (PEP). PEPS are used to evaluate the probability of a set of anticipations in a given context and thus reinforce the interpretability of the reasoning undertaken by the system in its environment [30]. Figure 12.10 shows rules generated by PEP-enhanced BACS in uncertain environments: each classifier is made of a Behavioral Sequence encoded as the set of all local perceptions of the system (here, an order

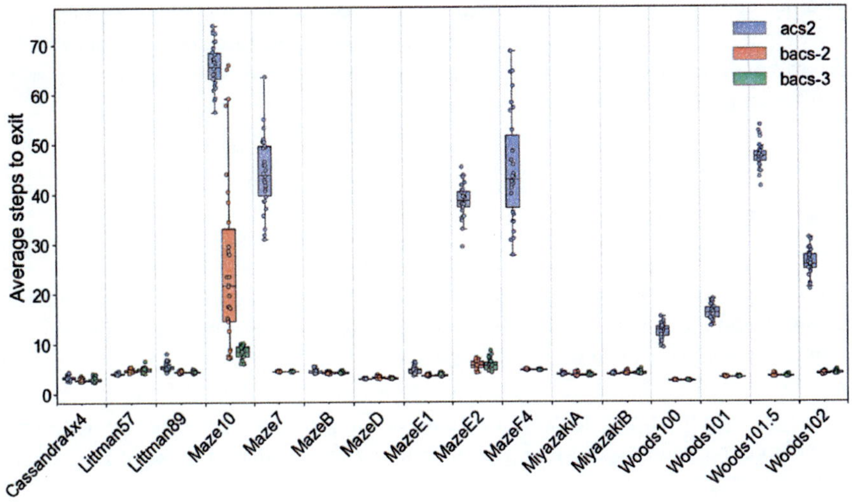

Fig. 12.9 Performance of BACS in uncertain environments: using Behavioral Sequences allows the systems to perform more efficiently than other ALCS in maze environments (the performance of the systems is measured as the average number of steps required by the systems to reach the exit over 30 runs) [29]

Fig. 12.10 Rules generated using both Behavioral Sequences and Probability Enhanced Prediction in uncertain environments [30]

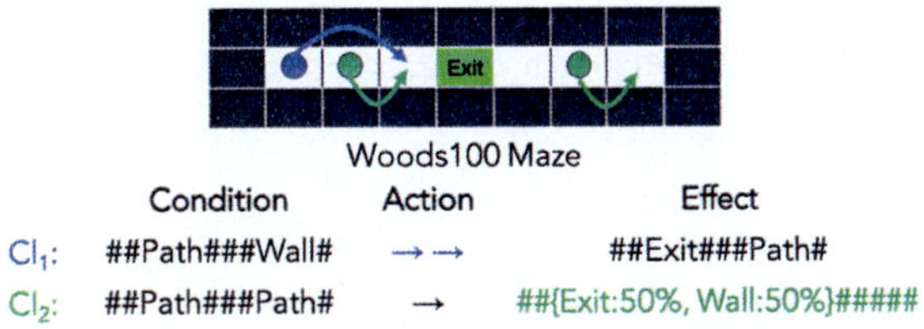

1 Moore neighborhood, that is, the eight adjacent cells to the agents in the environment). The actions that can be undertaken by the systems lead to the anticipation of the effect these actions have on the environment. When enhanced by PEP, several effects, associated with probabilities, are anticipated. This simple example shows the benefit of having a system with interpretable rules when it comes to decision or prediction making.

Such models can be put in perspective within a larger scheme integrating several paradigms aiming at providing more autonomy and more interpretability to machine learning systems [31]: instead of relying exclusively on deep model to make prediction and decisions, this DNN could merely be an entry point to feed facts to a reasoning system such as an ALCS. This approach is aligned with Spinoza's approach of human cognition [32]: while the *imagination* produces incomplete representations of the environment (non determinism), *reason* jointly uses these perceptions and the system's experience to infer decisions and/or actions, possibly anticipating the results of different possible outcomes.

12.4 Conclusions

In this chapter we have explored the joint use of formal knowledge models, symbolic reasoning and machine/deep learning to take a step towards a more explainable AI.

In particular, we have discussed different scenarios illustrating where to integrate knowledge models in a learning task. We have presented three different proposals carried out by our teams; mainly, a formal terminology for XAI, the use of formal models to induce the structure of a deep architecture and the joint use of DNNs and learning classifier systems.

Semantics for deep learning will also require to put an emphasis on Knowledge Representation and Transfer Learning if we want to achieve a coupling of semantic approaches to deep learning approaches. Indeed, DNN are somewhat limited in terms of input (as they can only process numerical vectors) and injecting semantic knowledge in such models requires a common representation of this knowledge that preserves it. This representation could be used both for feeding a DNN with knowledge, but also to understand predictions and decisions made by DNN by using their outputs jointly with an ontology. This area is the focus of our future research.

References

1. D. Gunning, Explainable artificial intelligence research at DARPA. https://www.darpa.mil/program/explainable-artificial-intelligence. Accessed 06 Jan 2020
2. C. Villani, Donner un sens à l'Intelligence Artificielle. Pour une stratégie nationale et européene. https://www.aiforhumanity.fr. Accessed 06 Jan 2020
3. J. Deng, W. Dong, R. Socher, L. Li, K. Li, L. Fei-Fei, ImageNet: a large-scale hierarchical image database, in *2009 IEEE Conference on Computer Vision and Pattern Recognition, Miami, FL, 2009*, pp. 248–255. https://doi.org/10.1109/CVPR.2009.5206848
4. 2018 in Review: 10 AI Failures. https://medium.com/syncedreview/2018-in-review-10-ai-failures-c18faadf5983. Accessed 06 Jan 2020
5. A. Adadi, M. Berrada, *Peeking Inside the Black-Box: A Survey on Explainable Artificial Intelligence (XAI) in IEEE Access*, vol. 6 (2018), pp. 52138–52160. https://doi.org/10.1109/ACCESS.2018.2870052
6. T. Miller, Explanation in artificial intelligence: insights from the social sciences (2017) https://arxiv.org/abs/1706.07269. Accessed 20 Jan 2020
7. I. Tiddi, M. d'Aquin, E. Motta, An ontology design pattern to define explanations, in *Proceedings of the 8th International Conference on Knowledge Capture* (ACM, 2015); Article no. 3

8. O. Brian, C. Cotton, Explanation and justification in machine learning: a survey, *IJCAI-17 Workshop on Explainable AI (XAI)* (2017)
9. M.T. Ribeiro, S. Singh, C. Guestrin, Why should i trust you? Explaining the predictions of any classifier, in *KDD* (2016)
10. J. Chen, F. Lecue, J. Pan, I. Horrocks, H. Chen, Knowledge-based transfer learning explanation, in *Principles of Knowledge Representation and Reasoning: Proceedings of the Sixteenth International Conference, Oct 2018, Tempe, United States*
11. F. Lecue, J. Chen, J.Z. Pan, H. Chen, Augmenting transfer learning with semantic reasoning, in *Proceedings of the Twenty-Eighth International Joint Conference on Artificial Intelligence (IJCAI-19)* (2019), pp. 1779–1885
12. M.A. Casteleiro, M.J.F. Prieto, G. Demetriou, N. Maroto, W.J. Read, D. Maseda-Fernandez, J.J. Des Diz, G. Nenadic, J.A. Keane, R. Stevens, Ontology learning with deep learning: a case study on patient safety using pubmed, in *SWAT4LS* (2016)
13. Patrick Hohenecker and Thomas Lukasiewicz. Deep learning for ontology reasoning (2017), arXiv:1705.10342
14. D. Dou, H. Wang, H. Liu, Semantic data mining: a survey of ontology-based approaches, in *Semantic Computing (ICSC), 2015 IEEE International Conference* (IEEE, 2015), pp. 244–251
15. A. Hotho, A. Maedche, S. Staab, Ontology-based text document clustering. KI **16**(4), 48–54 (2002)
16. A. Hotho, S. Staab, G. Stumme, Ontologies improve text document clustering, in *Third IEEE International Conference on Data Mining, 2003. ICDM 2003* (IEEE, 2003), pp 541–544
17. L. Jing, L. Zhou, M.K. Ng, J. Zhexue Huang, Ontology-based distance measure for text clustering, in *Proceedings of SIAM SDM Workshop on Text Mining, Bethesda, Maryland, USA, 2006*
18. N. Phan, D. Dou, H. Wang, D. Kil, B. Piniewski, Ontology-based deep learning for human behavior prediction with explanations in health social networks. Inf. Sci. **384**, 298–313 (2017)
19. H. Wang, D. Dou, D. Lowd, Ontology-based deep restricted boltzmann machine, in *International Conference on Database and Expert Systems Applications* (Springer, 2016), pp. 431–445
20. H.M.D. Kabir, A. Khosravi, M.A. Hosen, S. Nahavandi, Neural network-based uncertainty quantification: a survey of methodologies and applications. IEEE Access **6**, 36218–36234 (2018)
21. M. Bellucci, N. Delestre, N. Malandain, C. Zanni-Merk, Towards a terminology for a fully contextualized XAI. *Submitted to KES 2021 - 25th International Conference on Knowledge-Based and Intelligent Information & Engineering System*
22. X. Huang, C. Zanni-Merk, B. Crémilleux, Enhancing deep learning with semantics : an application to manufacturing time series analysis, in *KES 2019 - International Conference on Knowledge-Based and Intelligent Information & Engineering Systems. T. 159* (Elsevier, 2019), pp. 437–446. https://doi.org/10.1016/j.procs.2019.09.198
23. O. Sigaud, S.W. Wilson, Learning classifier systems: a survey. Soft Comput. **11**(11), 1065–1078 (2007)
24. W. Stolzmann, An introduction to anticipatory classifier systems. *International Workshop on Learning Classifier Systems* (Springer, Berlin, 1999)
25. M.V. Butz, W. Stolzmann, *An Algorithmic Description of ACS2, International Workshop on Learning Classifier Systems* (Springer, Berlin, 2001)
26. P. Gerard, W. Stolzmann, O. Sigaud, YACS: a new learning classifier system using anticipation. Soft Comput. **6**(3–4), 216–228 (2002)
27. M.V. Butz, D.E. Goldberg, Generalized state values in an anticipatory learning classifier system, in *Anticipatory Behavior in Adaptive Learning Systems* (Springer, Berlin, 2003), pp. 282–301
28. P. Gérard, J.-A. Meyer, O. Sigaud, Combining latent learning with dynamic programming in the modular anticipatory classifier system. Eur. J. Oper. Res. **160**(3), 614–637 (2005)
29. R. Orhand, A. Jeannin-Girardon, P. Parrend, P. Collet, BACS: a thorough study of using behavioral sequences in ACS2, in *International Conference on Parallel Problem Solving from Nature* (Springer, Cham, 2020), pp. 524–538

30. R. Orhand, A. Jeannin-Girardon, P. Parrend, P. Collet, PEPACS: integrating probability-enhanced predictions to ACS2, in *Proceedings of the 2020 Genetic and Evolutionary Computation Conference Companion* (2020), pp. 1774–1781
31. R. Orhand, A. Jeannin-Girardon, P. Parrend, P. Collet, DeepExpert: vers une Intelligence Artificielle autonome et explicable. In Rencontres des Jeunes Chercheurs en Intelligence Artificielle **2019**, 63–65 (2019)
32. Baruch Spinoza, Ethique, III, prop.6

Cecilia Zanni-Merk is a full professor of Computer Science at INSA Rouen Normandie with the LITIS research laboratory (France). She is the head of the MIND (multi-agents, interaction, decision) research group of LITIS and the deputy head of the MIIS (mathematics, information science and technology, systems engineering science) doctoral school. Her main research interests are in Knowledge Engineering, and more particularly in conceptual representation and inference processes applied to problem solving. The main keywords associated to Professor Zanni-Merk's research works are conceptualisation, ontologies and formal models, rule? based (crisp, fuzzy, probabilistic, spatio-temporal) reasoning, case-based reasoning, knowledge and experience capitalisation.

Anne Jeannin-Girardon is an associate professor of Computer Science at the University of Strasbourg (France). Her research focuses on machine learning and artificial intelligence. She is interested in both fundamental artificial intelligence (explainability, transfer learning, knowledge representation) and applications, in particular in the domain of bio-informatics and health ecosystems. Her recent work includes the area of the social integration of AI and particularly how AI can become more ethical and explainable.

Chapter 13
Linguistic Intelligence As a Root for Computing Reasoning

Daniela López De Luise

Abstract From ancient times it was the *vocal Language* an interesting tool for interchanging thoughts, the outcomes of person's reasoning. But it also constitutes the golden casket where encoding sets deep information clues about the internal and external environment of the individual. The exact nature of the reasoning performed by humans is not yet fully understood. But this chapter revisits part of the author's research effort to provide envision on how it would be technically evaluated, parameterized, and reapplied to other contexts. This chapter considers language processing as a tool for reasoning, understanding and mimics. The approach to do that is not the classical semantics but an exploration of alternate combinations of Morphosyntactics, Berwick's verbal behavior, entropy, and fractals. Along the way, emerge several useful tools to asses and evaluate the reasoning dynamics and the information flow thru diverse language pragmatics. Some of those outcomes are revisited here: *Morphosyntactic Linguistic Wavelets* (***MLW***), phrasal auto-expansion, learning profiling tools in Video Games (VG), reasoning and understanding of patients with Autistic Spectrum Disorders (ASD), evaluation of specific pedagogic Science Technology Engineering Art & Mathematics (STEAM) activities, consciousness for help building autonomous mobile robotics displacement and prediction tools for precision crops and pedestrian/vehicular risk management. From the author's perspective, there is still a long road to follow. Many pending questions must be answered before being able to overcome the issue of mimic natural language. She is currently working on a theory of language communication that may explain the underlying reasoning process.

D. L. De Luise (✉)
CI2S Labs (Computational Intelligence and Information Systems Labs), Buenos Aires, Argentina
e-mail: daniela_ldl@ieee.org

Center of High Studies in Information Technology), CAETI (Centro de Altos Estudios En Tecnología Informática, Universidad Abierta Interamericana (Inter-American Open University), Buenos Aires, Argentina

Laboratorio de Investigación Y Desarrollo en Tecnologías Informáticas, Research and Developing in Informatics Technologies Labs), IDTI Labs, Universidad Autónoma de Entre Ríos, UADER (Autonomous University of Entre Ríos), Buenos Aires, Argentina

M. Virvou et al. (eds.), *Advances in Selected Artificial Intelligence Areas*,
Learning and Analytics in Intelligent Systems 24,
https://doi.org/10.1007/978-3-030-93052-3_13

Keywords Natural language processing · Reasoning · Morphosyntactic linguistic wavelets · Harmonic systems · Robotic consciousness · Bacteria consciousness · STEAM metrics · Video games · Autistic spectrum disorder

13.1 Introduction

Although Natural Language Processing is usually focused in textual data, part of traditional methods includes some kind of statistical and heuristic processing of utterances and discourse analyses. It provides certain quality of linguistic reasoning to an automated system.

From language perspective every linguistic level relays in some way on morphology. Morphosyntactics describes the set of rules that govern linguistic units whose properties are definable by both morphological and syntactic paradigms. It may be though as a basis for spoken and written language that guides the process of externally encoding ideas produced in the mind. But semantic and ontological elements of speech become apparent though linguistic pragmatics. In fact, words distribution has been proven to be related to the semantics and ontology. Thus words become a powerful lens into human thought and other manifestations of collective human dynamics [1].

Perhaps one of the first big contributions in this perspective is Zipf's Empirical Law. It was formulated using mathematical statistics. It refers to the fact that many types of data studied in the physical and social sciences can be approximated with a Zipfian distribution, one of a family of related discrete power law probability distributions as in Fig. 13.1.

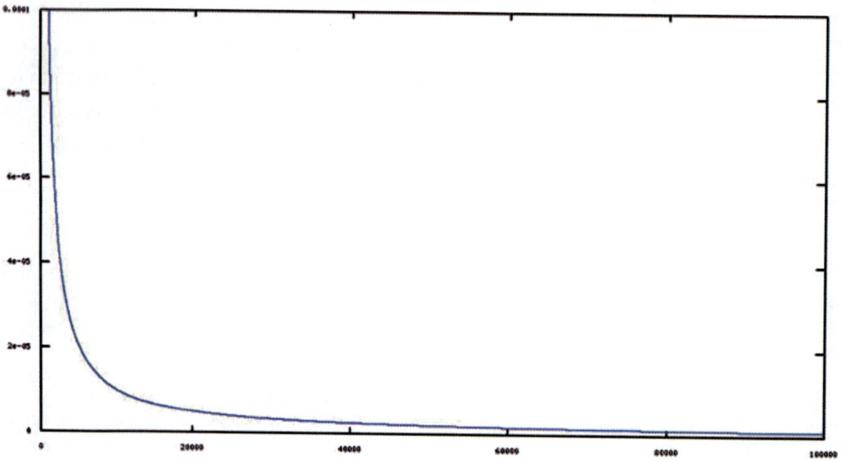

Fig. 13.1 Zipf law for texts in English

Zipf's law applies to most languages, that given some corpus of natural language utterances, the frequency of any word is inversely proportional to its rank in the frequency table. Equation 13.1 describes the behavior.

$$P_n \sim 1/n^a \tag{13.1}$$

Here $\boldsymbol{P_n}$ stands for the word positioned as the $\boldsymbol{n}$-th most frequent, and $\boldsymbol{a}$ is some real number. Experts explain this linguistic phenomenon as a natural conservation of effort in which speakers and hearers minimize the work needed to reach understanding, resulting in an approximately equal distribution of effort consistent with the observed Zipf distribution.

Following this concept, speech can be processed taking the utterances and its energy distribution. Part of the present article introduces other models like this. There is also an alternate approach for a systematic explanation of language production and managing, using heuristics and mining tools: The **M**orphosyntactic **L**inguistic **W**avelets (***MLW***). This provides a possibility to approximate abstraction process with reasonable precision during learning. It defines a process as a progressive sequence of filtering and selection steps. The approach is named ***MLW*** due to its management of granularity, analogous to traditional wavelets.

In the rest of this paper, a summary of the author's main work in this field is explained. Starting from language as a tool for communication (Sect. 13.2) which covers an introduction to ***MLW***, an explanation of how utterances in a dialog can be modeled by fractals and some applications. Then, there is an introduction on how language can be used in learning process to teach and to asses the process itself (Sect. 13.3). In the rest of this article, language is considered as an encoding of information, and introduces other ways that transmit information: In Sect. 13.4 there is an extension of seminal work from D. Hoftädter and S. Franklin with systems that model thru consciousness, considering elements in biology as a type of encoding similar to language. Section 13.5 presents information encoded as music harmony. Finally, in Sect. 13.6 are the conclusions, current and future works.

13.2 Language as a Tool for Communication

Language is one of the main encodings for human interactions. There are several channels to interchange language information. But it is usually reduced to visual or auditory representations. Many times the visual expression consists of a set of numbers, a specific set of sign, icons, or ideograms. Among them only words and numbers are linked to specific sounds there. They also have an agreed international meaning. That is the reason of their success and relevance as communication tool.

This section introduces ***MLW***, an automated approach to model its usage in current natural language pragmatics, and the relationship between fractals and oral dialog.

13.2.1 MLW

Since 2003, **C**omputational **I**ntelligence and **I**nformation **S**ystems **L**abs (CI2S Labs) has published a set of preliminary analysis [3] explaining how humans express itself using an encoding that is based on mathematics equations. The set of equations and steps to interrelate them were collected as a heuristic called **M**orphosyntactic **L**inguistic **W**avelets (***MLW***).

The brain can be compared with a unique device that can perform highly complex processing and at the same time is able to change itself with plasticity. This ability must be also modeled in order to generate an accurate representation.

Data driven learning is useful to determine the exact bias of natural language performance. But it has many problems to focus deep inside linguistics. That may happen due to the statistical bias, or because the heuristic bias is too much.

In any case, ***MLW*** is just an approach to heuristically convene a way to model semantics without using man-made ensembles of language productions like dictionaries or special ontological metadata. Nowadays, many data heuristics in deep learning claim to be suitable for linguistic processing, and are pretty similar to some part of the old fashioned ***MLW***.

This theory, constitutes a mimic of classical wavelets used for signal processing (typically sound or image data), that are very powerful since they can compress and successfully represent diverse and complex analogical data.

The first and main challenge for ***MLW*** is the systematic and automatic derivation of mathematical representations from something that is not numeric. Also to automatically represent subjectivity, context and derive common sense parameters. To overcome these tasks means to be able to model successfully not only physical and ontological bias of words, but also for sentences and context.

Using a wavelet like approach it is possible to process texts, expressions, and gain understanding of more subtle problems like early bilingualism and deaf effects on learning [2].

One interesting derivation of this technology is the symbolic language, an alternate representation of spoken words that entails the language flow of intended expressions in a universal way. Table 13.1 shows a few symbols and the rules of its usage for Spanish.

Table 13.1 *MLW* rules for symbolic representation in Spanish

ID	RULE OF APPLICATION	SYMBOLIC REPRESENTATION
1	A de B [y | o C[y |o ...]]	
4	A y|e B [XXX]	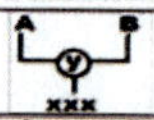
7	A o B [XXX]	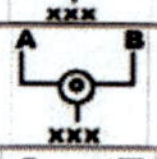
31	A se B	A→B

The translation from textual to symbolic representation can be performed automatically. When it was tested with native volunteers, they were able to understand and rebuild original texts almost with the same wording as the original text.

Most of ***MLW*** consists in a progressive parameterization of current textual expressions following simple heuristics. This way, it is possible to generate a set of "descriptors" that can be used for semantic clustering (called here as E_{ce}) like in Fig. 13.2, using structures of sentence ***MLW*** (E_{ci}). It is expected that future work can use them also to generate sentences as part of speech in natural language.

The underlying hypotheses is that the fewer hand made structures applied, the better language dynamics bias is taken into consideration and therefore the inherent subtle devices of the language production is part of the model.

One interesting finding is that the word dynamics and concept handling can be exhibited as graphical 3D representations. Figure 13.3 for instance, shows how words

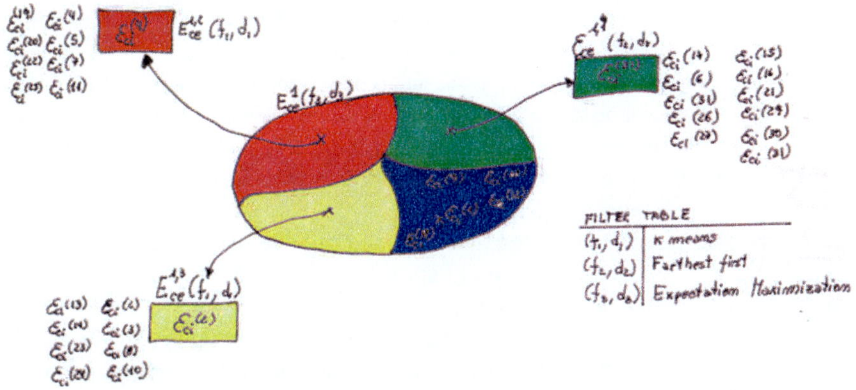

Fig. 13.2 ***MLW*** clustering using semantics

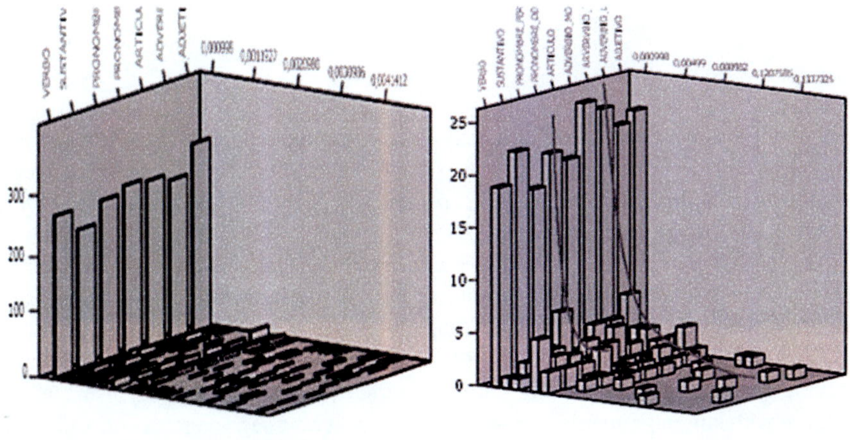

Fig. 13.3 ***MLW*** clustering using semantics as p_o in ***MLW***

elicited for sentences are not selected at random but with specific criteria (here denoted as po, one of the descriptors obtained with ***MLW***). The figure denotes on the left a random flat distribution. On the right side there is an asymptotic behavior, typical distribution in a natural language dialog (here in Spanish).

It is interesting to note that this confirms the Zipfian law, but in dialogs. Other interesting derivation is the ability to approximate the sound derivation of oral dialogs. It will be described in the following subsection.

13.2.2 Sounds and Utterances Behavior

This section describes the relevance in **N**atural **L**anguage **P**rocessing (NLP) of sound processing and phonetics and its role in human knowledge from the Verbal Behavior perspective as published in several papers. For more details see [4].

Verbal **B**ehavior (VB) has been proposed by B. F. Skinner [5]. He stated that language may be though as a set of functional units, with certain types of operant serving acting during language acquisition and production. An operant is a functional unit of language with a different and specific function. In fact he describes a group of verbal operants: Echoics, Mands, Tacts, and Intraverbals. These components of language are necessary for effective verbal communication and should be the focus of intervention meant to teach language.

It can be said that VB is shaped and sustained by a verbal environment. These practices and the resulting interaction of speaker and listener yield the phenomena which are considered here under the rubric of verbal behavior.

There is growing evidence on the onto-genetic sources of language and its development [6], and an interest in the theory that the verbal capability in humans is a result of evolution and onto-genetic development. It can be said that speech and sounds have an important relation to VB [7] and may serve as a way to identify certain disorders. An example is the case of patients suffering ASD (Autistic Spectrum Disorder).

13.2.2.1 L-Systems

In general, audible sounds are analogical signals ranging in the 20 Hz to 20 kHz. But it may vary for every individual and age (known as presbyaudia). This range covers 10 octaves (210 = 1024). Over audible spectrum there is ultrasound (acoustic waves with frequency over 20 kHz).

Approaches for sound processing range from the classical Fourier transform, spectrograms (time–frequency analysis) to very modern heuristics like dynamical spectrum, wavelets, Mel coefficients, Teager operator, signal intensity, etc. [8].

The relationship between sound, utterances and speech [7] can be modeled and become useful tools for tracking patients with several disorders. For instance, in children with language acquisition anomalies there is language degradation and even complete absence of speech.

Sound can be represented by curves and approximated by many recursive methods such as Lindenmayer systems, also known as L-Systems. An L-system is a parallel rewriting system and a type of formal grammar. It consists of an alphabet of symbols that can be used to make strings and a collection of production rules. Those grammar rules indicate that the way certain well defined symbols may expand into some larger string of symbols. There is also an initial *axiom* string from which to begin construction, and a mechanism for translating the generated strings into geometric structures.

L-systems have also been used to model the morphology of a variety of organisms and can be used to generate self-similar fractals [9] such as iterated function systems.

Let have symbols ***S* = *{X, Y, F, -, +, ε}***, the following grammar rules:

```
R1: X → X+YF+
R2: Y → -FX-Y
R3: F → ε
```

And axiom ***FX***.

Let symbol ***ε*** stands for the empty string, which means that we will delete any ***F*** from the result when the process is expanding the string using rules ***R1*** to ***R3***.

When symbols are evaluated with:

```
F: move forward one step
-: turn right 90°
+: turn left 90°
```

The resulting graph is the Dragon curve.

13.2.2.2 Dragon Curves

The dragon curve is a family of self-similar fractal curves. It is an **I**terated **F**unction **S**ystem (IFS) made up of the union of several copies of itself, each copy being transformed by function:

$$f_1(z) = (1 + i) \cdot z/2 \tag{13.2}$$

$$f_2(z) = 1--(1 - i) \cdot z/2 \tag{13.3}$$

With starting points {0, 1}, it can be graphically represented as its iteration 1, 2, 3, 4, 5, 6, 7 and 8 as in Fig. 13.4.

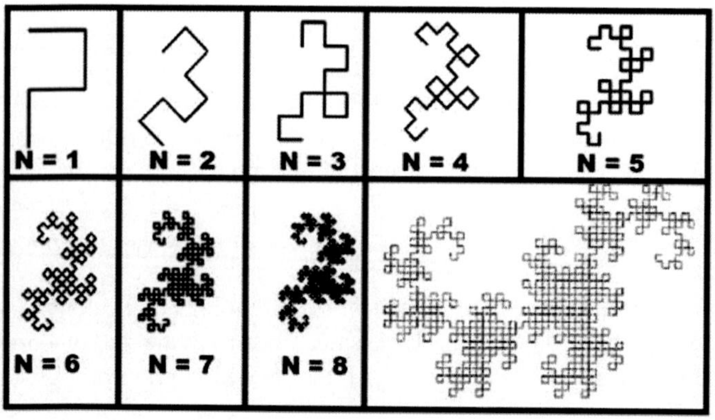

Fig. 13.4 Dragon Plot

13.2.2.3 Energy Distribution

Mining energy distribution of sound tracks leads to information that may be related to the statistical manner in which humans use spoken language. Time related features of the samples are closely related to simple changes in the signal energy. As a consequence, their changes may be used to distinguish audios. This type of analysis is a very simple way to characterize them but they lack of strong precision and must be combined with frequency information.

Let split sound $frame_i$ into $x_i(n)$; n = 1…N, with length N each one. Energy for each $frame_i$ is given by Eq. 13.4:

$$E(i) = \frac{1}{N}\sum_{n=1}^{N} |x_i(n)|^2 \tag{13.4}$$

This is used to automatically detect silences and to discriminate different types of sounds. Figure 13.5 presents both distributions for music and speech, making it evident the big different between them.

In general, voice productions have many silences between periods of high energy. As a consequence the convolution curve has many changes, and its standard deviation σ^2 may be used to detect speech. In order to get energy (σ^2/μ) is also used.

13.2.2.4 Spoken Language Analyses

Spoken language (mainly riddles, puzzles, locutions and dialogues) can be modeled using the law to produce utterances in natural language. Relevant information is contained in the Short Time Energy (the energy distribution framed values, in the following denoted as ***STE***).

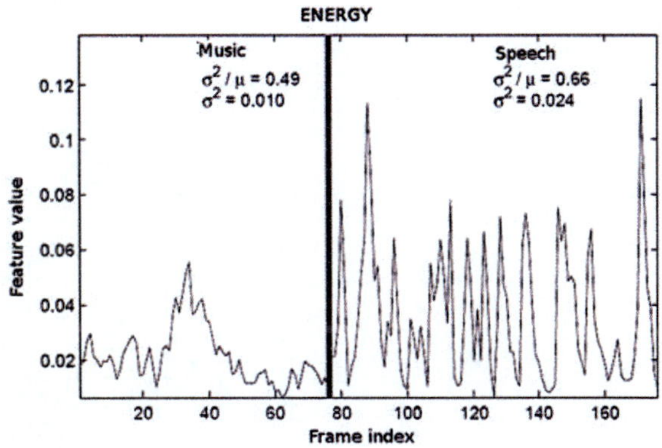

Fig. 13.5 .Comparison of energy distribution for Speech and Music

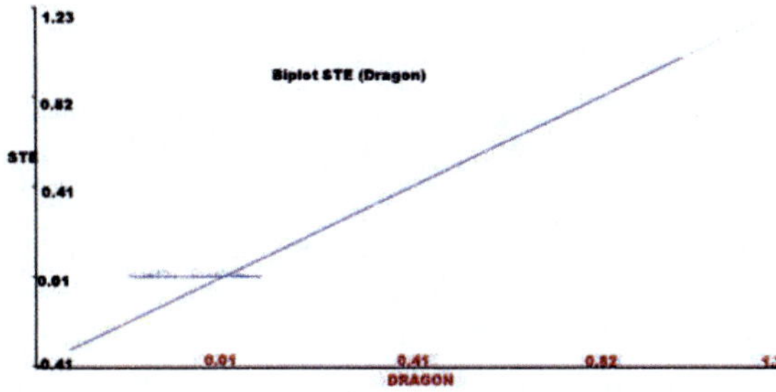

Fig. 13.6 STE versus Dragon plot

Coefficients from fractal Dragon and those derived from *STE* for sound records present a close relation. Figure 13.6 is the plot for both distributions (STE and DRAGON) in Infostat (c) [10], and a combination of linear correlation with some extra dots (it may be a sampling error).

Linear regression confirms that the correlation is statistically relevant. When each STE is projected on DRAGON the figure is suggesting a bit difference between them (see Fig. 13.7). The same happens with Zipf data.

This states that variation in data due to residual values is not significant. Therefore Dragon is a good model for STE.

This explains the relevance of sound processing for evaluating linguistics performance (from a fractal point of view) and relates it with written text. Fractal behavior has energy distribution according to Dragon curve. The author is currently working to adjust and explain the small and regular difference between them.

13.2.3 Semantics and Self-expansion

Other interesting perspective related to semantics is the ability to automatically rephrase a sentence or a question. In the field of **I**nformation **R**etrieval (IR) this is called self-expansion. This is a research topic since 2014 at IDTI Lab (Universidad

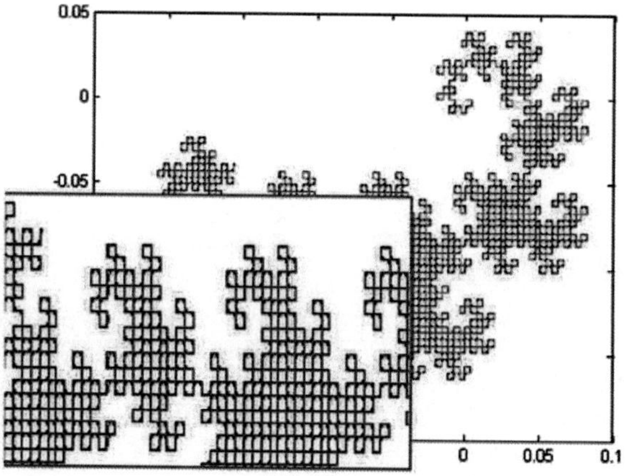

Fig. 13.7 STE versus Zipf plot

Autónoma de Entre Ríos, Department of Science and Technique (UADER-FCYT) at Concepción del Uruguay (Entre Ríos, Argentina). In collaboration with the Universidad Tecnológica Nacional branch located in the same province (UTN FRCU), a chatterbot was developed [11]. This project, called PTAH, implements a homonym for a restricted domain approach for handling information in the context of academy's regulation field. Among other challenges, it performs Natural Language **I**nformation **R**etrieval (IR) by means of a chatterbot. IR for Restricted Domains has no statistical error compensation. The slang in the documentation usually does not match the common language usage, and dialogs tend to be informal, making people tend to perform incomplete questions to the system. It represents an interesting challenge for the Natural Language Processing area, since many approaches tend to overcome only one of the previous aspects.

Among the advantages of PTAH's approach, is the lack of an explicit model of the knowledge [29]. Instead of this model, a straightforward linguistic inference is performed in a similar way to a **C**ase **B**ased **R**easoning (CBR). A set of patterns similar to regular expressions are extracted from a number of regulations. Expressions are used to address the content of the document. Patterns are used to perform questions about the main content of the regulation.

13.2.3.1 PTAH Chatterbot

Figure 13.8 depicts the main process starts in a remote PC where the user uses natural language (Spanish) to interact with the server where PTAH resides. Text feeds a subsystem (F.E., Fetch Evaluator) that infers a simple query from the dialog. The result is evaluated word by word against a Dictionary through a process denoted in the diagram as V.M. (Validation Management). It determines whether every word exists in the Dictionary or not. Then, an inference must be performed in order to properly evaluate the set of data to be answered (see I.M., Inference Management

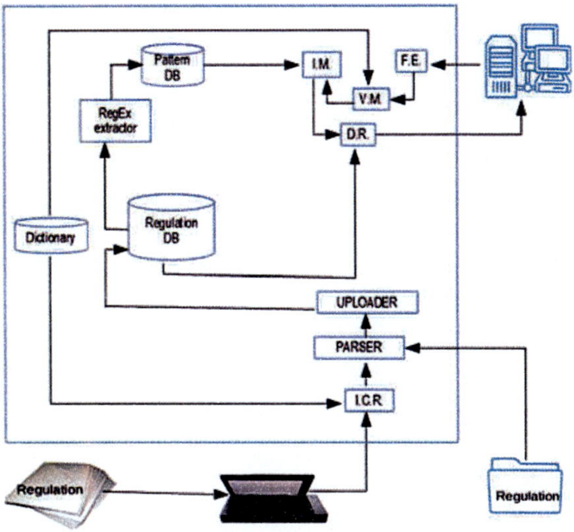

Fig. 13.8 PTAH workflow

in the figure). I.M. is the aggregate of algorithms for pattern matching and language processing. They are related with metrics for precision; recall and quality of the response (see Sect. 13.2.3 for further details). When the response is well scored, then D.R. (Data Response process) answers the query according to a certain quality threshold settled by I.M.

The main flow starts when a regulation is uploaded into the system. As it can be originally in a physical (paper) or digital medium, the information flow starts with a scanner or with a.docx file. When the regulation has to be scanned, an ICR (Intelligent Code Recognition) detects and corrects mistakes that may occur, using a specially tuned Neural Network that also performs certain steps to update the system's dictionary.

In both cases, the digital version is parsed (Parser module) to extract different sections and afterwards every section is stored in the Regulation Data Base (DB) with the Uploader. Then, a pattern (similar to regular expressions) is associated to every document's section in the database (in the figure denoted as Patterns DB).

13.2.3.2 Spoken Language Analyses

The documents are indexed by keywords. In every case the question is compared with the patterns in order to know if the related documents are the correct answers or not. The process is restricted just to Verbs, Nouns and Adjectives. This is performed since it has been demonstrated that semantics are mainly represented by these three syntactic categories. Similarity metrics considers the best **I**nverse **D**ocument **F**requency (IDF) terms. Therefore, words with high frequency commonly have less semantic relevance. The IDF is a numerical statistic that is intended to reflect how important a word is to a document in a collection or corpus. It is often used as a weighting factor in searches of information retrieval, text mining, and user modeling. It is a good metric of how good a word can discriminate a document within a set of documents. This retrieval process works with fuzzy logic and Morphosyntactic linguistic wavelets (***MLW***), adding not only the stems and IDF, but also typical ***MLW*** descriptors.

A pending task is the human feedback to improve the weighting process for reinforcement learning (as a complement of IDF). This should be the end of the retrieval process: the user interaction upon any system's response, to help the system learn and improve the hit and precision rates.

Currently the project is approaching the automatic conversion of documents into reduced patterns (up to now they were manually defined). Next section introduces a type of automatic linguistic indexing of the corpus, namely the semantic self-expansion.

13.2.3.3 Semantics in Automated Self-expansion

One key element for most of NLP concerning IR is the possibility of rephrasing questions or searches in order to be able to retrieve most relevant documents. To perform these tasks considering semantics is not simple and is part of current work under research as shown in [12].

The approach in PTAH is biased by morphosyntactic perspective with the following elements:

- query **H**.
- query for reference $\mathbf{M_j}$.
- pattern of the query $\mathbf{P_k}$.
- document $\mathbf{D_i}$.

Considering that any document $\mathbf{D_i}$ can be related to a pattern $\mathbf{P_k}$ and also rephrased as a $\mathbf{M_j}$. Figure 13.9 shows the relation between all these elements and the distance to be considered between them.

Here.

$d = f(d_1, d_2)$.

Fig. 13.9 PTAH elements

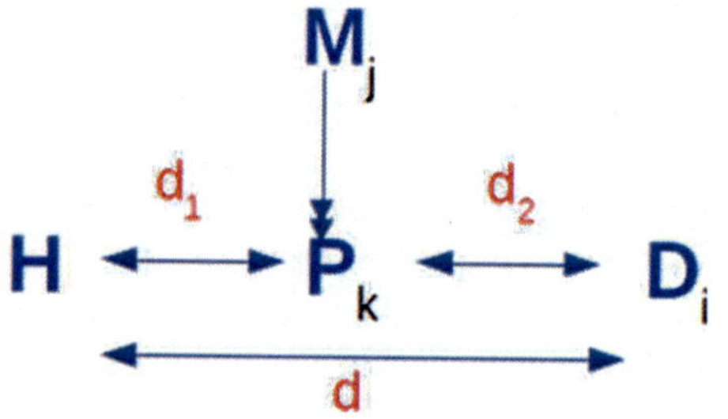

Fig. 13.10 PTAH extended elements

$d_1 = d(\mathbf{P_k},\mathbf{H})$, $d_2 = d(\mathbf{P_k}, \mathbf{D_i})$.

$\mathbf{P_k} = F(\mathbf{M_j})$, is derived from $\mathbf{M_j}$ using an accurate model.

To build the corresponding ***MLW***, it is mandatory to add certain structures with the following cardinality to entail the cardinalities shown in Fig. 13.10 [13].

With:

H: query performed by the user.

$\mathbf{P_H}$: pattern automatically derived from the query.

P: pattern automatically derived from a subset of documents in the corpus.

$\mathbf{D_i}$: document or part of it.

The pattern uses the ***MLW*** model and a heuristics that is currently under researching to expand the model by a pattern matching process between **H** and $\mathbf{P_H}$. This procedure is complemented with search by cluster proximity criteria based on **P**.

From the dynamic perspective, the managing of documents and queries can be summarized by Fig. 13.11.

With:

H: query performed by the user.

$\mathbf{P_H}$: pattern automatically derived from the query by algorithm **a.**

P: pattern derived from $\mathbf{P_H}$ using linguistic model **M**, improved by **G** learning process data driven by the set of words in **{W}.**

$\mathbf{D_i}$: document or part of it within the corpus.

$\mathbf{P_K}$: patterns automatically derived from element $\mathbf{D_i}$ by means of algorithm **a.**

{H}: log of queries previously performed by the user.

{W}: set of words extracted from patterns **P** and its metadata.

G: learning process and its self adaptation from **{H}.**

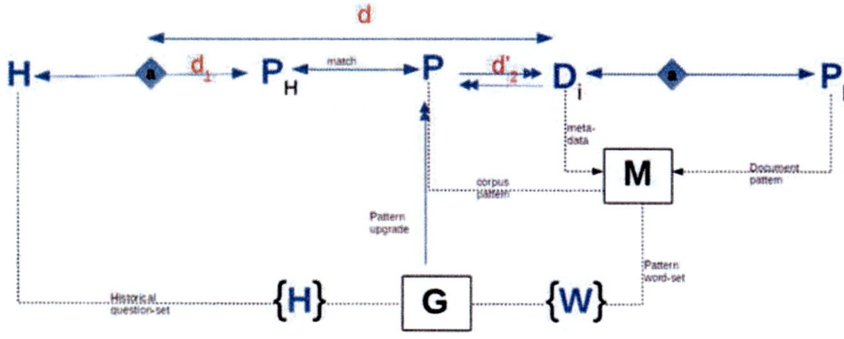

Fig. 13.11 PTAH dynamical ***MLW*** association

M: model that under the heuristic the builder of **P** based on *MLW.*

It is important to note that also there is a new semantic relevance conception that works with it and constitutes a mixed approach:

(a) **Accessing through pattern representing Eigen values of semantic regions of similar MLW descriptors.**

- *Level 1 metadata, denoting morphosyntactic features of the used word-set.*
- *level 2 metadata, derived from the context close to word* δ_i *belonging to* $\{D_l\}$.
- *Level 3 metadata, representing the usability of the sentence according to current words and universe of patterns* P_k.
- *level 4 metadata, expressing the semantic previous association expressed by dictionaries of the language.*

(b) **Accessing by simple pattern expansion.**

13.2.4 Semantic Drifted Off from Verbal Behavior

Autistic disorder can be of different degrees. There are patients who can communicate using spoken and written language, but many others can't. This section presents an approach that intends to perform the automatic management of Autistic communication patterns by processing audio and video from the therapy session of individuals suffering **A**utistic **S**pectrum **D**isorders (ASD).

The research is leaded by CI2S Labs in collaboration with IDTI Lab as the ***BIOTECH*** project, and with participation of students from Centro de Altos Estudios en Tecnología Informática of the Universidad Abierta Interamericana (CAETI–UAI, Center for Advanced Studies in Information Technology of the Inter-American Open University).

ASD patients usually have social and communication alterations that make it difficult to evaluate the meaning of those expressions. As their communication skills may vary, it is very hard to understand the semantics behind their verbal behavior. ***BIOTECH*** is based on previous work on Machine Learning for Individual Performance Evaluation. Statistics show that autistic verbal behavior are physically expressed by repetitive sounds and related movements that are evident and stereotyped. So it is possible to automatically detect patterns in audio and video recording of patient's performance, which is an interesting opportunity to communicate with ASD patients.

13.2.4.1 Autism Degrees

DSM-V from the American Psychiatric Association, defines a classification guide to allow scientific community to diagnose and detect many mental problems. People diagnosed with ASD can fall in one of the following categories:

Grade 1. Patient presents communication deficiencies that produce severe problems. There are difficulties to start social interactions, atypical responses or unsatisfactory activity for opening to other persons. It may seem that the individual lacks of interest to social interactions. Behavior is restricted and repetitive.

Grade 2. There are notable deficiencies in social, verbal and non-verbal communication. Patient has problems with socialization even with in situ *assistance. Limited initiatives for social interactions and reduced or lack of normal answer to social openness to other persons. As an example, a person produces simple sentences, with interactions guided by specific and concrete interests, with an eccentric non verbal communication. The behavior is restrictive and repetitive, with anxiety or difficulty to change attention focus.*

Grade 3. There are severe communication deficiencies (verbal and non verbal) that deeply affect functioning and very limited starting of social interactions. Patient also shows minimal response to social aperture to other persons. For example, a person with very few intelligible words, very occasionally starts any kind of interaction, and when it happens, performs in a very unusual way, only to meet basic needs. The individual reacts only to direct social approximations.

DSM-V mentions a verbal and a non-verbal social communication as one of the typical problems that presents ASD patients. B. F. Skinner (Skinner, 1981) performed several studies of verbal behavior. According to him, vocal language is just a part or a subset, while sound utterances and many other actions like gestures, are verbal even when they are not part of an organized language. This is so, since they produce a reaction in the listener or observer that is similar to a vocalized language, therefore they should be part of verbal behavior.

13.2.4.2 Movements and Patterns

Repetitive and stereotyped movements are a relevant symptom of ASD. Jumping, swinging and other rhythmic body movements were described informed in several publications. Ángel Riviere [15] classifies every probable behavior which, with all the detectable movements described by Leo Kanner.

BIOTECH intends to automatically process videos of therapy sessions with an ASD patient. A set of small and coordinated modules process them. In order to validate the processing steps, sounds and movements are manually processed and every action of the patient is identified and classified. Some parameters like Time, Type of movement, and specific features of the file are considered. This hand made processing is validated with patient's reactions and therapist's instructions.

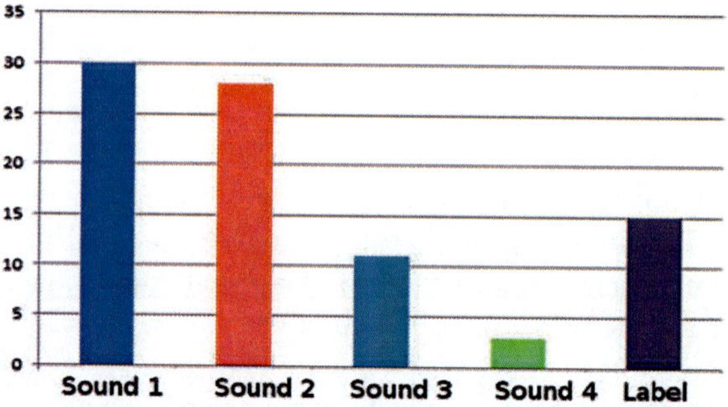

Fig. 13.12 Histogram of sound patterns

The steps to process multimedia records are the following:

- conversion of video from the original format.wav
- Creation of clippings (data samples)
- Analysis of every clipping
- Tagging and statistically evaluations

A careful protocol is mandatory prior to perform this activity. All the protocols must be validated with the patient's therapist. The recorded audio from the test is used to detect sounds and movements of the patients related to it. After an exhaustive manual processing, events were identified in which a sound the predefined movement takes place.

The summarized results are grouped by similarity to collapse them into main categories. Figure 13.12 presents the results from a real case.

Expressions produced by the patient upon stimuli are determined and related to them. ***BIOTECH*** aims to model that relationship that is different for every patient, using a set of parameters. In the Fig. 13.13 sounds with the same expression "iuuu" is identified as *Sound 1*; expression "aaaa" is *Sound 2*; expression "brrrr" is *Sound 3* and "Slapping" is *Sound 4*. It is important to note that might be other sounds (in this case besides Sound 1, Sound 2, Sound 3 and Sound 4). For instance happiness sounds and many others that are very complex and different between them. Many times background sounds are related to patient's sounds or movements. That happens due to the fact that in ASD the patient is unable to categorize stimuli, and every one deserves an "answer" (see [4] for more details). In other opportunities the patient performs a slapping with a registered sound.

13.2.4.3 Analysis of Every Clipping

Fortunately, sounds that act as an "answer" are associated to a movement. For instance sound 1 in the figure came every time with an angry gesture. Considering that, it is possible to discriminate between noise and intended articulation.

Clippings of audio have a time frame with $\pm$ 2 extra seconds. They are processed with FFT (**F**ast **F**ourier **T**ransform) and derived some other parameters mainly using

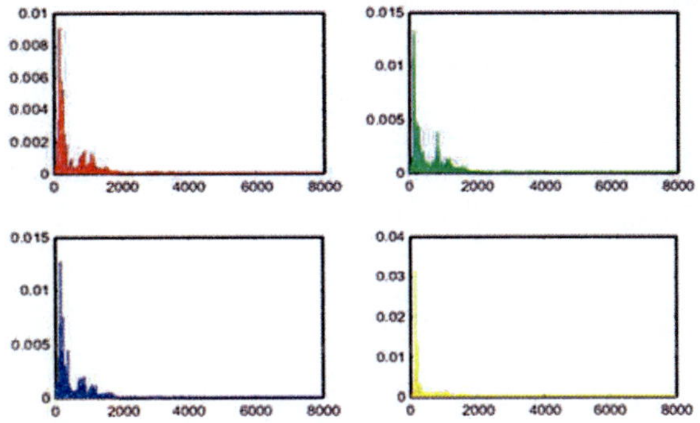

Fig. 13.13 Spectrum of the patterns

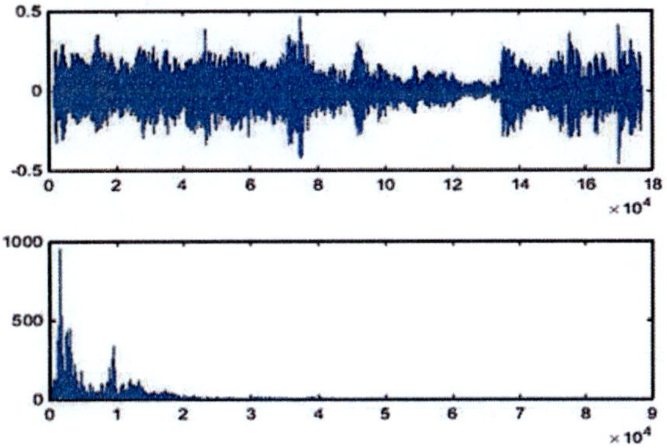

Fig. 13.14 Time/frequency for the first pattern

a combination of Octave and Python language. Figure 13.13 shows spectral representations of clippings related to the testing previously presented, and Fig. 13.14 the time/frequency decomposition of the first sound pattern.

To train a **N**eural **N**et (NN), the database for the patient's case must contain all the video and sound parameters. Some of them are: maximum amplitude, minimum amplitude, frequency, standard deviation, variance, etc. Every column has the derived information of previous analysis on every clipping. Parameters were processed using a **M**ulti **L**ayered **P**erceptron (MLP) NN. Most of the tests were performed with RapidMiner (c), the rest with WEKA (c) both platforms useful to help mining data.

The parameters for NN are "-" for the number of hidden units. That let the algorithm decide the number and architecture of hidden layers. The number of epochs are 1100 and the speed of learning ($\alpha = 0.1$).

After several tests the model presented a bias reduced with extra parameters: the coefficient of the polynomial approximating the convolution curve of each one can be used as one of the features to discriminate them. The polynomials corresponding to the current test case are in Table 13.2, as long its accuracy (R^2).

Those convolutions present a notable difference between them. Figures 13.15 and 13.16 show that graphically.

Table 13.2 Sounds convolution

Sound	y	R^2
Sound 1	y = −E-Q7 × 4 + 8E-05 × 3 - 0,0146 × 2 + 0,4077x - 38,044	$R^2 = 0{,}8649$
Sound 2	y = −7E-08 × 4 + 7E-05 × 3 - 0,0155 × 2 + 0,7622x - 46,378	$R^2 = 0{,}8361$
Sound 3	y = −E-08 × 4 + 5E-05 × 3 - 0,0096 × 2 + 0,1744x - 39,547	$R^2 = 0{,}8291$
Sound 4	y = 6E-07 × 4 - 0,0002 × 3 + 0,027 × 2 - 1,3408x - 29,891	$R^2 = 0{,}9016$

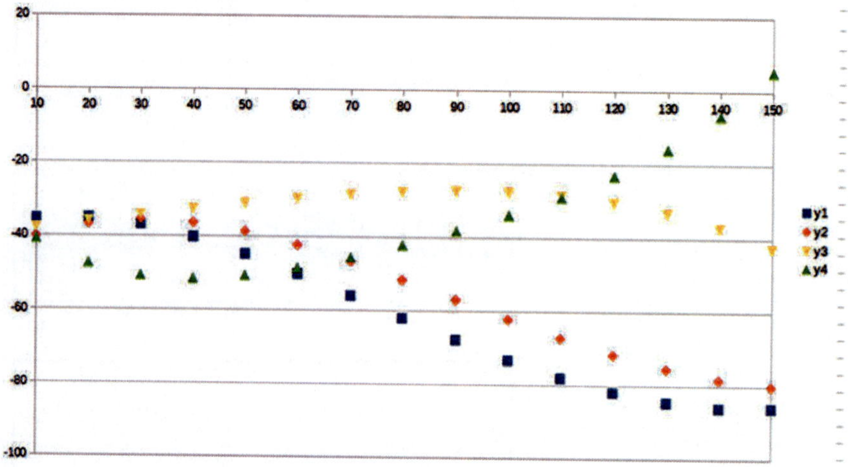

Fig. 13.15 Low range convolution

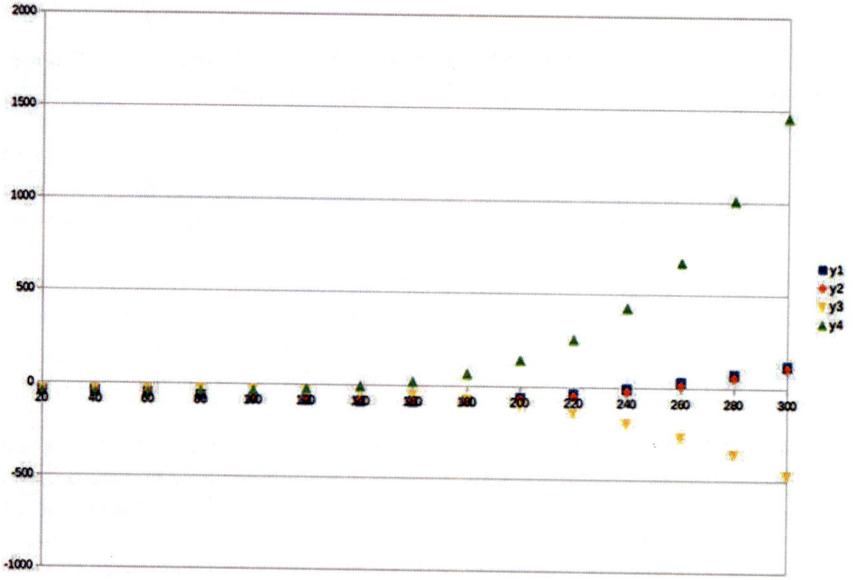

Fig. 13.16 High range convolution

ASD has many typical corporal expressions that may be considered as a verbal communication. But every patient performs a specific use of a subset of them. Thus there is the possibility to consider it to understand its internal status and, eventually some of its intended expressions.

The current status of this research can be summarized as follows:

- Pitches and sounds produced are limited to a specific set that is being reused as a tool of expression/communication, along with other verbal manifestations.
- It is possible to automatically detect each of the main verbal expressions using a MLP, trained with the small set of the following parameters: maximum and minimum amplitude of the spectrum, its average, a_DEVEST, standard deviation, Frequency variance, minimum FFT value, maximum FFT value, average FFT and standard deviation of FFT.
- The detection rate and precision could be improved using extra parameters
- There is enough indication that produced sounds and movements correspond to specific messages (moods of annoyance, happiness, etc.).
- The set of expressions, are specific for the patients
- Sounds are produced in specific frequencies (not by random)
- Sounds can be represented by certain parameters with customized characteristics:

  (a) Spectrum is always tetra-modal, with a first peak that sublimates the rest.
  (b) Spectrums of sounds have specific and differentiated peaks with similar proportion relationship among peaks.
  (c) Time spectrum is a descriptive probabilistic stamp of every sound.
  (d) Phase spectrum can be used as a complementary property of the sound.

13.2.5 Semantics and Augmented Reality

As shown previously, autistic disorder can be of different degrees. Severe cases are unable to perform abstract associations between objects in the environment and their related ontology.

ARA is a research project intended to make it explicit the semantics to ASD patients in a visual way [16]. This research started in 2014 in CI2S Labs, is now almost finished with a working prototype developed by a student from CAETI as part of his thesis.

An interaction model is automatically derived from the natural interaction between a patient and his environment. This project differs from current treatments and tools, in that the individual is not trained by imposing pictures or semantic patterns, but ideograms built from the patient's preferences and environment.

By using Augmented Reality, the autistic is being treated in an innovative way: the model gathers the environment variables and through communication by exchange of images (**P**icture **E**xchange **C**ommunication **S**ystem, PECS) the treatment becomes an agile, continuous and flexible process. The procedure is expedited since the patient does not have to select PECS (see Fig. 13.17 for some typical examples),

Fig. 13.17 PECS

but they appear to him. The activity recording is them processed in such a way to control and describe the cognitive and social profile of the patient. It also, performs a customize performance statistics. Likewise, the special administration of these statistics is intended to lay the groundwork for more representative future work that could allow the derivation of unbiased patient's evolution [17].

ARA collects and records activity of the patient and it can be used to perform statistics and help improve therapeutic strategies. To understand ***ARA*** [17], it is necessary to introduce the concept of Augmented Reality, and its usage within this prototype. It provides the patient with an expanded vision of his environment, just through a daily device. Physical world's data are combined with virtual elements to create a mixed reality in real time. Through this controlled environment, it presents a simplified visual and auditory customized access to the environment. It also provides the basis to generate specific training activities for ASD, focus on those areas where they have limitations. The proposed model can:

- Provide a simple and customized representation of the environment.
- Use information to foresee situations that are likely to happen to the patient in order to help him.
- Simplify the real world for the autistic.
- Highlight relevant part aspects of the environment for the patient.
- Strengthen the visual interaction (images, photos, videos, etc.) for patients that easily communicate that way.
- Strengthen the auditory interaction (sounds, spoken voice) for patients with this communication channel.
- Enhance the patient's process of generalization / abstraction.
- Allow a better customization of the therapy.
- Extend therapeutic and home context with technology.
- Track and measure patient's evolution using analysis of certain metrics and logs.
- Profile the patient using Computational Intelligence.

ARA prototype, aims to provide a customized treatment, handle innovative variables and monitor the model's metrics, in order to define how methodologies are best applied to each patient. It is expected that it also constitutes an excellent tool to:

- Improve the patient's social communication.
- Incorporate new concepts through reinforcement learning.
- Generate, stimulate and determine interest of the patient for his environment.
- Help the patient to understand his environment.
- Generalize similar objects and make evident that generalization.
- Associate sounds and activities to objects.
- Provide the use of Augmented Reality, as a tool to assist the therapist, for a focused therapy in specific topics.
- Evaluate metrics on LOGS files, to be able to determine specific autistic verbal behavior.

Traditional therapy manages the patient to make him integrate himself into his environment. It is important since autistic persons isolate themselves in an inner world. The approach of this proposal reverses the problem: it works with an alternate communication process, supported by new technologies. The medium is adapted to the patient's mental process thus entering into the reality of the autistic. This way the patient can better imitate concepts shown by the application.

13.2.5.1 Architecture of ARA

One of the main contributions of this project is the new shifted paradigm: dropping out the traditional focus on getting the autistic adapts to the environment. PECS is a traditional communication system, combined with Augmented Reality to provide semantics in the patient's environment.

ARA has twofold architecture with desktop and a mobile application. The therapist is the user of the desktop module. He first registers the patient's data into the system, then, a treatment with certain sessions. Treatments can be settled in advance. There is also the possibility to add new treatments. Every treatment can have one or more sessions. When a session is added, a set of PECS must be attached with sounds associated to it and a vocal utterance of a mono or bi-syllabic word, working as a sound icon.

It is pending to work with aromas and textures as extensions of the original icon. Figure 13.18 shows the global architecture.

After selection of PECS, sounds and videos, the next step is to generate the context where the patient will perform his treatment. To do that, the generated information has to be sent to the mobile device or tablet.

Despite of Augmented Reality framework do not require it, the current implementation of ***ARA*** works with QR tokens. Consequently, the QR codes associated with the generated PECS must be printed. These codes should be placed in any space where it is expected the patient would be working with the device.

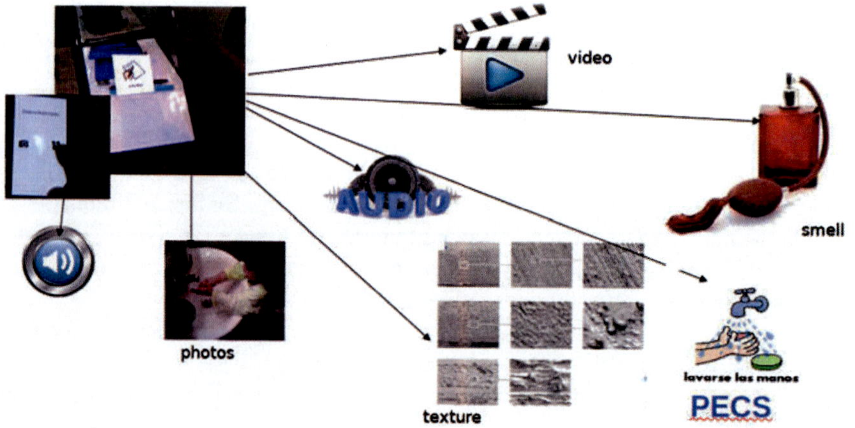

Fig. 13.18 *ARA*

The Augmented Reality application is rooted in the mobile device or tablet. The screen shows anytime the camera captures in real time. At a QR code loaded for the session, the application recognizes it and displays on the screen the PECS related to the QR code as the current camera capture. (See Fig. 13.19).

When the PECS appears on the screen, the sound loaded and the word will be reproduced as well. It is expected to favor vocalization of patient. When the PECS is activated, two large and characteristic icons are presented, one for displaying images and another for inspecting videos. The photo camera icon allows accessing a gallery of images related to situations to which the PECS refers. The video-camera icon stands for showing one or more videos of related situations.

Fig. 13.19 *ARA* recognition of QR

ARA combines traditional ASD therapy with Augmented Reality. From its logs it is possible to use the data set reflecting the activity of the patient. It allows summarizing the information related to sessions, from several treatments of the same patient.

It also can be used to define new patterns, evaluation metrics, etc. It remains to use the relevant information that may arise from this process, in order to provide feedback to the application.

An important point to be considered is that the characteristics of ***ARA*** allow extending the activity that can be performed in the therapist's environment. Under expert supervision, patient's environment may become a closer environment.

13.3 Language in the Learning Process

Cognitive processes are influenced by several factors. They have been studied by several authors. Particularly in the lasts years TIC (**T**echnology for **I**nformation & **C**ommunication) emergence enable, now more than ever, to affirm that cognitive processes are complex, variable, and of difficult characterization. Thus, it is not enough to perform practical tests under certain circumstances. Linguistic skills bias the results. Everything seems to show that, far from being a closed and stable process, it has evolutionary features.

This section introduces some basic concepts of projects MIDA and LEARNITRON, that aim to establish how to asses and characterize the cognitive process by using educational videos instead of traditional questionnaires that depend on the individual ability to understand the meaning of the questions.

As a counterpart, it is presented a research leaded by SCA in Argentina about STEAM effects on students, to derive a suitable model using approaches from the field of Machine Learning. This later project lies with current STEM and STEAM experiences.

13.3.1 Modeling Learning Profiles

LEARNITRON is a model that explains learning profiles using different strategies. The architecture and functioning is closely related to a virtual museum called MIDA, which is being introduced in the following subsection. Both aim to promote engineering careers using video games (VG) as a tool. It also is used by SCA and GTC to develop strategies that may tackle the problem of university desertion in engineering careers. The percentage of students dropping out of engineering degrees is very high, reaching approximately 70%.

The MIDA prototype, through the collection of information to obtain usage statistics, can help students to improve their understanding of engineering subjects, all through a playful aspect perspective. Gamification constitutes the use of game

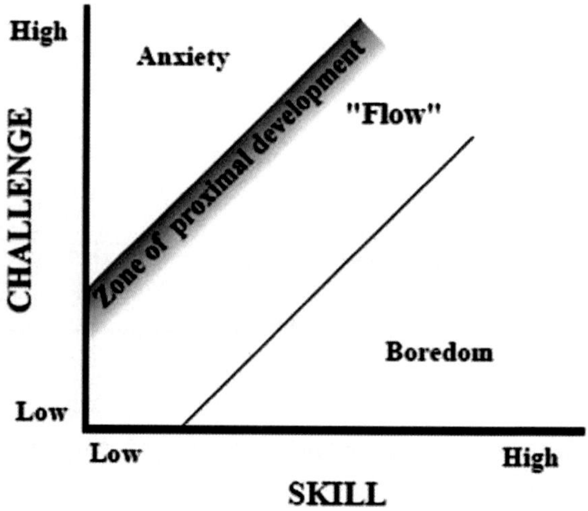

Fig. 13.20 Balance between challenge and skill in VG

mechanics in non-play environments in order to enhance motivation, concentration, effort, loyalty and other positive values common to all games. But VG are effective in the teaching–learning process, if they have certain specific characteristics: a proper balance between the challenge and the skills in the presented tasks. You can see in Fig. 13.20, how Csíkszentmihályi represented it [21].

It is possible to determine a four step mental processes in VG: exploration, hypothesis, re-exploration and rethinking. It is possible to build mathematical models to explain the quality of that cycle and consider it as a measurement of the whole process. This way the flow of knowledge between the system and players can be quantified.

13.3.2 Looking for Additional Teaching Tools in Academy

Education is a big step towards development in a country. According to UNESCO, if every child has the same access to education, the average incoming per capita would increase a 23% in the next 40 years [19].

Perhaps the simplest way to promote careers is to present contents in the current language of young people: video-games. That concept grounds much effort in that way in the world.

MIDA (**M**useo de la **I**ngeniería **D**esde la **A**ntigüedad, Museum from ancient engineering) is one effort in this way. Started by CI2S Labs in 2013, it had several collaboration entities and is now being developed jointly with CAETI and IDTI Lab. It aims to make more attractive and to demystify Engineering careers.

The main goal behind MIDA is to learn by using ancient devices. At the same time its activity is used to feed a set of interconnected intelligent modules, also known as LEARNITRON project. The next section presents more information about it.

13.3.2.1 MIDA Architecture and Features

In order to provide solutions to the problem of students' desertion, MIDA provides a platform that introduces Gamification tips in engineering careers [18]. It is video game, enhanced with a small kernel that implements an intelligent learning model. Such combination is intended to help teaching-and learning engineering concepts going beyond traditional linguistic communication. It has a dynamic and playful interface able to evaluate how the learning process evolves during the user interaction. Figure 13.21 depicts main MIDA global architecture.

The figure shows how the video-game (enclosing black rectangle) is connected to other components that are mounted out of the UNITY (c) platform, mainly a data server or a terminal of user.

Among others, MIDA has the following characteristics:

- Traces behavioral information to be able to discriminate junior from senior visitors and their abilities.
- Shows ancient devices as a way to introduce complex concepts from the physics, mathematics, chemical, hydraulics, mechanics, etc.
- Provides a confident and efficient tool to assess the impact of certain multimedia artifacts to learning and conceptualization processes. Among other usages of such model, could be custom advice for a specific classroom about which are the best suited sequence of concepts, multimedia approaches, learning speed and reinforcement requirements.
- Shows and explains real world applications of abstract concepts that are hard to visualize.
- Promotes engineering careers, as it can be visited by anyone that takes time to play with devices, read/listen information and play in the sandbox.
- Tracks and evaluates certain factors in the dramatic decrease of the engineering students and high desertion in first levels. This is done by indirect tracking the persistence, motivation and other clues.

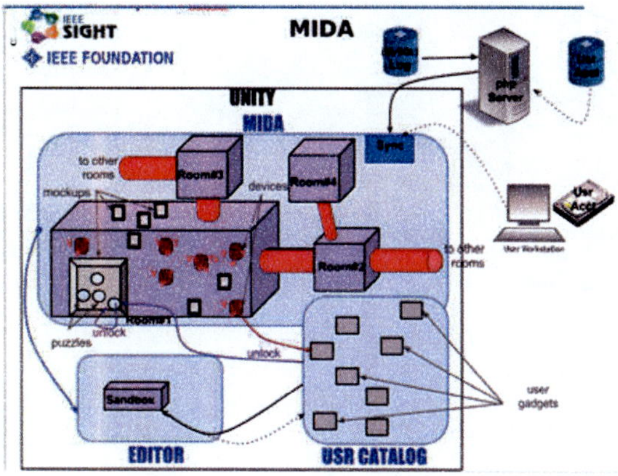

Fig. 13.21 MIDA architecture

13.3.2.2 How MIDA Works

MIDA is organized in several rooms according to the type of concepts exposed in the showroom. But all of them have the following two components:

Mock-ups: with many puzzles that may or may not be unlocked according to the learning stage of the user (see Fig. 13.22). The more puzzles are successfully completed; the more ones will be unlocked. This is because a puzzle is a mini test of certain tip or concept within a topic.

Devices: ancient objects selected by its characteristics (most of them simple machines that use simple physics, mathematics, mechanics, etc., see Fig. 13.23). Every object in the showroom has videos (introductory or function's demo), texts (introductory or technical) and images or photos. Besides, it is possible to assemble and disassemble the device to understand its architecture.

In despite of the fact that the museum shows antiques and simple devices, it also serves to teach basic engineering principles whenever the user must learn how the exposed objects operate. Every room associates objects to challenges that evaluate the level of understanding of the user. As the user plays within the Museum a specific intelligent activity is performed to derive the user learning preferences. This would be used to bias the Museum behavior, improving the visitors' experience.

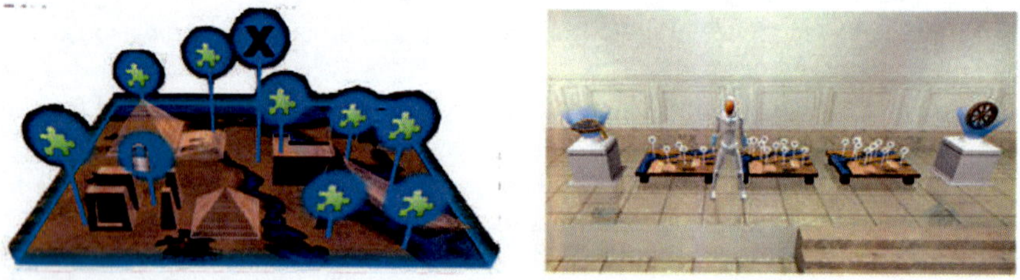

Fig. 13.22 MIDA Mock-up

Fig. 13.23 MIDA devices

13.3.3 *LEARNITRON for Learning Profiles*

The project named LEARNITRON, a branch of the research project MIDA. It aims to add a set of tools based on machine learning and other approaches from the Computational Intelligence. All those modules intend to provide self-adaptation of MIDA's behavior and to enhance the experience of learning certain very abstract concepts in engineering. As a spring off, there are also a set of metrics and models to describe users and their performance.

As the student population trend to change due to many factors, the main goal of this work is to automatically derive a model describing typical learning profiles through the usage of a video game (MIDA museum). The model is able to process real-time information from the MIDA repository in order to detect the current user profile. Data are collected, pre-processed and feed to the model proposed here, obtained mainly with data mining.

That way it is possible to determine the impact of the stimuli generated by the museum during the learning process of the players In order to derive the current learning profiles, the visitor activity is registered in log files such as the one shown in Fig. 13.24.

Keeping track of the activity not only includes the exact timing of each action, but also the component (objectID) and the action (value). After PCA (**P**rincipal **C**omponent **A**nalyses) two axes CP1 and CP2 were processed with Cluster analysis (it is a multivariate technique that seeks to group elements trying to achieve maximum homogeneity in each group and the largest difference between groups).

Results indicate that there are four clusters that matches Dendogram at similarity 84,70 (see dotted red line in Fig. 13.25) obtained shows that there are four identifiable clusters (Fig. 13.26).

Clusters correspond to four main profiles:

Cluster 1 CP1 y CP2 negatives, profiles with small variation of activities and short timings. It is a precise (no exploration of the solution) and efficient solver.

Cluster 2 Both positive, profile with big activity variation and longer timings. It is an explorer and less systemic solver.

←T→				id	userId	date	actionId	objectId	value
☐	Editar	Copiar	Borrar	1	10	2014-03-17 00:36:30	0	10000	1
☐	Editar	Copiar	Borrar	2	10	2014-03-17 00:36:31	0	20001	1
☐	Editar	Copiar	Borrar	3	10	2014-03-17 00:36:32	0	40001	1
☐	Editar	Copiar	Borrar	4	10	2014-03-17 00:36:37	0	50001	1
☐	Editar	Copiar	Borrar	5	10	2014-03-17 00:36:39	0	50002	1
☐	Editar	Copiar	Borrar	6	10	2014-03-17 00:36:40	0	50001	1
☐	Editar	Copiar	Borrar	7	10	2014-03-17 00:36:41	0	60001	1
☐	Editar	Copiar	Borrar	8	10	2014-03-17 00:36:44	0	50001	1
☐	Editar	Copiar	Borrar	9	10	2014-03-17 00:36:51	0	40001	1
☐	Editar	Copiar	Borrar	10	10	2014-03-17 00:36:55	0	40002	1

Fig. 13.24 Log file example

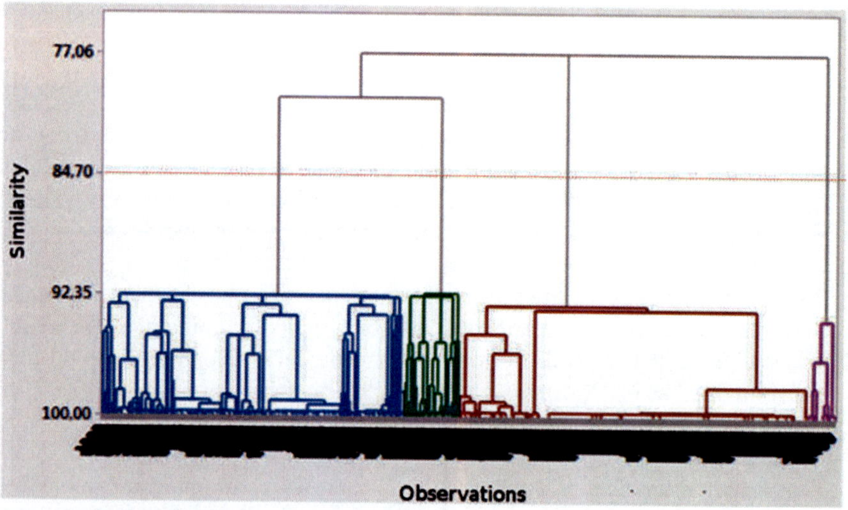

Fig. 13.25 *Dendogram*

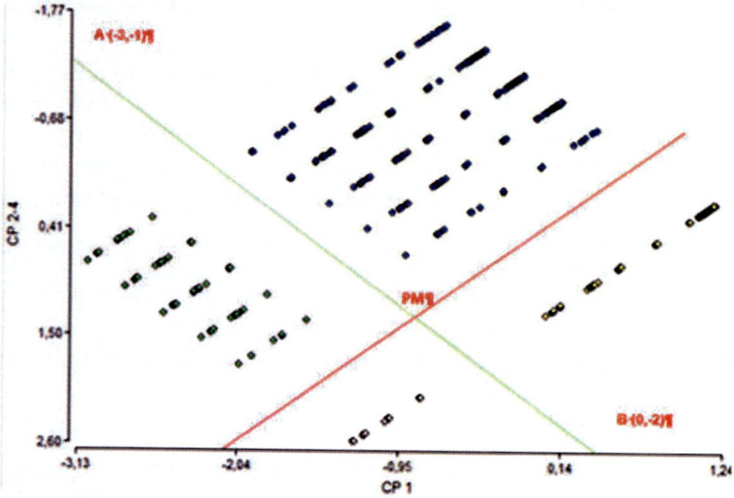

Fig. 13.26 Model

Cluster 3 CP1 negative with high variations. CP2 positive. It is an explorer and systemic with high efficiency.

Cluster 4 CP1 negative with less variation. CP2 positive (much higher value). Profile of an explorer, systemic but with low efficiency.

From the previous results, a simple two-line model: collecting the log data, and passing it thru the model, it is possible to find out the cluster that represents the user, and therefore the learning profile (see in Fig. 13.26 the model as represented graphically).

It is highly probable that efficiency may be due to background knowledge that would indicate that cluster 1 differs from cluster 3 by the degree of previous knowledge. But they have a similar approach for learning.

Cluster 2 differs from cluster 4 in that the approach for learning is less systemic, that would indicate the trend to not organize knowledge and presumably not to detect in advance generalities.

The topic immersion is more intuitive, with less previous structures on concepts: it is learning from details to generality. It is the opposite of cluster 4.

It is expected that a user from cluster 2 does not like to read definitions and global concepts before testing them by itself. The opposite happens with clusters 3 and 4.

13.3.4 Profiling the Learning Process: Tracking Mouse and Keyboard

Learning is a complex and multi-factor process. Stress is one of he known parameters affecting the productivity. This section describes the analysis of mouse and keyboard to track individual's feeling and its impact in performance [22]. The goal is to be able to asses stress by simple and cheap devices. A set of metrics and equations are derived by mining logs collected during several tests.

Three moods were considered: relax, neutral and stress. The environment was conditioned to induce the desired feeling prior to perform the test (see Fig. 13.27).

An interesting finding is the statistical confirmation of Berwick's verbal Behavior [22]:

(a) **For Stressed individuals manifestations are:**

- Defensive body position, usually with crossed arms.
- Easily lost focus.
- General discomfort.
- Less manipulation of hardware.
- Rigid position, sometimes rub his hands, face tense.
- More difficulty in understanding the questions.

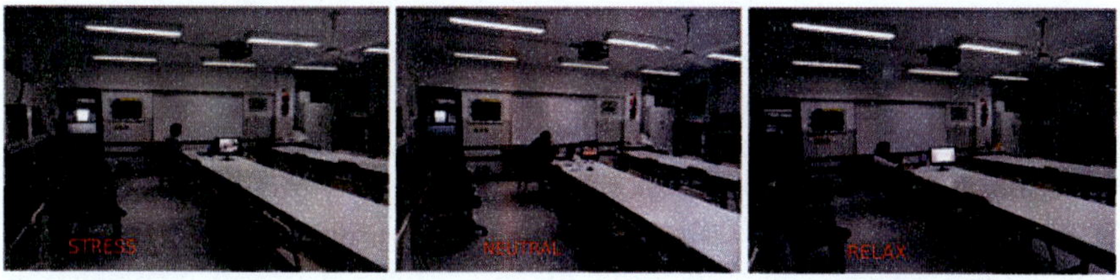

Fig. 13.27 Status: stress, neutral, relax

(b) **For Neutral mood:**

- Very well focused with almost permanent sight on the screen.
- Body indicates a slight interest in the activity: leaning forward, slow and permanent contact with devices, soft movements.
- Takes care of his understanding asking extra questions.
- Reads quiet, at his own time, with calm.
- Continuous management of the hardware.
- Some short times of relaxing.

(c) For Relaxed individuals:
- Sits in a comfortable position.
- Body and face expression of satisfaction, sometimes smiling.
- Eats and drinks while he is working.
- Deep concentration in the reading and tasks.
- Sometimes much contact with hardware.

Some other findings are related to the performance:

–Stress leads to longer time to answer.

–People with Stress have more probability to make errors in the typing and clicking.

–Neutral mood indicate people that don't matter about the relevance of performing tasks.

–Under Neutral and Relax status, people are quicker.

–Typing is faster for Neutral than Relax and Stress.

–Mouse will remain more time inactive for Neutral than for Relaxed, and Relaxed than for Stressed people.

–The number of clicks with the mouse (for both right-handed and left-handed users) is higher for relaxed people than for Neutral, and Neutral is higher than for Stress.

–The frequency of the DELETE and BACKSPACE for Relaxed people is the highest, followed by Neutral and then by Stressed individuals.

As shown, there are several manifestations for the inner condition that can be evaluated in forms of verbal behavior and thus considered to evaluate when the individual is able to acquire new knowledge. Understanding the status or the student would help provide the conditions required to improve his performance, reduce a possible disappointment and improve the rate of continuity in the careers.

13.3.5 Profiling the Learning Process: Tracking Eyes

VG are presented as an interesting tool for teaching due to its flexibility and attractiveness, but they also constitute a kind of interface useful to track the mood of the individual in diverse ways. This section presents a model based on eye tracking for inferring the degree of focus of the user [23]. Eye behavior is considered in

LEARNITRON project to be one more encoding of language according to Berwick's perspective.

The cognitive process requires of a proper inner state for its best performance. The goal behind this model is to be able to automatically detect and drive the activity of the VG in order to present itself in a suitable way that makes, in other words, a real time reduction of the stress level and increase the focus.

It also can be used as a metric to evaluate the whole learning process using the peaks and valley of attention as a reference for organizing the contents and extra material about certain topic related to the specific moment.

This module intends to model the contraction and opening of pupil, and the gaze direction. The main task here is to detect, tune and infer the proper equation that relates them to inner status. The model can determine the level of fatigue and stress, and in some way the interest degree of the individual.

13.3.5.1 A Set of Parameters for Eye Behavior Encoding

Tracking eyes requires a set of calibration marks, namely a set of reference points that guarantee the right interpretation of the changes in the image being processed (see Fig. 13.28). The model obtained uses the point where the gaze focuses, denoted as PoR in the figure.

Then it makes use of the coefficient NDAC in Eq. 13.5, to asses the current level of Depression, Anxiety and Engagement of the individual during his activity.

$$NDAC(x) = f(EE, ED, SE, NC, ZD, TTA) \tag{13.5}$$

Equation 13.6 describes EE.

$$EE = B(O) + B(M) + B(K) + B(U) \tag{13.6}$$

ED: Individual age.
SE: Gender of the individual.
NC: Nationality of the individual.
ZD: Right-handed or Left-handed.
TTA: Total Time for performing the Activity in seconds.

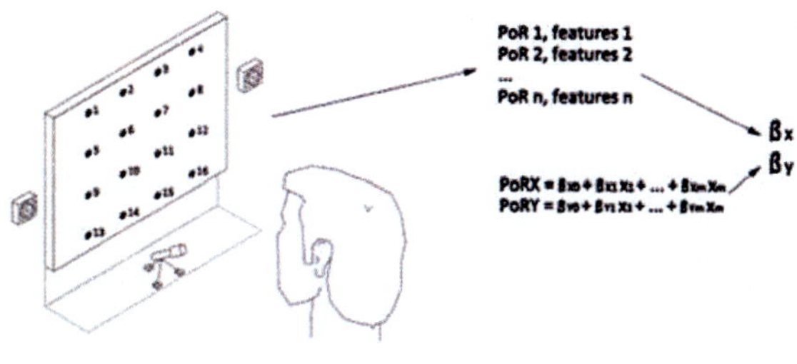

Fig. 13.28 Calibration marks

And B(O) corresponds to Ocular variations described in Eq. 13.7.

$$B(O) = f(DPI, DPE, API, APE, EM, DP, TE) \tag{13.7}$$

where:

DPI: Diameter of Left Pupil (pixels).
DPE: Diameter of Right Pupil (pixels).
API: Area of Left Pupil (pixels).
APE: Area of Left Pupil (pixels).
EM: Eye Movement (Focus).
DP: Distance in Pixels.
TE: Time elapsed (milliseconds).

The activity itself is measured also by the mouse performance with the Eq. 13.8.

$$B(M) = f(MNE, MVS, MTS, MTT) \tag{13.8}$$

With the following parameters:
MNE: Mouse environment.
MVS: Selected value.
MTS: Time elapsed.
MTT: Total Time with mouse.

When used with other devices (Mouse, keyboard and camera) the accuracy of the predicted Stress and focus levels increase. But the model stands as a good approximation for user activity and is still under research how to improve its performance with any type of camera and light conditions. Furthermore, some other parameters are under research like blinking and global face expressions. Preliminary tests allow to state that these types of models are useful for user mood profiling.

13.3.6 STEAM Metrics

This is a short presentation of a new initiative leaded by SCA, with the active collaboration of IDTI, CI2S Labs, Centro de Altos Estudios en Tecnología Informática (CAETI), Universidad Católica de Santiago del Estero DASS (UCSE DASS) in Jujuy province, Escuela Manuel Dorrego (EMD) in Buenos Aires province, UADER FCyT-CdelU in Entre Ríos province, IEEE Games Technical Committee (IEEE GTC), and Proyecto Escuela Gamificada Inmersiva (PEGI–LR) in La Rioja province [20, 24]

There is a world trend to apply new educational approaches, well known as STEM y STEAM. In general this implies a cross-cutting of main topics, usually involving multimedia and VG. This movement started in 2001, as Ramaley was the director of a human educational resources at the National Science Foundation, working to develop career plans to improve science, mathematics, engineering and technology.

The change aims to relate science and mathematics as a required addendum of technologies and engineering, and a step to understand the universe. Past US president Obama started an official program in that country referring to it as the new way education should be considered. STEAM is an extension of the former STEM initiative (Science, Technology, Engineering, and Mathematics) with "Art" as an extra perspective to be considered.

Such interconnections between science and other fields were formerly highlighted by technological leaders like Apple's founder Steve Jobs who mentioned it as "useful and beautiful" products. Many efforts try to keep rates of formal education in all countries.

In 2018 SCA founded LINCIEVIS (**L**aboratorio de **IN**vestigaciones **CIE**ntíficas **VI**deo juegos y **S**TEAM) a laboratory focused on topics related to science and VG. The first project started there is the one being presented in this subsection: how to determine metrics for proper tracking STEAM activities, and its relationship with human resources development.

13.3.6.1 Standardize the Diversity of STEAM Procedures

As STEAM, as well as STEM experiences are diverse and depend on the current context of application, there are too many parameters to be exhaustive in the tracking of the experience performance. In order to overcome that, and obtain a valid data-set to perform systematic analyses of the experiences, it is mandatory to collect in s standard way.

The SCA solution is to determine a set of WEB Forms to collect a set of parameters in the same way. Figure 13.29 shows one of them.

Those parameters represent the information about the institutions performing the activity, the students (or public) and the event itself. Some of those variables are also a current state of the art previously detected by other practitioners of such activities in the past, and eventually reported in some congresses and articles.

By applying Machine Learning, some preliminary results indicate that it is possible to determine an adaptive parametric model to model and predict student's continuity in the academy, and also to keep track of the STEAM activities and its effects on the institution and students.

Many previous publications are aligned with these findings and the success cases are valuable references to tune up the model. But it is important to say that a proper control over this type of changes in the transference of knowledge will help to gain experience more quickly and to obtain better quality processes of transference.

Current events are highly biased towards technology tools and may be carefully driven since many other aspects of education experience must also be preserved like literature and artistic perspective as well as psychological and interpersonal skills.

The linguistic performance is one of the challenges since there is an increasing barrier due to the globalization (different languages) and the digitization (much more time spent in activities that do not induce reading and writing). A proper balance

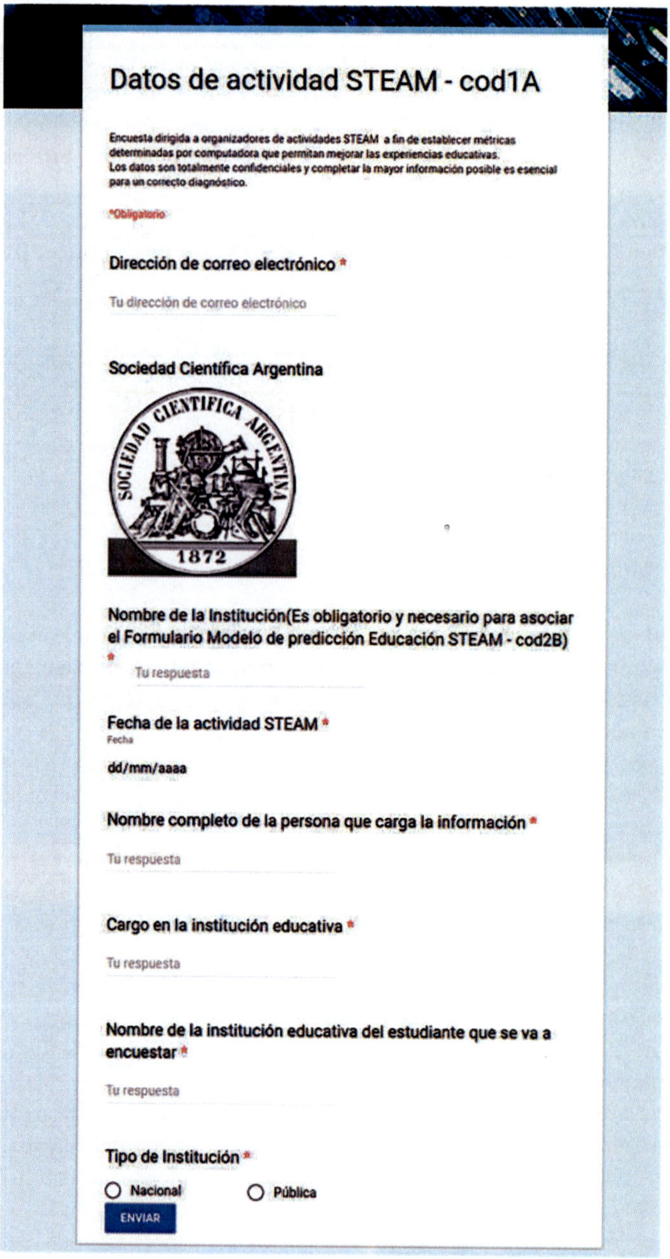

Datos de actividad STEAM - cod1A

Encuesta dirigida a organizadores de actividades STEAM a fin de establecer métricas determinadas por computadora que permitan mejorar las experiencias educativas.
Los datos son totalmente confidenciales y completar la mayor información posible es esencial para un correcto diagnóstico.

*Obligatorio

Dirección de correo electrónico *

Tu dirección de correo electrónico

Sociedad Científica Argentina

SOCIEDAD CIENTIFICA ARGENTINA 1872

Nombre de la Institución(Es obligatorio y necesario para asociar el Formulario Modelo de predicción Educación STEAM - cod2B) *

Tu respuesta

Fecha de la actividad STEAM *
Fecha
dd/mm/aaaa

Nombre completo de la persona que carga la información *

Tu respuesta

Cargo en la institución educativa *

Tu respuesta

Nombre de la institución educativa del estudiante que se va a encuestar *

Tu respuesta

Tipo de Institución *

Nacional Pública

ENVIAR

Fig. 13.29 Heading WEB form

Table 13.3 First predictive model

IF	THEN
Mother_profession='House mother' AND year in(1988,1989, 1990) AND gender=M AND neighborhood in('Obrero', 'Policarpa', 'Pueblo nuevo', 'Chinita') AND number<> 5 AND city<>'Closed neighborhood' AND department= 'Antioquia'	EARLY-SCHOOL-DROPOUT
Mother_profession='House mother' AND year in(1988,1989, 1990) AND gender=M AND number =9 AND city<>'Closed neighborhood' AND department= 'Antioquia'	EARLY-SCHOOL-DROPOUT
Mother_profession<>'House mother' AND father_profession in('Professional' , 'Salesman','Policeman') AND year in(1988,1989, 1990) AND gender=F AND neighborhood in('Velez', 'Ortiz', 'villa del rio') AND number<>5 AND city<>'Closed neighborhood' AND department='Antioquia'	NON-EARLY-SCHOOL-DROPOUT

between those ingredients is subject to be researched and the final equation is still pending in the whole community.

13.3.6.2 How STEAM Works

According to age group and biography, it is possible to predict the possibility that a student. The first attempt to perform this task in this project is the model presented in [25]. It was obtained from data collected in the Urabá region (Colombia, Antioquia) considering the variables reported in previous studies reported in Latin America as relevant to determine the continuity in education.

The information corresponds to real cases from year 2000 to 2005. The results are fully aligned with other works in the field and introduced in a rule-based model as summarized in Table 13.3.

First findings indicate that:

– The age of the student is not relevant.
– The neighborhood is relevant.
– The mother and father occupations are relevant.
– Religion and region (department) are relevant.

As can be seen the model has a few variables. Currently the research goal is to add the new parameters obtained from real-classroom experience in order to gain sensibility and more accuracy.

13.4 Language of Consciousness to Understand Environments

According to ECHO (European Commission for Humanitarian aid Operations), Natural disasters affect more than 300 million of persons per year. Many countries have considerable lost recovering from natural disasters.

In the case of Argentina, the problem may appear mainly as a flooding, earthquake, or fire. Figure 13.30 shows the risk distribution for each one.

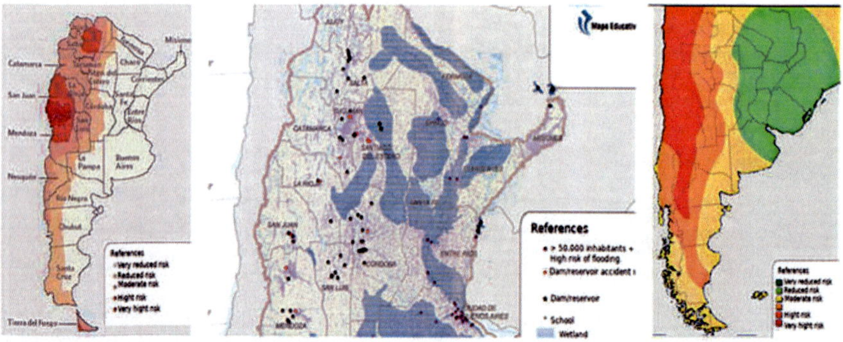

Fig. 13.30 Risk zones in Argentina

13.4.1 COFRAM Framework

COFRAM is a prototype implemented by CI2S Labs as a lightweight version of consciousness theory [3] for autonomous mobile robot for disaster zones. Currently, Escuela Superior Técnica del Ejército (EST, Argentina), UCSE DASS, and IDTI are in collaboration for this project.

The grounding steps in this field were performed by D. Hofstädter, under the philosophy that a system can understand itself and therefore perform certain reasoning using the cognition derived from that process. Certain processes serve as the factory of concepts and knowledge in a predetermined life-cycle that lets the system function in a neural-like performance. But the main difference with Neuron Networks is the learning by reinforcement more than a convolution process. It has been compared to ant colonies. Those "ants", have simple and repetitive tasks. They are represented as "*codelets*" in this context, and are cataloged in a stock called *Coderack*. They become activated from time to time under a probabilistic principle. *Codelets* help build *concept*s that can be promoted to a *Slipnet* (a long term memory) or destroyed if the *concept* does not worth enough. This process is performed in a kind of blackboard named *Workspace*. There, *codelets* are responsible for building relevant or irrelevant *concept*s (denoted as nodes in Fig. 13.31 [31][31][31]) and their relations (the links between nodes).

The main contribution of this type of intelligent systems is the plasticity for learning. Many *Workspaces*, lead to parallel searching in different solution spaces, something similar to the approach taken by *Genetic Algorithms*.

Most of first applications of this theory were performed by Stan Franklin at University of Memphis [26], who collaborated also in the start up process of ***COFRAM***.

Figure 13.32 shows the ***COFRAM***'s framework GUI (**G**rafical **U**ser **I**nterface) used to configure the platform for a specific robot, and Fig. 13.33 shows one of the several robots used to test the prototype. It is important to say that the project is now

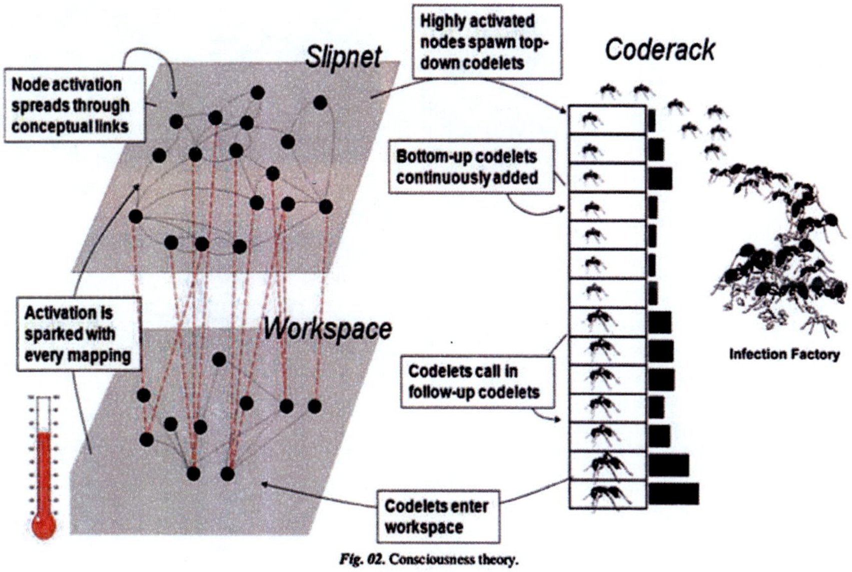

Fig. 13.31 Consciousness main architecture (credits S. Franklin et al.)

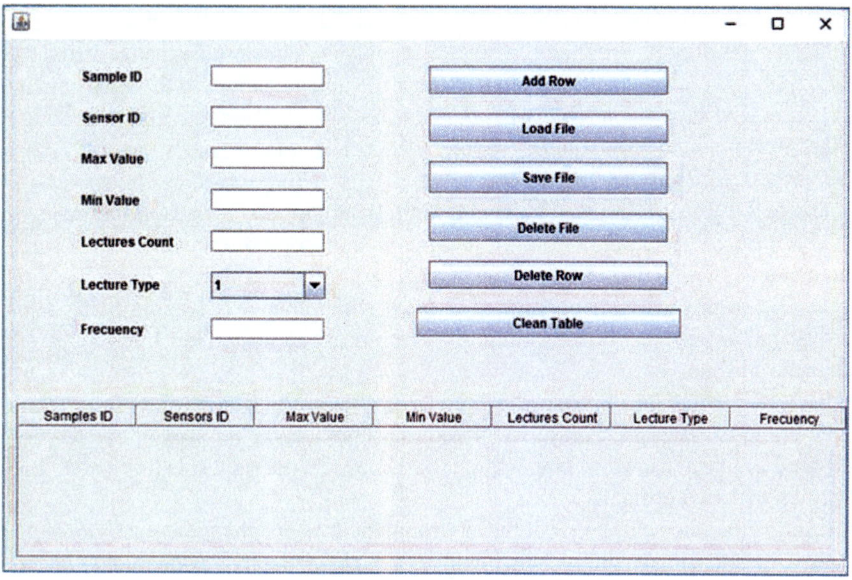

Fig. 13.32 Consciousness main architecture (credits S.Franklin et al.)

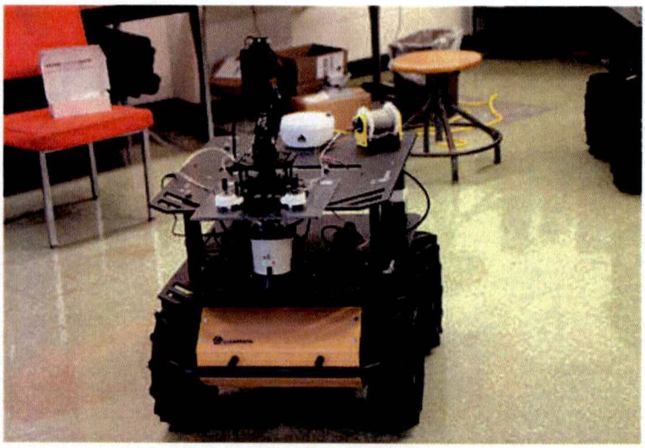

Fig. 13.33 One of the robots used for testing

working with a free version of *MQTT* (c) to be able to control the robots that are in different places.

13.4.2 Bacteria Infecting the Consciousness

The model implemented in ***COFRAM***'s kernel is called *ALGOC*, and was tested in controlled out-door environment. After a first analysis, a key problem arises from the tests: *concepts* are not diverse enough to track the quick changes in the environment [27]. That is to say the encoding of *codelets* is not able to manage certain variability in the information to be handled by robotics. As a consequence, a new type of *codelet* was introduced: an infected *codelet.*

The new type of *codelet* is built in a kind of factory able to adapt and behaves as a bacteria during an infection. Time domain and scope of domain are just some parameters of the bacteria that will perform its activity in a mimic of bacteria's litic cycle. That results in new generation of *codelets* with a change of paradigm that helps to vary certain features of the solution space out of the *workspace* and its current status at run-time.

Bacteria help thus improve the bias of problem solution adding some extra creativity to the process. Figure 13.34 shows on the left the number of created and activated codelets in a world map without obstacles. On the right side this is the same analyses with two obstacles.

It shows that the complexity of the environment does not change the activity level. That is, the conscious explores it using almost the same type of evaluation *concepts* without chances to explore more diverse alternates.

Results in further testing show high variability in the *codelets*, and solutions are more diverse. More complex solutions arise and original alternates.

It is interesting to note that bacteria introduce an extra flexibility in the encoding. It is like considering an extension in the language for expressing new things. For instance the word "empanada" represents a typical food in Argentina. There is

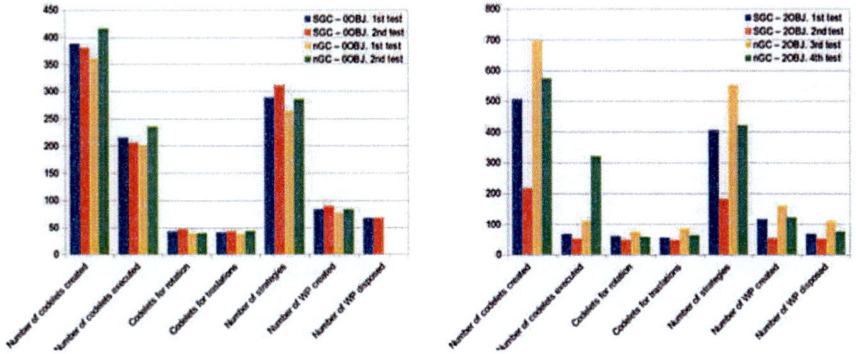

Fig. 13.34 Codelets with zero (left) and two (right) obstacles

no word in other languages to represent it. Something similar happens with word "throughput": there is not a word like this in Spanish, but a sentence. So, the best way to speak about it in Spanish is to use the word itself. That is how languages have continuous changes by adopting new terms (sometimes adapting it to the local phonetics). In several cases the original vocabulary (the official dictionary) has to be extended.

So the model obtained by ***COFRAM*** suffers some changes leaded by external factors out of the conscious process of the environment (see Fig. 13.35).

Virus in this context has two types of input: *codelets* or *concepts*. The outputs are just infected *codelet*s. The "organism" is the current conscious model, represented by its status and its abilities. An initial colony grows at the beginning with certain probabilistic changes. The infection can be selective or massive. At the end of the cycle, the individual is evaluated to determinate the viability of the changed organism to solve the problem.

The figure shows all the process where squares represent status, *codelet*s are circles, and polygons are the selectively infected organism. An implementation for robotics can be:

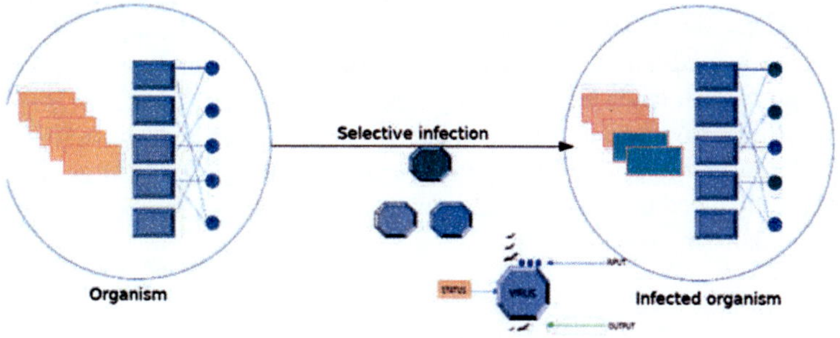

Fig. 13.35 Selective infection by a virus

```
STATUSi (PARAMETER1, PARAMETER2... PARAMETERn)
PARAMETER1(codelet that can start a process in the current work-
space)
PARAMETER2= cn (number of images)
PARAMETER3=n (number of interest points)
PARAMETER4=vector [PERCEPT1, ..., PERCEPTcn] list of images collected
by sensorial concepts called PERCEPT that reflect camera activity
PARAMETER5= vector [CONCEPT1, ..., CONCEPTn], determined by Slipnet
PARAMETER3. CONCEPTi is the sensed interest points
PARAMETER6= u [min threshold β1 and max threshold [β2]
PARAMETER7= sensed distance, size, direction, etc. as CONCEPTj
PARAMETER8= clp (number of polygons for tracking), clp ∈ [3...M]
PROCESS i= (FUNCTION1, FUNCTION2,..., FUNCTIONn)
FUNCTION1= {select interest points}
FUNCTION2= {eval. Optic flow}
FUNCTION3= {filter vectors}
FUNCTION4= {create areas}
FUNCTION5= {create polygon describing the object}
FUNCTION6= {create blob according selection criteria}
Codeleti-1: there is object or not
Codeleti+1: n interest points in current object
Codeleti+2: detect n pixels as interest points for Codeleti+1 ac-
cording strategyi-1
Codeleti+(n): idem according strategy i+n
Codeleti+(n+1): filter vectors with PARAMETER5
Codeleti+(n+2): determine clustering criteria for PARAMETER7 (this
proposal makes use of closeness criteria)
Codelet i+(n+3): build polygon with XX sides (XX=PARAMETER8)
```

13.5 Harmonics Systems: A Mimic of Acoustic Language

Although humans perform a great work working and encoding several objects' information, it is also useful to refer to other ways the nature uses to perform the same task. From this perspective one interesting very powerful encoding is the wave and acoustic language: it is named ***Harmonic Systems*** (***HS***). For further details see [27].The original theory was also extended by adding fuzziness. The goal of ***Fuzzy Harmonic Systems*** (***FHS***) is to extend the model to data that may require high subjectivity [30].

Both ***HS*** and ***FHS*** can be applied to problems with time evaluation of events. This implies the need to include a "timestamp" of the events, in order to be under the scope of these models.

Characteristics of FHS

The flexibility of ***FHS*** models reduces its computational complexity, and its fuzzy inference engine makes it simpler and appropriate to apply the benefits of HS without affecting its performance:

Firing subroutines: Detection of certain characteristics of an event during resonance can trigger subroutines. That is the case of ***Kronos***, when a risk case is detected, when the user is in a risky zone, the risk level can influence in the resulting level calculated by the system. For instance, if the risk level is "Low" and the zone risk is "High", the subroutine weights the level as "Medium".

Filters: it is possible to apply different filters to reduce the number of comparisons and improve the velocity of data processing. Furthermore it is possible to apply filters on certain patterns. For instance, a pattern could select pedestrians or drivers, or determined weather conditions.

Unsupervised: Unlike other prediction approaches, HS does not require a training step in order to start functioning. It adjusts automatically its parameters and adapts by self to the context of the problem.

Data model very strong: As it has been said in the previous item, it does not need training in order to get in work. It is possible to change data model in a very simple way letting at the same time the system keep working. That makes the system very strong, allowing data addition very quick.

To use fuzzy logic ***FHS*** model it is important to perform a fuzzification step of every variable in the model. To do that, just take the relevant properties of the problem [17]. Although it is possible to fuzzify any parameter, it does not make any sense to work on all of them.

13.5.1 HS for Traffic's Risk Predictions

Project ***KRONOS*** is focused in this type of encoding process to manage complex information that can't be contextualized to perform interpretation (that is, semantics is not a feature since it requires too much time to interpret the information). But some of the context should be considered in some way in real time.

This is the topic of research headed in ***IDTI LAB*** since 2015. The first use of this type of models was the ***KRONOS*** for traffic risks.

In this context an Android device collects data that is filtered and feed ***KRONOS*** with information like temperature, atmospheric pressure, outlook, traffic information, gender of the driver, etc.

This prototype can work for pedestrian, cyclist, or any type of driver. Figure 13.36 shows the Android GUI.

It performs a matching against patterns using a process similar to physical resonance, with harmonics being alternate events with minimal variations. The key concept here is to perform the whole process just focusing on timings and giving the rest of the problem a category of parameter no matter its meaning. The way to

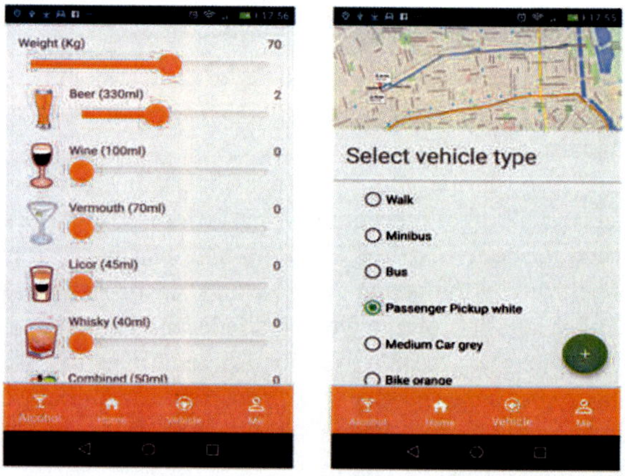

Fig. 13.36 Android GUI in KRONOS for traffic

Fig. 13.37 A Pattern in ***HS*** for any generic problem

T	PARÁMETRO-1	PARÁMETRO-2
t1=λ_1	PROC=A	USR=034
t2=λ_2	PROC=C	USR=035
t3=λ_3	PROC=A	USR=035

$$U=U+\eta_u\,(U-P_o(t_1|pattern).P_o(t_2|pattern).P_o(t_3|pattern)$$

$$\lambda_1=\lambda_1+\eta\,(t_1-\lambda_1)$$

$$\lambda_2=\lambda_2+\eta\,(t_2-\lambda_2)$$

$$\lambda_3=\lambda_3+\eta\,(t_3-\lambda_3)$$

Fig. 13.38 Updating process in ***HS*** with a Hebbian like heuristic

represent dynamical processes or any other information is to determine patterns as a sequence and alternate sequences. Figures 13.37 and 13.38 show a typical pattern and an update process according to HS.

Tests show the ability of this type of models to perform accurate predictions in real time. Figure 13.40 shows an example of the output of KRONOS.

There, ***M*** stands for **M**aximum risk, ***N*** for **N**eutral and ***A*** for **A**bsence of risk.

13.5.2 HS Application to Precision Farming

Precision farming requires a good understanding of the requirements for every type of crop to be considered. IDTI Lab started in 2018 an application of ***HS*** in the field

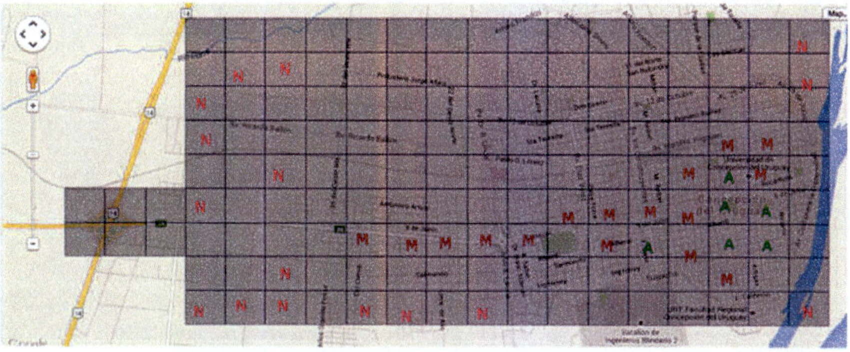

Fig. 13.39 Output screen with risks level in the map

of rice crops. The current knowledge in the zone allowed starting with a first version of a weather station (see Fig. 13.40).

This version collects data about humidity in the air and ground, ultraviolet radiation, temperature and wind. It is still pending the implementation of a Ph-meter.

Patterns were derived from the relationship between data variables collected and previously analyzed with Data Mining.

Early findings indicate that humidity behaves in a different way according to the month, and affects in some way but not linearly to the water level. One interesting finding is the minimum sampling rate differs from the traditional dairy control. That is, the time elapsed between two consecutive controls is usually one day at least. But Data Mining indicates that this timing is too long.

Other important finding is the relation between humidity in the environment and the water level on at ground level is relevant for predicting the behavior in the following hours.

Fig. 13.40 Output screen with risks level in the map

Another important result is the determination that the battery should not be loaded with solar panels since the zone can have several raining days and it cannot be guaranteed the same performance. Several patterns have been determined for water level control. Currently the research focuses on testing those patterns.

13.6 Conclusions and Future Work

Language is a key part of dairy performance. It not only can be a set of vocalizations or texts. Its main goal is to transport information from a sender to a receiver, functioning as an encoding. But there are many other ways to encode information, other "languages" that may be very diverse. Those languages sometimes can be easily related to objects and events, but sometimes it is hard to find the link between them. For instance, R. Berwick described the verbal behavior as body manifestations as alternate encoding of language. In fact music itself is an encoding that can be described by a special type of mathematics named harmony.

Some works performed in CI2S Labs have been focused on how to transform visual information to acoustic information. As part of the project named ***HOLOTECH*** [28] a prototype in Android was developed to help blind people outdoors with a sound language. The goal is to prevent the individual against cars, obstacles or any other object in the surrounding. This type of translation between a visual to an auditory encoding is very hard since sight receives information in a very efficient, synthetic and compact way, but ear process sequentially the information.

To summarize, information is everywhere and encoded in very diverse ways: movements, music, speech, noise, drawings, shapes, colors, dog barking, whale songs, etc. Sometimes, its expressions are more subtle than others. Also, information flows constantly but just a few encodings are considered by formal linguistic.

Furthermore, there are many very well known approaches to quantify and track spoken and written information, but there is still a long way to fully understand entropy, real world and language dynamics. For instance, Natural Language is still hard to produce automatically. Many other pending works are how information can flow, new transmission channels, how reasoning manages and creates it, how entropy relates to reasoning with information, etc. this is an exciting world that is not property of humans but nature.

References

1. Piantadosi S. T., Zipf's word frequency law in natural language: A critical review and future directions. Psychonomic Bulletin and Review. 2015.
2. López De Luise D., MLW and bilingualism. Adv. Research and Trends in New Technologies, Software, Human-Computer Interaction, and Communicability. IGI Global. 2013.
3. López De Luise D., El uso de Soft Computing para el modelado del razonamiento. Argentina Scientific Society (SCA) Trans. Sociedad Científica Argentina. In press. 2019.

4. López De Luise D., Azor R., Sound Model for Dialogue Profiling. International Journal of Advanced Intelligence Paradigms archive. Volume 9 Issue 5–6. Pages 623–640. 2017.
5. Sundberg M.L., A Brief Overview of a Behavioral Approach to Language Assessment and Intervention for Children with ASD. Assoc. for Behavior Analysis Newsletter, 30 (3). 2007.
6. Douglas Greer D., The Ontogenetic Selection of Verbal Capabilities.Int. journal of psychology and psychological therapy, 8 (3), 363–386. 2008.
7. Bustamante P., Lafalla A., López De Luise D., Párraga C., Azor R., Moya J., Cuesta J. L., Evaluación sistemática del comportamiento somático y oral como respuesta a estímulos sonoros en pacientes con TA. CLAIB. 2014.
8. J. Alm, J. Walker, Time-Frequency Analysis of Musical Instruments. SIAM **44**(3), 457–476 (2002)
9. Azor Montoya J. R., Perez J. R., Análisis del comportamiento autosimilar con distribución Pareto del tráfico Ethernet. Anales del XVI Congreso Argentino de Ciencias de la Computación. 2010.
10. Di Rienzo J.A., Casanoves F., Balzarini M.G., Gonzalez L., Tablada M., Robledo C.W., InfoStat, Córdoba University, Argentina. URL http://www.infostat.com.ar. 2013.
11. López De Luise D. et al., Information Retrieval in Restricted Domain for ChatterBots. 2018.
12. López De Luise D., Hisgen D., Language Modeling with Morphosyntactic Linguistic Wavelets. In Automatic content extraction on the web with intelligent algorithms. CIDM. 2013.
13. López De Luise D., Morphosyntactic Linguistic Wavelets for Knowledge Management. Book: Intelligent Systems, ISBN 979–953–307–593–7. InTech OPEN BOOK. 2011.
14. López de Luise D., Raúl Saad B., Pescio P., Saliwonczyk C., Autistic Language Processing by Patterns Detection. Int. Journal of Artificial Life Research (IJALR) 8(1). 2018
15. Riviere A., Martos J., El niño pequeño con autismo. ISBN: 9788460702610. 2000.
16. López De Luise D., Azor R.J., Párraga C., Conducta verbal autista: Modelo automatizado del perfil paciente-audio. Ed.OmniScriptum GmbH & Co. ISBN 978–3–59–07183–6. 2015.
17. Menendez E., López De Luise D., Augmented Reality as Visual Communication for People with ASD. Journal of Systemics, Cybernetics and Informatics (JSCI). 2018.
18. López De Luise D., Videojuegos como entidades del desafío cognitivo. Ludology Magazine. In Press. 2019.
19. UNESCO , El desarrollo sostenible. Report. La Habana. 2015.
20. Ariel Zambrano, Daniela López De Luise , Impact of multimedia in learning profiles. Int. J. Of Advanced Intelligence Paradigms. ISSN online: 1755–0394, ISSN print: 1755–0386. 2018.
21. Csíkszentmihályi M., Beyond Boredom and Anxiety. Jossey Bass Publishers. 1975.
22. Vargas Ligarreto D. E., Diseño de métricas derivadas de teclado y mouse para la medición del nivel de aprendizaje. Information Technology Thesis. UAI. Argentina. 2019.
23. Silva F. G., López De Luise D., Modelado de la Conducta de Usuario en Video Juegos. IJAIP. 2018.
24. Ruiz Tabarez E.A., López De Luise D., Modelo automático de evaluación en experiencias STEAM. COINCOM2019, Colombia. 2019.
25. Ruiz Tabarez E.A., López De Luise D., Modelo de predicción de deserción de alumnos. COINCOM. 2018.
26. Franklin S., et. al., A LIDA cognitive model tutorial (2016). Biologically Inspired Cognitive Architectures. 2016.
27. Rancez L., Maciel M., López De Luise D., De Elía B., Menditto J.P., Carro Verdia J., Pagano M., Temporalidad Aplicada a Sistemas de Conciencia. JAIIO. 2018.
28. Park N., López De Luise D., Rivera D., Bustamante L., Hemanth J., Multi-Neural Networks Object Identification.SOFA2018. Advances in Intelligent Systems and Computing. 2018.
29. López De Luise D., Carrillero P., Tournoud M., Pascal A., Alvarez C., Pescio P., Code Recognition by means of Improved Distance and tuned Dictionary. IEEE URUCON 2017.
30. Bel W., López De Luise D., Costantini F., Antoff I., Alvarez L., Fravega L., Fuzzy Logic in Predictive Harmonic Systems. SOFA2018. Adv. in Intelligent Systems and Computing. 2018.

31. Geoperspectivas. http://geoperspectivas.blogspot.com/2015/03. 2015.
32. Cinabrio. http://cinabrio.over-blog.es. 2019.
33. Noqueremosinundarnos. http://noqueremosinundarnos.blogspot.com/2011/11. 2011.

Daniela López De Luise Dipl. in System Analysis (Buenos Aires University, UBA), Computer Science Eng. (Spain write: 2004/H04041), Expert System Eng. (Buenos Aires Technological Inst. ITBA), PhD in Computer Sciences (National University at La Plata. UNLP) and Major in Public communication of Science and Technology (UBA). Main work: Morphosyntactic Linguistic Wavelets (MLW), Harmonics Systems (HS), Bacteria consciousness (BC) for robotics, Currently working on Caos & Language Theory (CLT), to produce and process Natural language. Currently, leading member of CETI (Center for Intelligent Technologies) at National Acedemy of Sciences in Buenos Aires (Academia Nacional de Ciencias de Buenos Aires, ANC BA - CETI), member of de Scientific Society of Argentina (Sociedad Científica Argentina, SCA ICD), Outreach project leader for STEAM activities. Associated Editor of IEEE Latin America Transaction. Museum of History Sarmiento: steering committee and leader of Museum-Lab LINCIEVIS for STEAM activities Director of Specialization for Computer Science. Universidad Autónoma de Entre Ríos (UADER). Director of CI2S Lab (Computational Intelligence & Information Systems Lab), since June 2013. Founder of the local branch and leading member of IEEE CIS Game Technical Committee. Director of IDTI Lab (Institute of Information Technology) since March 2017. Undergraduate and graduate teacher of Universidad Autónoma de Entre Ríos (UADER). Professor in National University of Technology (Universidad Tecnológica Nacional, UTN), Branch Concepción del Uruguay. Member of team GIBD (Data Base Research Group). Undergraduate and graduate teacher of Universidad Abierta Interamericana UAI). Consultant in Intelligent Systems (Computational Intelligence), soft computing (Since 1997). Member of WCI (Woman in Computational Intelligence) Argentina since 2015. Member of WIE (Woman In Engineering) Argentina since 2018. Member of IEEE SIGHT Argentina since 2013. Founder of Argentina chapter of IEEE Computational Intelligence Society, and current treasurer. Member of Local Lecturer Program (IEEE AR) since 2010. Science Journalist: As a Specialist in Public Communication of Science I founded and lead "Journalist by a day". Editor Dr. Mario Diván. Prize IEEE Foundation Funds for projet MIDA (2014). IEEE Computational Intelligence Society recognition for Notable Services in 2011–2012, Relevant Engineer recognition in LA Region (IEEE) among others like Sadosky prize, Bank Rio, etc.

Chapter 14
Collaboration in the Machine Age: Trustworthy Human-AI Collaboration

Liana Razmerita, Armelle Brun, and Thierry Nabeth

Abstract Collaboration in the machine age will increasingly involve collaboration with Artificial Intelligence (AI) technologies. This chapter aims to provide insights in the state of the art of AI developments in relation to **human-AI collaboration**. It presents a brief historic overview of developments in AI, three different forms of human-AI collaboration (e.g. conversational agents) and introduces the main areas of research in relation to human-AI collaboration and potential pitfalls. The chapter discusses the emergent multifaceted role of AI for collaboration in organizations and introduces the concept of trustworthy human-AI collaboration.

Keywords Collaboration · Trustworthy AI · Personalization · Organizational analytics · Behavioural analytics · Agent design · Organizational design · Chatbots · Cultural analytics · Attention management

14.1 Introduction

Artificial Intelligence (AI) has been a field of study for some years now. It is concerned with embedding intelligent behavior in artifacts and how intelligent behavior is generated and learned. AI has as one of its long-term objectives the development of machines that do the same things as humans, possibly even better (Nilsson 1998).

Building such AI systems has been a daunting, complex and controversial task as its main goal has been to understand intelligent behavior, emotions and cognition

L. Razmerita (✉)
Copenhagen Business School, Copenhagen, Denmark
e-mail: lra.msc@cbs.dk

A. Brun
Université de Lorraine, CNRS, Lorraine, France
e-mail: armelle.brun@loria.fr

T. Nabeth
P-Val Conseil, Paris, France
e-mail: thierry.nabeth@pval.com

M. Virvou et al. (eds.), *Advances in Selected Artificial Intelligence Areas*,
Learning and Analytics in Intelligent Systems 24,
https://doi.org/10.1007/978-3-030-93052-3_14

in order to instill it in machines. This ambitious scientific goal was emphasized by James Albus paraphrased by Nilson (1998) in his introductory chapter:

> understanding intelligence involves understanding how knowledge is acquired, represented, and stored; how intelligent behavior is generated and learned; how motives, and emotions, and priorities are developed and used; how sensory signals are transformed into symbols; how symbols are manipulated to perform logic, to reason about the past, and plan for the future; and how the mechanisms of intelligence produce the phenomena of illusion, belief, hope, fear, and dreams and yes even kindness and love.

AI was traditionally a science and engineering domain "concerned with the theory and practice of developing systems that exhibit the characteristics associated with intelligence in human behavior, such as perception, natural language processing, problem solving and planning, learning and adaptation, and acting on the environment". AI systems were developed as components of other more complex applications by adding intelligence in various ways (e.g. reasoning, learning, adapting) [49].

However, in the last decades new AI developments towards a more numerical or data-driven approach of AI have been made possible due to:

- The availability of a large amount of data (big data) that can be used to discover patterns (e.g. digital traces of user activities available from social media platforms, Google or other digital platforms).
- The availability of huge and affordable computing power (e.g. graphical processors) and storage capabilities (e.g. cloud).
- New advances in machine learning algorithms, associated tools and data science that can be used to collect and analyze data. (various tools and open source libraries support these processes: e.g. IBM Watson, PowerBI, TensorFlow, Weka, Matlab facilitate developing AI systems of various complexity, etc.)
- Progress in the creation of agents or chatbots.

Such recent developments of AI open up new opportunities to integrate AI systems and data technologies in various types of applications, for example, natural language processing, human–machine interaction, information retrieval, graphics and image processing or robotics and provide new opportunities for businesses to innovate and derive value in new ways. As a result, in the past decade we have witnessed an explosion of the use of AI in a wide range of sectors such as healthcare, services, education, mobility, or commerce, which promises to revolutionize these sectors in the future by automating the existing process and enabling or inventing totally new applications.

Furthermore, leveraging AI systems can make data smart by developing new ways to process data going beyond the use of analytics in organizations. Conversely, AI systems develop AI technologies integrating (e.g. machine learning algorithms) that learn from data and inform decisions. The use of AI within a business context has led to the development of a new set of terminologies such as business intelligence, cognitive computing or computational intelligence [14]. AI technologies offer new possibilities for the relationship between humans and machines with respect to performing work tasks on digital platforms, and for the effective design and governance of platforms [40].

AI plays an increasing role in knowledge collaboration, through facilitation of human-AI interaction: conversational agents in the form of assistants, chatbots, and personalization using algorithms or recommender systems. The emergence of big data and data science will yield to unprecedented data-driven decisions and business insights driven by artificial intelligence, algorithms and machine learning techniques.

According to [14] AI will support three important business needs such as:

1. automating business processes,
2. gaining insights through data analysis and
3. engaging with customers and employees.

The use of AI in a business context is also associated with the term business intelligence, in particular when AI systems are used to gain insights and support decisions based on data analysis. This data-driven intelligence can be used to build informed decisions and business strategies. We provide some examples of how artificial intelligence collaborates with humans in the different cases below:

- In the service sector, AI is used to assist call centers by proposing personalized conversational agents that are able to answer 24/7 the more basic questions and alleviating the workload of the call center agents. They are considered to be a next generation paradigm for customer services: arguably this technology allows employees to focus their time and energy on more complex customer problems and help eliminate rote work that might be repetitive and tedious.
- In the mobility sector, artificial intelligence is used to provide automatic driving assistance, and in the future will be used to assist driving automatic cars or drones, thus enabling goods to be delivered easily.
- In the healthcare sector, AI is already used on a large scale for analyzing radiological images and diagnosing cancer in collaboration with expert doctors. In the future, AI will be used to monitor health and provide a continuous, personalized and just in time healthcare assistance to every citizen, preventing the development of diseases.
- In the e-commerce sector, AI intervenes in the analysis of customer behavior, anticipates their needs and provides recommendations in order to manage their attention toward items of interest and eventually persuade them to buy them. Using data from different sources, AI systems may also support the process of managing inventories.
- In e-education, AI is used to capture and understand students' learning strategies, level of knowledge, etc. and can be used to move them to personalized learning goals. AI can also be used to encourage learners to interact with peers, through group collaboration, to improve the learning outcome, increase motivation, attendance, etc.

Recent pop culture has also developed new science fiction scenarios of AI use. Movies like "The operating system" have popularized the idea that AI can become human and even able to develop relationships with other humans. Other TV series like"Black mirror" create a rather dystopian view of how algorithms and AI technology may impact human relationships and our civilization. The Social dilemma documentary showcases how AI and user profiling based on social media data (Facebook) have been used to influence elections. The documentary also highlights the danger of misusing personal data to influence users and manipulating them through targeted interventions. At this level of AI development, it becomes particularly important to address trustworthy collaboration with AI and the ethics of AI.

Within this chapter, we focus on applications to areas where AI can support collaboration with humans (e.g. by engaging with customers and employees) in different forms *of personalized interaction, attention management* and *persuasion.*

Collaboration in this context refers to the process of humans and AI systems working together to pursue a common objective. Indeed, in the machine era, AI goes well beyond mere ICT as passive tools (e.g. word processors) that are controlled by the user, and actually help the user in the realization of a task. More specifically, AI systems have certain cognitive capabilities (perception, interpretation, plan making, execution and learning), as well as a level of autonomy, and are driven by goals that can be conducted without direct human supervision. An agent may play the role of an assistant at the service of this user, but it can also be autonomous and take initiative, and even serve the goals of other actors (such as the company offering this agent as a service).

The objective of this paper is to look at the use of AI to support collaboration, and future development. This is increasingly important as recent developments of AI are opening up new avenues for developing new capabilities that will impact behaviors, organizing and work in general [16, 25]. As organizations integrate AI systems, collaboration with AI is emerging in different scenarios of knowledge work. Trustworthy AI looks at the different elements that intervene in collaboration and the associated challenges.

We distinguish between three types of collaboration with AI:

1. Human—computer collaboration where AI is embedded
2. Human—AI collaboration (or conversational AI)
3. Human–Human collaboration where AI can intervene.

These three types of collaboration will be discussed in detail in Sect. 3. This chapter will consist of the following sections: the second part presents a brief introduction and historical overview of artificial Intelligence. The third part presents how AI ca be applied to support collaboration at the tree levels: enhancing the collaborative process and making it more fluid; providing more advanced and proactive mechanisms supporting the collaborative & social processes (trust, motivation, stimulation); in informing the organization design of more collaborative organizations. The fourth part briefly overviews challenges and ethical issues of the use of AI (e.g. privacy, impact on the society) followed by a conclusion.

14.2 Artificial Intelligence: An Overview

14.2.1 The Role of AI—Definitions and a Short Historic Overview

The term AI is difficult to define in a unified way to reach consensus among different fields and application domains. As already mentioned above in the introduction, AI encompasses a cluster of computing technologies including intelligent agents, machine learning, natural language processing and decision-making supported by algorithms [51].

The term "Artificial Intelligence" was first coined by John McCarthy together with other AI influential scholars, including (e.g. Allen Newell, Marvin Minsky, Herbert Simon) in a 1956 workshop held at Dartmouth.

Although the discipline of Artificial Intelligence AI was created more than 60 years ago, its exact definition has been the subject of numerous debates and has embraced a number of concepts and objectives [21].

Early AI systems were designed in symbolic programming languages (e.g. Lisp, Prolog, Scheme) or using agent-based architectures. In the early stages, AI was divided into different isolated subfields such as: natural language processing, knowledge representation, problem solving and planning, machine learning, robotics and computer vision. Given the expectation that AI can help create machines that think and learn and even surpass human intelligence, it was not surprising that many of the early AI systems failed to scale up and solve complex real-world problems or to exhibit real intelligence [49].

According to [21], AI is defined as "a system's ability to interpret external data correctly, to learn from such data, and to use those learnings to achieve specific goals and tasks through flexible adaptation". This definition seems to relate to the recent Machine Learning perspective, since "learning" is not a necessary characteristic of all artificial intelligence systems.

In this chapter, we define artificial intelligence informally as systems (e.g. algorithms, robots) with a high level of autonomy, aiming at assisting, guiding or automating human tasks. The authors of this article argue that the term AI is overused and not all simple systems can be classified as AI. Hence, not all algorithms are AI but all AI systems are based on algorithms.

Historically AI has been associated with the design of an "artificial general intelligence" aiming at replicating human intelligence, and its ability to solve a broad range of problems. Modern AI dates back to the Turing test, which was originally developed in 1950 by Alan Turing. It was one of the first attempts to embed human intelligence in a system. The challenge was to create a system that "could think", that answered questions similar to the way a human would, and ideally could not be differentiated from a human. This was dubbed the "imitation game".

Turing's test is fundamental but also controversial as it reduced intelligence to a conversation. The test is considered passed by the machine if the machine answers could not be distinguished from answers by humans. Later in 1963, Allen

Newell and Herbert A. Simon developed the idea that the mind can be viewed as a system that manipulates bits of information according to formal rules. They proposed the idea that "symbol manipulation" was the essence of both human and machine intelligence. AI consisted of programs based on symbolic reasoning or was based on rules (if … then) in earlier phases.

A further step in AI development beyond: "Can machines think?" is the problem of consciousness or intentionality. A mind usually has thoughts, ideas, plans or goals. Several questions have been addressed by AI researchers and philosophers: "Can machines have a mind?" or "Can machines have consciousness?" or If we assume that AI will become similar to humans and thus imitate human characteristics, other questions can be further developed, e.g., "Can machines have emotions?", "Can machines be self-aware?" or "Can machines be creative?", "Can machines have a soul or can machines be hostile?".

It has become important to address these questions as the idea that AI can become self-sufficient, autonomous, and make its own decisions has become popular in the last years. The concept of "singularity" was introduced to present a vision of technological developments and "intelligence explosion", or super intelligence embodied in machines, which could become dangerous for humanity. It has been promoted by both science-fiction writers such as [54] as well as famous scientists and celebrities (Stephen Hawkings and Elon Musk).

14.2.2 AI and Agents

An agent is defined as a knowledge-based system that perceives its environment (which may be the physical world or other agents, or a complex environment); reasons to interpret perceptions, draw inferences, solve problems, and determine actions; and acts upon that environment to realize a set of goals or tasks for which it has been designed. Agents may continuously improve their knowledge and performance through learning based on the interaction with other agents, users, or based on other type of data [49].

In the past, various types of agents have been designed. Eliza was one of the first computer-enabled technologies designed to build some sort of human-technology interaction [55].

The development of AI is linked to the development of different types of intelligent agents that perform different functions in a Society of Mind [29]. The society of mind theory views the human mind and any other naturally evolved *cognitive* systems as relying on individually simple processes known as *agents*. This theory has developed as a collection of essays and ideas that Minsky started writing in the early 1970. These agents cooperate in a similar way with people in a society.

Historically agents have been part of the AI endeavor to converse (chatbots) or provide information, to entertain, support humans in various tasks (e.g. learning), or guide or persuade users. Rosalind Picard has introduced the concept of affective computing as an additional endeavor for achieving genuine intelligence, by considering the role of emotions. Affective computing recognizes the role of emotions and try to give computers the ability to recognize, express and understand emotions [37].

14.2.3 *Beyond Modern AI*

AI currently primarily relies on a data-driven approach (e.g. machine learning) however in the past many systems have been developed based on a symbolic approach (e.g. expert systems or rule-based systems). In some other cases, it consisted of the automation by a machine of the reasoning of human experts in specific domains (cf. the expert systems), the realization of some complex planification tasks (cf. constraint programming), or the modelling of knowledge and associated cognitive processes. It has integrated sophisticated mechanisms aiming at solving complex problems in novel ways (e.g. genetic algorithms inspired by evolutionary principles, or fuzzy logics) based on mechanisms such as emergence and adaptation.

More recently, artificial intelligence has been associated with machine learning. However, machine learning is just a subset of AI. Machine learning AI is able to learn and to adapt by processing sizeable amounts of data, and to identify automatically patterns that the systems will be able to use to solve problems in similar situations. Machine learning algorithms have a rather narrow scope and limited capabilities. Nowadays deep learning allows for use of both numerical and symbolic data, such as in recommender systems [57].

From a usage perspective, artificial intelligence has been considered both as a means (1) to augment human cognitive capabilities (e.g. helping them sensing, interpreting the context, making plans, and implementing decision), or (2) to automate the human process completely (replacing intellectual operations conducted by humans by machines). In the first case, AI maintains the role of humans at the center of the decision loops, whereas in the second case it makes the human superfluous.

14.3 The Role of AI for Collaboration

Collaboration is an integral part of organizational working and learning practices. In the last years, new forms of collaboration have emerged due to new forms of collaboration technology, a continuing trend in globalization and global scale adoption of hybrid or remote work. Digital collaboration can be defined as the articulation of personal knowledge into collective wisdom made possible via a diversity of digital platforms, including enterprise social media (e.g. blogs, micro-blogs and wikis) [43], collaborative platforms (e.g. GoogleDocs, Dropbox) or more recently even using AI technologies (e.g. enterprise AI platforms Grace[1]).

[1] https://2021.ai/offerings/grace-enterprise-ai-platform/.

Collaboration can be established at different levels. According to Wikipedia, collaboration is the process of two or more people, entities or *organizations* working together to complete a task or achieve a goal.

Collaboration can consist in the interaction of two individuals that interact (share information, contribute to the production of something) in order to produce an output more effectively. Collaboration can also extend to the interaction of a group of people aiming at the production of a common good belonging to this group. Collaboration may also be considered at a more global level (societal level), in which the member of a society may contribute to the realization of a common good. An example of this form of collaborative innovation for the common good is the production of a new protocol of care or development of the Covid 19 vaccine. In fact, AI has been used for the development of the Covid 19 vaccine.

Different factors can foster or hamper collaboration. AI can be used to build predictive models that assess the likelihood that a user pursues certain actions (e.g. dropout from a course) and user's intentions such as the user's intention to engage or not in collaboration. Predictive models can be built taking into account various factors (e.g. independent variables that influence a dependent variable, individual and communal factors). Prior research has outlined important factors that contribute to the intention and decision to engage in collaboration in a digital environment. Engagement in collaboration may be influenced by goals that are set and expectations from collaboration arising from previous experiences, but also the perceived ability to work in groups and the peers' attitudes towards collaboration [42].

Trust represents one of the most important factors, among those acknowledged in the literature, for enabling collaboration. Trust is the expectation individuals have of behaviors of others and in particular cooperative behaviors of others. Virtual agents can play a key role in a value creation process and the establishment of engagement and online trust [9]. AI can help to construct trust through reputation mechanisms or recommendations [24].

14.3.1 Human–Computer Collaboration Where AI is Embedded

Collaboration with AI can be seamless or "automated" by making the content customized or personalized through algorithms (e.g. recommender systems). This area has been the subject of research for many years for scholars in different fields, including artificial intelligence, computer science, focusing on user modelling, recommender systems and more recently data science and digital marketing. Personalization aims to give users, customers, and employees a web experience with the highest relevancy possible but also to achieve socially intelligent behavior. Personalization is achieved through programs or algorithms that take into account individual users' preferences, behaviors as well as context.

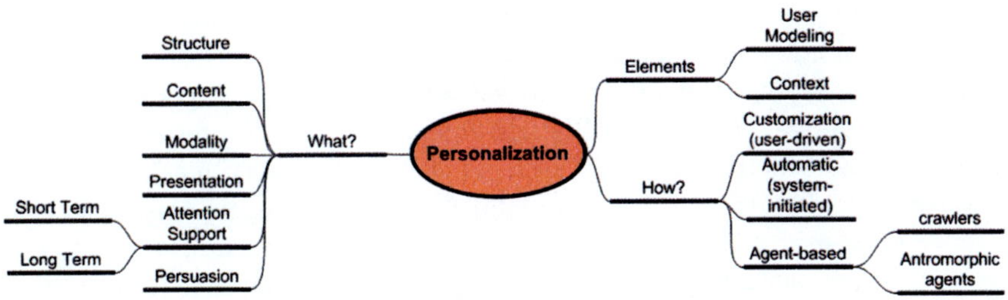

Fig. 14.1 A taxonomy of personalization techniques according to [45]

The objective of personalization is to improve the efficiency of the interaction with users for the sake of simplification and to make complex systems more usable. Personalization is particularly important in e-business. On the one hand, consumers expect personalized interaction with online retailers, but on the other hand, personalization is perceived to conflict with the desire for privacy. Personalization relies on the collection and the exploitation of personal data which may raise serious privacy concerns and lead to a "personalization-privacy paradox" [23].

Figure 14.1 presents a taxonomy of personalization techniques. This taxonomy summarizes different forms of personalized human–computer interaction (what can be personalized?) and the elements that contribute to supporting it (how to achieve this). Personalization relies on different elements such as the user's preferences, or other characteristics of the users (e.g. age, gender, culture, demographic profiles, personality traits that could be captured from available data or through user profiling) and/or context (e.g. location). Users' profile data may also include biometric information (e.g. fingerprints, iris scans) or medical data that can easily be captured and stored using different apps and devices (e.g. cell phones, smart watches). Such big data sets may be stored and mined for different purposes (e.g. personalization, predictions, or even unexpressed desires). They may be used to create a digital identity (e.g. create digitized human clones) and offer a variety of personalized services for the users, but access to these sets raises privacy concerns for the users.

Personalization can be implemented in different forms that include: personalization of structure, content, modality, presentation, attention support, and persuasion. User profiling is a form of user modeling based on the users' data or digital traces of interaction with the system using a multitude of methods as presented in [7]. It is a way to bypass the lack of information provided by the users that allows one to personalize the interaction with applications that adapt to their user needs and accommodate their preferences [45]. Context awareness allows AI systems (e.g. agents or other intelligent systems) to adapt to the environment and to the users' characteristics and needs. Furthermore, it is an important element for integrating intelligence or intelligent behavior.

Forms of personalization can be implemented as a form of social information access. Social information access is defined as a stream of research that explores "methods of organizing the past interactions of users community in order to provide better access to information for future users" [8].

Personalization of structure refers to the way in which the hypermedia space is structured and presented to the different groups of users. "Personalized" can be agent-based interaction or automatic (system-initiated using algorithms). Agents can be conversational or embodied agents (with or without anthropomorphic features). Their physical aspect can also be selected for embodied agents (for instance, it can be close to a human agent to provide a certain "human touch"). The design of personalized interaction within computer-based systems (e.g. learning environments) is often designed based on agents or even multi-agent systems that can interact autonomously.

Agent architectures are implemented to collaborate with users or carry out tasks on their behalf, based on knowledge about users, their goals or desires. Agents can intervene in supporting different forms of human–computer collaboration through personalized interventions. For example, pedagogical agents have been designed to support learning processes taking into account students characteristics, perceived needs in relation to learning objectives and emotions [6]. Some early prototypes of the agents have been implemented in the KInCA system (Knowledge Intelligent Conversational Agents) described in [2, 5]. KInCA is designed as an agent-based architecture to support the adoption of knowledge-sharing practices within organizations. Several expert agents are assigned different roles such as: interacting with the user for diagnosing the user state, implementing persuasion strategies such as storytelling. KInCA relies on the idea of offering **personalized user support**. The system observes the user's actions and, whenever appropriate, makes suggestions, introduces concepts, proposes activities that support the adoption of the desired behaviors. Conversational agents aim at providing personalized guidance through the whole adoption process; from the introduction of the behaviors to the user (e.g. explaining what the desired behaviors are and why they should be adopted) to their practice within the community. Such conversational agents are designed to fulfill the role of change agents as they motivate people to learn and adopt new behaviors [46] using different strategies at different stages based on users' activity. Change agents implement different forms of persuasion. Persuasion strategies are associated with interventions that may include tracking and monitoring users' activity, sharing, social support, but also gamification strategies (e.g. competition, comparison or rankings). Persuasion technologies are currently used in many different domains including education, commerce, healthcare and well-being [35, 36].

Agents have also been designed to better manage attention of the users in a social context through various interventions. Based on observation of users, agents' intervention includes: the guidance on the use of the platform, reminder about the completion of the user's profile, notification of approaching deadlines, tracking inattention, identifying bursts of attention by the community as a whole, or suggesting the adoption of practice (e.g. open up to others) or encouraging pro-social behaviors [47].

Different forms of personalization, including attention support and persuasion, are increasingly used in data-driven digital marketing. Personalization of content is effective in capturing the attention of customers and tailor communication to specific customers. It can be combined with predictive models to find the best time to send communication to specific customers (termed "mechanical AI"). Personalized communication can leverage the effect of persuasion and increase the chance of cross-selling opportunities. For example, companies delivering flowers can remind users of birthdates and special events that can be accompanied with a nice bouquet of flowers. Furthermore, other associated items (e.g. chocolates or champagne) may be recommended to them. The analysis of the card texts that accompany the flowers are used to detect emotions and make specific recommendations for gifts that fit the specific occasion. However, these algorithms require handling of big amounts of data. AI use in e-commerce is associated with new flavors of AI: "feeling AI", "thinking AI" and the implementation of emotional commerce. AI can thus contribute not only to competitive advantage but also to an enhanced customer experience [56].

14.3.2 Human—AI Collaboration (Or Conversational AI)

Human-AI collaboration can take place through the use of conversational agents or AI-based digital assistants. Digital assistants have a specific extent of interactivity and intelligence in order to help users perform tasks. AI-based assistants rely on a conversational user interface (e.g. speech-based, text-based or visual input) for receiving input and delivering output to the users on the basis of natural language processing or machine-learning algorithms. AI-based assistants have representation of the domain knowledge and the capability to acquire new knowledge using machine learning algorithms [26].

They are not only a next-generation paradigm for customer service and in everyday domestic interactions. AI-assistants are increasingly integrated in workflows and interactions using enterprise platforms (e.g. Teams, Slack) and thus shape the future of work and collaboration.

Different platforms incorporate AI in the form of assistants: Siri by Apple, Alexa by Amazon, Google Home by Google or Cortana by Microsoft. These assistants combine natural language processing capabilities with AI-powered search and Internet of Things (IoT). Conversational AI coupled with AI powered search assist users in various way including finding favorite music tunes, informing them on demand about the weather forecasts, activating smart home features or searching information on the web [48].

Chatbots represent an increasingly popular technology for supporting customer service. "A chatbot is a virtual person who can effectively talk to any human being using interactive textual as well as verbal skills" [52]. A chatbot can be designed and integrated in a system and fulfil different purposes e.g., for customer support, or to augment the software development process in open-source communities (see

Github[2]). More recently chatbots have even been used to support well-being or digitize dead persons and assist the grieving process [6].

Agents have started to take part in basic interactions on social media platforms. Agents, also or chatbots, can be given different tasks, such as: coordination of knowledge work, brokerage of knowledge, communication and knowledge collaboration. These assistants are designed to act or react as a human would in a dialogue session. Natural language processing, including dialogue processing, is at the core of these assistants [34]. One remaining challenge to make the collaboration more intuitive and useful is to design adaptive scenarios [10].

Furthermore, in order to behave in a human-like manner different forms of human intelligence need to be implemented. Among different forms of intelligence (e.g. abstract, practical, kinesthetic), social intelligence is particularly important as it relates to "the ability to get along with others and get them to cooperate with you". The authenticity of the agents represents another important element to consider, as it is a crucial element in preventing manipulation, and establishing cooperation and trust [33].

14.3.3 Human–Human Collaboration Where AI Can Intervene

In this section we provide an overview of the use of AI as systems designed to support collaboration and how they can be used at the tree levels: for enhancing the collaborative process and making it more fluid; providing more advanced and proactive mechanisms supporting collaborative social processes (e.g. trust, motivation, stimulation); and in informing the organization design of more collaborative organizations.

AI can also be used to change the workplace culture (e.g. for helping to design organizations that are collaborative or help organizations to become more collaborative). The integration of AI into workflows and business processes also contributes to the emergence of analytics driven organizations and it transforms organizational culture towards a culture of analytics. Such organizational culture relies on data, rather than intuition and experience of human managers. An analytics of culture fosters data literacy and uses insights from data for supporting business decisions in various activities, including organizational design. We will outline below the **emergent role of AI and Analytics for Collaboration and Organizational Design**.

14.3.3.1 Data Science and Analytics for Organizational Design

The availability of data, and progress in machine learning have augmented considerably the possibility to study and analyze the functioning of organizations and use

[2] Making a Github bot (https://www.geeksforgeeks.org/making-a-github-bot/).

tools to inform the design of organizations in a more objective or scientific manner. A scientific manner involves relying on factual data analysis rather than only on mere human intuitions and experience. The term organizational design analytics has been introduced as a branch of computational social science. It combines data science and analytics methods for organizational design. Computational social science is defined as the study of society and human behavior through the prism of computational analyses and the sensing of human signals. It is becoming more and more of a reality. It aims to be applied to the design and the operation of more effective organizations. The use of "scientific method" such as psychometric analysis (i.e. personality tests) has been used for a long time, notably in helping organizations to recruit the right profiles. They have been limited by the effort required to use them, originating from the necessity for respondents to fill out questionnaires, and for the recruiters to employ staff possessing specific qualifications to analyze the results. Besides, employee surveys and questionnaires have significant shortcomings since employee self- reports are often tainted with cognitive bias [11]. The values and beliefs that people proclaim may be significantly different from how they behave in real life and who they really are. The validity of psychological analysis methods has also been questioned for its inability to validate the method in real world settings, and more specifically afterward to assess the reality of its predictive power.

Scholars and practitioners have started to make the link between AI and organizational design [30, 38] by adopting a data science approach to design more effective organizations. This approach is referred to as "organizational analytics" [30] or "organizational design analytics" [38]. It relies on the use of algorithms to analyze data from organizations and use the result of this analysis as input to be used to build a better performing organization. More specifically, [38] propose the use of a suite of methodologies and tools helping to better perceive the organizational environments of a company, make predictions and experiment quickly with new practices.

The term "Org 2.0" that they suggest refers to a radical evolution of organizational design, and involves enabling designers to make much more sophisticated design decisions than in the past, and moving away from merely copying other designs in favor of "haute couture", specifically adapted to the situation. Their analytics driven approach can be conducted at three levels:

- perception, which is based on the combination of big data and traditional statistical methods to capture the current situation;
- prediction, which is based on the application of machine learning and AI to Big Data, to forecast what is going to happen;
- prototyping, which is based on agent-based modelling for testing the hypothesis.

Morrison [30] proposes a number of analytical tools to organization design teams and HR to provide them with a new and better way to design, transform and operate their organizations.

A cultural analysis can be derived from the digital traces of people interaction [11]. Previous researchers have mined millions of e-mail messages exchanged among employees in a high-technology firm for assessing the cultural fit of its employees and monitoring its evolution [20].

Using means of artificial intelligence such as natural language understanding (NLU) applied to the information that employees provide in electronic communication (email, Slack messages, and Glassdoor comments) offers new ways to provide insights into the culture of an organization and how people behave effectively, rather than what people claim they do [11].

AI provides the means to collect and analyze information much more effectively than humans. AI also guides organizations in the formation of teams that are likely to collaborate optimally by encapsulating and making easily available the expertise of the members of the team in the composition of teams. Effective teams should be composed of people with different profiles that complement each other's strengths. More specifically, AI can help:

- To collect data: for instance, web agents are used to scrape social media data (e.g. LinkedIn, Twitter, Glassdoor) that will be available for the analysis.
- To analyze and make sense of this data (profiling): for instance, machine learning algorithms (supervised or non-supervised learning; classical machine learning or deep learning) are used to analyze data that can be available in a large variety of forms (e.g. number, natural language). Dimension reduction, natural language understanding, or clustering represent examples of means that are now available for data analysis.
- To provide guidance. AI systems, including analytics recommender systems or expert systems, can help organizational managers to design more effective organizations or help the transformation of organizations (fix misfunctioning organizations, or the fusion of organizations).

14.3.3.2 Team Members' Personality, Team Composition, Sociology and Collaborative Culture

Exploiting information of people personality, group sociology or culture represent means that can be exploited for guiding the design or the transformation of more collaboration effective organizations.

First, the personality of a team member as an individual has some impact on the likeliness of this team member to collaborate with others. Good team players are often defined in terms of traits like being dependable, flexible, or cooperative [15]. Different researchers have explored the link between the personality of team members and teamwork effectiveness [12]. For instance, they have looked, using the big five personality model, how extraversion, agreeableness and conscientiousness can be associated with peer-rated contributions to teamwork. For instance, a good level of extraversion can be associated positively to collaboration since it favors the establishment of social interaction. Agreeableness is particularly useful in situations involving interpersonal conflict by alleviating conflicts that may arise in collaborative endeavors and interactions. And very conscientious members are likely to be highly committed to group tasks. However, these traits should not be in excess, in which case they may have a disruptive effect [12]. An excessive level of extraversion may originate from a dominant personality resulting in interaction based on power relationships and competition rather than cooperation. Members very high on conscientiousness may be perfectionists, being overly focused on their individual goals leading to relationship tensions, whereas too agreeable members may be too risk-averse to conflicts, leading to the reduction in the quality of the interaction.

Second, at a collective level, the composition of a team will also impact the quality of collaboration. The creation of highly effective teams therefore involves assembling different members with expertise and competencies required to tackle the problems for which this team was created, but also to fill different roles that are necessary when solving problems. Belbin's [3] work on the composition of teams has proposed that highly effective teams should be built by assembling in the same team a combination of members having preferences for fulfilling certain roles. For example a *plants role* is a behavioral attitude about creative, unorthodox and generators of ideas; A *teamworker role* is about acting as the "oil" between the cogs that keep the machine that is the team running smoothly) [1, 31].

Finally, organizational design can take into consideration the social and cultural levels to make organizations more collaborative. Social roles and social norms, which refer to the mostly unwritten rules that manage how human agents interact with one another or in groups, also have a strong influence on the way people collaborate. Some societies and organizations that rely in an important manner on status impose strong constraints on the level of interaction, for instance by limiting the expression of dissident views based on factors like seniority, position, and gender. The lack of trust between the members of an organization may also limit the willingness to engage in interaction and take risks. Sociological theories such as the work of Boltansky and Thevenot [4] on people justifying their actions in a social context can then be used to describe the functioning of organizations from a sociological point of view [19] and provide guidance about how to improve them. For instance, one of their theories identifies six categories of "worlds" to which one can associate different sets of justification of social action, such as the domestic world that is driven by values of tradition or family, or the civic world that relies on democratic value and consensus.

Organizational culture can be defined as a collection of shared values, expectations, and internalized practices that guide behaviors of the members. Some organizational cultures will favor collaboration, whereas others will make it very difficult. An important stream of literature exists on this subject. Previous research on cross-cultural communication has shown how cultural differences can create barriers to interactions [53].

14.3.3.3 Data Science, Organizational Design and Collaboration

Personality-related information is not something new in the design of more collaborative teams (e.g. Belbin teamwork inventory was used well before the generalization of the use of data science technique), but the advent of artificial intelligence promises to considerably augment its utilization in the design of more collaboration-effective organizations. AI and machine learning (ML) enable us to derive insights from different types of data (e.g. social media, enterprise data, or other digital communications). Social Media Analytics allow us to infer individuals' personality characteristics from digital data such as emails, text messages, tweets, or forum posts. However, user profiling, or the inference of personality traits based on data, online interactions or digital text raises ethical concerns. In particular, such algorithms may embed biases or may give rise to discrimination. An example of such a tool is IBM Watson personality insights[3] although this has been recently discontinued.

AI may support groups or teams' formation in organizations. Creating heterogeneous, diverse groups is important for performance, creativity and learning. Collaborative outcomes (e.g. quality of learning) depend on characteristics of the group and the group composition. Algorithms can support the formation of groups taking into account specific characteristics and certain criteria (e.g. culture, gender, personality). Previous work has been done on the use of AI to support groups in classrooms and to support diverse or heterogeneous teams e.g., [1, 41].

In relation to the data collection, recent research, based on the processing of 25,104 job advertisements published on the online job platforms, proposes a text mining approach combining topic modeling, clustering, and expert assessment in order to identify and characterize six job roles in data science. Monster, and Glassdoor [28].

Consulting companies have designed a set of analytical tools that can help the design or the transformation of organizations:

Crystal[4] offers a set of tools that integrate the DISC personality insights, and that can be used to identify team strengths and weaknesses. These tools make use of machine learning techniques, which have been employed in the profiling of people personality based on their LinkedIn profile [15]. For instance the tool "Crystal for Teams" offers teams personality-based insights to navigate important conversations, including one-on-ones, performance reviews, group meetings, or conflicts.

Talentoday[5] offers people an analytics platform based on the analysis of personalities used both at the individual and collective levels. If it seems to be primarily used by human resource professionals; it is also used by organizational designers to guide the fusion of organizations. One of the tools that this company offers to its clients, "MyPrint Collaboration report" is aimed at improving one-to-one collaboration, and consists in 10 pages report automatically generated that includes an analysis of the areas of synergy and risk between two individuals in terms of personality.

P-Val conseil[6] has developed "la méthode monde", a method based on the work of the sociologists Boltansky and Thevenot [4] on people justifications of their actions in a social context. Boltansky and Thévenot have identified six worlds ('Inspiration', 'Merchant', 'Industrial', 'Civic', 'Domestic' and Opinion) that are governed by different values. Based on these, they have derived a set of sociological profiles. P-Val has derived from this work a set of sociological profiles that are used both at the individual and collective levels to help decode and anticipate resistance to change

[3] IBM Watson personality insight (https://cloud.ibm.com/apidocs/personality-insights).

[4] Crystal (https://www.crystalknows.com/).

[5] Talentoday (https://www.talentoday.com/).

[6] P-Val conseil (https://pval.com/).

in organization transformation. P-Val uses this approach to facilitate the merging of organizations by assessing cultural proximity of the organizations to be merged. P-Val uses artificial intelligence as a means to profile and analyze sociological data, using methods such as clustering or natural language understanding. More specifically, a 'civic'-oriented person may receive in this personalized report some advices about how to deal the more effectively with a 'merchant'-oriented person.

Assessment tools can be used to identify different ways in which people are inclined to make an impact and a contribution. For example, the firm GC Index[7] proposes five categories of individuals among which the **Game Changers** are the individuals who generate the ideas and possibilities that have the potential to be transformational. The **Play Makers** individuals focus on getting the best from others, individually and collectively, in support of agreed objectives.

14.3.3.4 AI Collaboration and the Management of Attention

The advent of social media has put a strong emphasis on online social interaction. With the Web 2.0 technology, the Internet is no longer used only to access a massive amount of information, but also to interact and collaborate with others on a global scale. At an organizational level, the technology is also increasingly used to share and collaborate with others, a trend that has been considerably reinforced with the Covid 19 pandemic which has forced people to work at home (part time, or even full time during the more difficult times of the pandemic). Thus, a variety of tools (e.g. email, collaborative platforms, video-conferencing systems) are now used on a regular basis to work and "collaborate with others". This phenomenon has created the conditions of a massive **social interaction overload** where people are overwhelmed by solicitations and opportunities to engage in social exchanges, but they have little means to deal effectively with this new level of interaction [32]. Different studies have shown that instant messaging notification on the desktop creates distraction and disruption, which is detrimental to productivity [13].

More generally, the individualization of work has made it more difficult for knowledge workers to manage their time, with the risk of having a substantial part of their time consumed in the interaction, and the difficulty to separate work life from private life. The solution to this problem has been proposed in the form of systems that are able to help users to manage their attention, e.g., the time that they dedicate in each category of tasks (e.g. how much time they work on their own on a text editor, how much time they spend writing emails, or how much they interact or collaborate). An example of such a system is Microsoft MyAnalytics Office 365, which "allows people to work smarter with productivity insights". Such tools even though dedicated to self-monitor and provide individual feedback it may also be used in a concerning way for individuals.

[7] GC Index (https://www.thegcindex.com/).

Attention aware systems rely largely on advanced analytics, which are implemented based on a variety of AI techniques, such as the collection of attention data and their analysis using machine learning or deep learning techniques.

14.3.4 Challenges of Using AI: Toward a Trustworthy AI

AI is becoming embedded in different forms of human–computer interaction. Furthermore, big data associated with the use of AI leads to new opportunities for businesses, but also raises ethical and privacy concerns. However, important challenges are associated with the use of AI on a large scale. Among these challenges the most important ones are: lack of transparency of algorithms, systems fragmentation, data privacy regulations, but also data quantity and quality.

Most AI technologies rely on data, but these technologies can also make data smart and can be used in a range of scenarios. People are increasingly aware and concerned about the power of big data, its potential use and misuse in both organizations and society. Our digital traces expose everything we do and like in the digital world. Hence transparency and management of "visibilities" become crucial to consider in a digital world [18]. As new technologies, including artificial intelligence and robotics, "have the potential to infringe on human dignity and compromise core values of being human" [22], ethics and privacy concerns are important to be considered a truly trustworthy human-AI collaboration.

Recent research has emphasized the fact that learning algorithms are distinguished by four consequential aspects: 'black-boxed performance, comprehensive digitization, anticipatory *quantification*, and hidden politics" [16]. **Trustworthy AI** needs to overcome such negative consequential aspects in which humans are reluctant to engage with AI or regard AI as a competitor rather than an assistant.

Trustworthy AI is an emerging topic for both academics and industry [50]. Trustworthy AI aims to contribute to the well-being of individuals as well as the prosperity and advancement of organizations and societies [50], while avoiding risks of infringing individuals' privacy, discriminating part of the population and being unfair. To reach this goal, legal and ethical dimensions are at the core of work in Trustworthy AI. We can for example cite beneficence, non-maleficence, autonomy, justice, and explicability, proposed in [17].

More generally, **trustworthy AI** is fully human-centered and aims to offer high levels of human control, with the goal to lead to wider adoption and increase human performance, while supporting human self-efficacy, mastery, creativity, and responsibility. Notice here that there is a twofold danger: excessive human control and excessive computer control. One challenge lies in the trade-off between both controls. From this point of view, trustworthy AI has a significant role to play. Indeed, not only does AI has to foster collaboration, it also has to guarantee that this collaboration will be beneficial for both parties. Specifically, it may not only focus on increasing collaboration when AI is used but may foster a long-term collaboration regardless of AI use. Indeed, designing an AI that is temporally myopic can be counterproductive in the long term. Considering that a positive benefit is obtained in the short term, if no control or knowledge about the long-term impact may come down to designing an AI that only compensates the consequences of its passed decisions. In addition, AI has to be fair: each person in the collaboration deserves to be equally considered so that there is no discrimination between those persons.

The adoption of AI by society, both at the individual and global levels, depends to a large extent to the trust that people can develop toward AI technologies and their application. People should believe in the ability of AI to deliver value without at the same time representing a risk that would lessen this value. In management and organizational science trust can be approached from different perspectives.

First different factors of perceived trust in AI intervene in the formation of trust [27, 39]:

(1) the *ability* of the system to fulfill its role. The lack of explainability of AI can for instance generate suspicion of the workability of an AI solution.
(2) the *benevolence* of the system. People have to be convinced that AI is used for their own good, and not for the benefit of tierce parties. The belief that AI systems are controlled by the small group of organizations such as the GAFAM (Google, Apple, Facebook, Amazon and Microsoft) or government agencies (even in democratic countries) may create rejection.
(3) the *integrity* of the system, and in particular the perception that the rules on which it is based are clear. For instance, if the rules that explain the acceptance of a loan by a bank using AI to support their decisions are too obscure to the citizens, there is little possibility that AI will appeal.

Second, trust can also be based on rational reasoning, i.e. cold and rational calculation that have been formalized with the agency theory in economy [39] versus gut feeling. In the latter case, people are subject to biases and may develop an irrational negative feeling that is disproportionate with reality. For instance, they may fear that the AI machines take over the world (as in the movie Terminator), even if this is for now only a very distant possibility.

14.4 Conclusion

The focus of our chapter has been to discuss the emergent role of AI in shaping collaboration in different forms. We have highlighted the benefits and pitfalls of human-AI collaboration and introduced the concept of trustworthy AI. This chapter has presented a vision of the role of ICT in the machine age that has shifted from the traditional ICT tools totally dedicated to the realization of specific functions (controlled by humans, or automating processes), to AI systems with some cognitive capabilities, autonomy and individual goals, and which support or collaborate with humans to complete certain tasks. These AI entities may be at the direct service of the humans that they serve, but also controlled and serve the goal of other entities (e.g. GAFAM or tech savvy organizations).

We have seen in this chapter that this vision can be observed at different levels:

- At the individual level, with personalized services that have developed a certain understanding of the users (e.g. user profile) and that can be used by an agent or chatbot, personalized web services (e.g. in e-commerce), and also via cell phones, becoming in the latter case a cognitive extension to the human being.
- At the organizational level, with the augmented / smart organization, that is constantly monitoring its environment (analyzing data), and acting with some level of autonomy to solve problems and adapt, or in helping the design of more effective organizations, thanks to the new AI-powered advanced organization analytics.

Yet this chapter has also raised the concerns of this machine age such as: loss of control and expandability; risk of these AI systems to be controlled by tierce entities that may not be benevolent to the users (e.g. authoritarian regimes, business-oriented companies firstly driven by their own commercial goals).

In sum, we distinguish between three types of collaboration with AI. First, it can be used to support personalized interaction where AI is in the background. Second, Artificial Intelligence can intervene in the collaborative process in knowledge & social platforms to support the social mechanisms associated with collaboration. It can provide feedback on the collaborators. It can help in the construction of trust in knowledge and social platforms. It can provide recommendations contributing in team formation by suggesting who to connect to (e.g. people that have more affinity). Thirdly, Artificial Intelligence can be used more indirectly by informing the design and transformation of organizations that are more likely to be collaborative. AI may be used to suggest the formation of highly functioning teams or to guide the cultural transformation of organization towards a more collaborative or knowledge-sharing oriented culture.

Acknowledgements We would like to thank to Inger Mees for readproofing and Daniel Hardt for providing feedback and comments on this chapter.

References

1. J.M. Alberola, E. Del Val, V. Sanchez-Anguix, A. Palomares, M. Dolores Teruel, An artificial intelligence tool for heterogeneous team formation in the classroom. Knowl.-Based Syst. (2016). https://doi.org/10.1016/j.knosys.2016.02.010
2. Angehrn, A., Nabeth, T., Razmerita, L., & Roda, C. (2001). K-InCA: Using artificial agents to help people learn and adopt new behaviours. In *Proceedings - IEEE International Conference on Advanced Learning Technologies, ICALT 2001*. https://doi.org/10.1109/ICALT.2001.943906
3. M. Belbin, *Management Teams* (London, Heinemann, 1981), ISBN 978-0-470-27172-8
4. L. Boltanski, L. Thévenot, Les économies de la grandeur. Cahiers du Centre d'études de l'emploi, Paris, PUF, Czerwinski (1987)
5. Brna, P., Cooper, B., Razmerita, L. (2001). Marching to the wrong distant drum: pedagogic agents, emotion and student modeling. In *Proceedings of Workshop on Attitude, Personality and Emotions in User-Adapted Interaction*. Sonthofen, Germany.

6. Brown, D. (2021). AI chat bots can bring you back from the dead. *The Washington Post*. Retrieved from https://www.washingtonpost.com/technology/2021/02/04/chat-bots-reincarnation-dead/
7. Brun, A., Boyer, A., & Razmerita, L. (2010). Compass to locate the user model I need: Building the bridge between researchers and practitioners in user modeling. In *Lecture Notes in Computer Science (including subseries Lecture Notes in Artificial Intelligence and Lecture Notes in Bioinformatics)*. https://doi.org/10.1007/978-3-642-13470-8_28
8. Brusilovsky, P., & He, D. (2018). Introduction to social information access. In Lecture Notes in Computer Science (including subseries Lecture Notes in Artificial Intelligence and Lecture Notes in Bioinformatics). https://doi.org/10.1007/978-3-319-90092-6_1
9. S. Castellano, I. Khelladi, J. Charlemagne, J.-P. Susini, Uncovering the role of virtual agents in co-creation contexts. Manag. Decis. **56**(6), 1232–1246 (2018). https://doi.org/10.1108/MD-04-2017-0444
10. Colace, F., de Santo, M., Lombardi, M., & Santaniello, D. (2019). Chars: A cultural heritage adaptive recommender system. In *TESCA 2019 - Proceedings of the 2019 1st ACM International Workshop on Technology Enablers and Innovative Applications for Smart Cities and Communities, co-located with the 6th ACM International Conference on Systems for Energy-Efficient Buildings, Cities* (pp. 58–61). https://doi.org/10.1145/3364544.3364830
11. Corritore, M., Goldberg, A., & Srivastava, S. (2020). The new analytics of culture. *Harvard Business Review*.
12. Curşeu, P. L., Ilies, R., Vîrgă, D., Maricuţoiu, L., & Sava, F. A. (2019). Personality characteristics that are valued in teams: Not always "more is better"? *International Journal of Psychology, 54*(638–649). https://doi.org/10.1002/ijop.12511
13. M. Czerwinski, E. Cutrell, E. Horvitz, Instant messaging: Effects of relevance and time, in *People and Computers XIV: Proceedings of HCI 2000*, ed. by S. Turner, P. Turner (British Computer Society, 2000), pp. 71–76
14. Davenport, T. H., & Ronanki, R. (2019). Artificial Intelligence for the Real World. In *On AI, Analytics and the New Machine Age* (Harvard, pp. 1–17). Boston, MA.
15. Driskell, J. E., Goodwin, G. F., Salas, E., & O'Shea, P. G. (2006). What makes a good team player? Personality and team effectiveness. *Group Dynamics*. https://doi.org/10.1037/1089-2699.10.4.249
16. Faraj, S., Pachidi, S., & Sayegh, K. (2018). Working and organizing in the age of the learning algorithm. *Information and Organization, 28*(1), 62–70. https://doi.org/10.1016/j.infoandorg.2018.02.005
17. L. Floridi, J. Cowls, A Unified Framework of Five Principles for AI in Society. Harvard Data Science Review **1**(1), 1–15 (2019). https://doi.org/10.1162/99608f92.8cd550d1
18. Flyverbom, M. (2019). The digital prism transparency and managed visibilities in a datafied world. *The Digital Prism: Transparency and Managed Visibilities in a Datafied World*. https://doi.org/10.1017/9781316442692
19. Fridenson, P. (1989). Luc Boltanski et Laurent Thévenot, Les économies de la grandeur, Paris, Presses Universitaires de France, « Cahiers du Centre d'études de l'emploi », 1987, XVI- 367 p. *Annales. Histoire, Sciences Sociales*. https://doi.org/10.1017/s0395264900063204
20. A. Goldberg, S.B. Srivastava, V. Manian, W. Monroe, C. Potts, Fitting in or standing out? The tradeoffs of structural and cultural embeddedness. Am. Sociol. Rev. **81**, 1190–1222 (2015)
21. M. Haenlein, A. Kaplan, A brief history of artificial intelligence: On the past, present, and future of artificial intelligence. Calif. Manage. Rev. (2019). https://doi.org/10.1177/0008125619864925
22. Jasanoff, S. (2016). The Power of Technology. In *The Ethics of Invention: Technology and the Human Future*. (pp. 4–58). W.W.Norton.
23. A. Kobsa, H. Cho, B.P. Knijnenburg, The effect of personalization provider characteristics on privacy attitudes and behaviors: An Elaboration Likelihood Model approach. J. Am. Soc. Inf. Sci. (2016). https://doi.org/10.1002/asi.23629
24. Kunkel, J., Donkers, T., Michael, L., Barbu, C. M., & Ziegler, J. (2019). Let me explain: Impact of personal and impersonal explanations on trust in recommender systems. In *Conference on*

Human Factors in Computing Systems - Proceedings (pp. 1–12). https://doi.org/10.1145/3290605.3300717
25. P.M. Leonardi, COVID-19 and the New Technologies of Organizing: Digital Exhaust, Digital Footprints, and Artificial Intelligence in the Wake of Remote Work. J. Manage. Stud. (2021). https://doi.org/10.1111/joms.12648
26. A. Maedche, C. Legner, A. Benlian, B. Berger, H. Gimpel, T. Hess, M. Söllner, AI-Based Digital Assistants. Bus. Inf. Syst. Eng. (2019). https://doi.org/10.1007/s12599-019-00600-8
27. R.C. Mayer, J.H. Davis, F.D. Schoorman, AN INTEGRATIVE MODEL OF ORGANIZATIONAL TRUST. Acad. Manag. Rev. (1995). https://doi.org/10.5465/amr.1995.9508080335
28. Michalczyk, S., Nadj, M., Maedche, A., & Gröger, C. (2021). "Demystifying Job Roles in Data Science: A Text Mining Approach." In *ECIS* (p. 115). Retrieved from https://aisel.aisnet.org/ecis2021_rp/115/
29. Minsky, M. (1986). *The Society of Mind.*
30. Morrison, R. (2015). *Data-driven Organization Design: Sustaining the Competitive Edge Through Organizational Analytics.*
31. Mostert, N. M. (2015). Belbin-the way forward for innovation teams. *Journal of Creativity and Business Innovation.*
32. Nabeth, T., & Maisonneuve, N. (2011). Managing attention in the social web: the AtGentNet approach. In *Human Attention in Digital Environments* (pp. 281–310). Cambridge University Press. https://doi.org/10.1017/cbo9780511974519.012
33. M. Neururer, S. Schlögl, L. Brinkschulte, A. Groth, Perceptions on authenticity in chat bots. Multimodal Technologies and Interaction **2**(60), 2–19 (2018). https://doi.org/10.3390/mti2030060
34. Nuruzzaman, M., & Hussain, O. K. (2018). A Survey on Chatbot Implementation in Customer Service Industry through Deep Neural Networks. In *Proceedings - 2018 IEEE 15th International Conference on e-Business Engineering, ICEBE 2018* (pp. 54–61). https://doi.org/10.1109/ICEBE.2018.00019
35. Orji, F. A., Oyibo, K., Greer, J., & Vassileva, J. (2019). Drivers of competitive behavior in persuasive technology in education. In *ACM UMAP 2019 Adjunct - Adjunct Publication of the 27th Conference on User Modeling, Adaptation and Personalization* (pp. 127–134). https://doi.org/10.1145/3314183.3323850
36. R. Orji, K. Moffatt, Persuasive technology for health and wellness: State-of-the-art and emerging trends. Health Informatics J. (2018). https://doi.org/10.1177/1460458216650979
37. Picard, R. W. (1997). *Affective Computing.* MIT press.
38. Puranam, P., & Clément, J. (2020). The Organisational Analytics eBook: A Guide to Data-Driven Organisation Design. Version 1.0. Retrieved from https://knowledge.insead.edu/blog/insead-blog/organisational-data-the-silver-lining-in-the-covid-19-cloud-15516
39. P. Puranam, B. Vanneste, Artificial Intelligence, Trust, and Perceptions of Agency. SSRN Electron. J. (2021). https://doi.org/10.2139/ssrn.3897704
40. A. Rai, P. Constantinides, S. Sarker, Editor's Comments: Next-Generation Digital Platforms: Toward Human–AI Hybrids. Manag. Inf. Syst. Q. **43**(1), 9 (2019)
41. Razmerita, L., & Brun, A. (2011). Collaborative learning in heterogeneous classes: Towards a group formation methodology. In *CSEDU 2011 - Proceedings of the 3rd International Conference on Computer Supported Education* (Vol. 2).
42. L. Razmerita, K. Kirchner, K. Hockerts, C.-W. Tan, Modeling collaborative intentions and behavior in Digital Environments: The case of a Massive Open Online Course (MOOC). Academy of Management Learning & Education **19**(423), 469–502 (2020)
43. L. Razmerita, K. Kirchner, T. Nabeth, Social media in organizations: leveraging personal and collective knowledge processes. J. Organ. Comput. Electron. Commer. **24**(1), 74–93 (2014)
44. Razmerita, Liana, Nabeth, T., Angehrn, A., & Roda, C. (2004). *Inca: An Intelligent Cognitive Agent-Based Framework for Adaptive and Interactive Learning.*
45. Razmerita, Liana, Nabeth, T., & Kirchner, K. (2012). User Modeling and Attention Support. In *Centric- The fith International Conference on Advances of Human Oriented and Personalized Mechanisms* (pp. 27–33). Lisbon.

46. C. Roda, A. Angehrn, T. Nabeth, L. Razmerita, Using conversational agents to support the adoption of knowledge sharing practices. Interact. Comput. **15**(1), 57–89 (2003). https://doi.org/10.1016/S0953-5438(02)00029-2
47. Roda, Claudia, & Nabeth, T. (2008). Attention management in organizations: Four levels of support in information systems. In *Organisational Capital: Modelling, Measuring and Contextualising*. https://doi.org/10.4324/9780203885215
48. Schmidt, R., Alt, R., & Zimmermann, A. (2021). A conceptual model for assistant platforms. In *Proceedings of the Annual Hawaii International Conference on System Sciences* (pp. 4024–4033). https://doi.org/10.24251/hicss.2021.490
49. G. Tecuci, Artificial intelligence. Wiley Interdisciplinary Reviews: Computational Statistics. (2012). https://doi.org/10.1002/wics.200
50. S. Thiebes, S. Lins, A. Sunyaev, Trustworthy artificial intelligence. Electron. Mark. (2021). https://doi.org/10.1007/s12525-020-00441-4
51. L. Tredinnick, Artificial intelligence and professional roles. Bus. Inf. Rev. **34**(1), 37–41 (2017). https://doi.org/10.1177/0266382117692621
52. A. Trivedi, Z. Thakkar, Chatbot generation and integration: A review. International Journal of Advance Research **5**(2), 1308–1312 (2019)
53. Trompenaars, F. (2006). *Managing people across cultures. Proceedings of 20th IPMA World Congress on Project Management.*
54. V. Venge, *Technology Singularity* (2021)
55. J. Weizenbaum, ELIZA-A computer program for the study of natural language communication between man and machine. *Communications of the ACM* (1966). https://doi.org/10.1145/365153.365168
56. A. M. Williamson & K. M. Akeren, *Artificial Intelligence in Digital Marketing.* Copenhagen Business School (2021)
57. S. Zhang, L. Yao, A. Sun, Y. Tay, Deep learning based recommender system: a survey and new perspectives. ACM Comput. Surv. **52**(1), 1–38 (2019). https://doi.org/10.1145/3285029

Liana Razmerita is associate professor in learning technologies at Copenhagen Business School. Her research investigates new ways of organizing, collaborating and learning in the digital age. She is interested in how emerging technologies (such as AI) and ICT shape new ways of working, learning and co-creating value for organizational and social change or innovation. She holds a PhD from University of Toulouse, France and an engineering degree in automation and computer science from University of Galati, Romania. She has previously worked at INSEAD Fontainebleau, INRIA Sophia-Antipolis, France and University of Leeds, UK. She has published over 100 scholarly written articles in refereed journals, conference proceedings and book chapters. Her work has been published in journals such as: Academy of Management Learning & Education, Journal of Knowledge Management, Online Information Review, Journal of Organizational Computing and Electronic Commerce, Journal of Applied Artificial Intelligence, Interacting with Computers and IEEE Systems, Man and Cybernetics.

Armelle Brun (PhD HDR) works on recommender systems, data mining, explainable algorithms, user privacy and ethics. She is involved in several European, national and regional projects, where she is in charge of work packages. She leads a work package in the French eFran METAL project (2016–2021) dedicated to the mining of logs of learners' activities and the recommendation of resources. She coordinates the National PEACE Numerilab project (for the French Ministry of Education) (2019–2022). She currently holds the scientific excellence distinction for research and doctoral supervision. She has published over 90 articles and is regularly involved in national and international conference and workshops organisation (including chairing). She was awarded the "best paper" at the ASONAM 2009 conference. Her recent paper about grey sheep users modeling has been nominated "outstanding paper" at the ACM UMAP 2016 conference.

Thierry Nabeth is a senior A.I. scientist at P-Val Conseil working on a variety of Artificial Intelligent projects such as advanced organizational analytics, personalized chatbots, or natural language generation of reports for the banking sector. He has previously worked at INSEAD Centre for Advanced Learning Technologies in Fontainebleau, France as a senior research fellow, in the domain of advanced knowledge management systems, advanced social platforms, or learning technologies.